Geological Perspectives of Global Climate Change

Edited by
Lee C. Gerhard
William E. Harrison
and
Bernold M. Hanson

AAPG Studies in Geology No. 47

Published by
The American Association of Petroleum Geologists
Tulsa, Oklahoma, U.S.A.
In collaboration with the Kansas Geological Survey
and
The AAPG Division of Environmental Geosciences
Printed in the U.S.A.

Association Editor: Neil F. Hurley
Geoscience Director: Robert C. Millspaugh
Publications Manager: Kenneth M. Wolgemuth
Managing Editor, Publications: Anne H. Thomas
Special Publications Editor: Hazel Rowena Mills
Cover Design: Rusty Johnson
Production: Custom Editorial Productions, Inc., Cincinnati, Ohio
Front Cover: Tana Glacier in the Chugach Mountains, Alaska. Photograph by Lee C. Gerhard.
Back Cover: Synthetic satellite image by Rudolf B. Husar, Washington University, Saint Louis,
Missouri; land layer from the SEAWIFS project; fire maps from European Space Agency; sea-
surface temperature from Visualization Laboratory, Naval Oceanographic Office, U.S. Navy;
cloud cover from Space Science and Engineering Center, University of Wisconsin, Madison.

This and other AAPG publications are available from:

The AAPG Bookstore
P.O. Box 979
Tulsa, OK 74101-0979
Telephone: 1-918-584-2555 or 1-800-364-AAPG (USA)
Fax: 1-918-560-2652 or 1-800-898-2274 (USA)
www.aapg.org
http://bookstore.aapg.org

Australian Mineral Foundation
AMF Bookshop
63 Conyngham Street
Glenside, South Australia 5065
Australia
Tel. +61-8-8379-0444
Fax +61-8-8379-4634
www.amf.com.aux/amf

Geological Society Publishing House
Unit 7, Brassmill Enterprise Centre
Brassmill Lane, Bath, U.K.
BA1 3JN
Tel +44-1225-445046
Fax +44-1225-442836
www.geolsoc.org.uk

Affiliated East-West Press Private Ltd.
G-1/16 Ansari Road   Darya Ganj
New Delhi   110 002
India
Tel +91 11 3279113
Fax +91 11 3260538
e-mail: affiliat@nda.vsnl.n

Preface

Climate science is in a state of flux. Scientists from several disciplines are working diligently to show the effects of anthropogenic activities on global climate. As this work continues, however, many are finding that natural processes and variability effectively mask or obscure anthropogenic contributions. The major problem, in our view, revolves around the fact that current scientific activities related to climate-change issues do not adequately consider pertinent evidence available in the geologic record. These records serve as archives for evaluating such critical issues as the natural range of temperature conditions and the variation in greenhouse-gas concentration levels through geologic time.

Geologic science has a unique role in assessing the phenomena of climate change. Geology is the only discipline that routinely works backward in time to unravel facts and interpretations from materials that are hundreds of thousands, millions, or even billions of years old. Climate changes must be viewed over the same spans of time that create and drive them. Some climate changes (e.g., small-scale solar cycles) may be as short as a human life span, while others span tens of thousands to hundreds of millions of years.

This book was designed to bring into careful consideration the geologic parameters and measurements that illustrate the full range of past climate changes against which anthropogenic effects may be compared. Such a documented background is essential if the human consequences on climate are to be truly distinguished from naturally occurring variation. Most of the papers collected here are new research results that add to the already solid literature in geologic science that deals with the variability of climate. It is important that the full spectrum of science be brought to bear on this matter. It is especially important if significant national policies (e.g., U.S. adoption of the provisions of the Kyoto Protocol) are based on flawed or selective consideration of scientific investigations.

If the costs associated with significant reductions in greenhouse gases were trivial, we would have fewer reservations about developing and deploying such measures. However, the projected costs are extremely high ($315 billion per year [based on the U.S. Department of Energy's Information Administration estimated carbon tax of $348/ton of carbon, Murphy Oil Company estimated carbon content per barrel of oil, and an assumed U.S. daily oil consumption rate of 19.5 million barrels per day] in new taxes) and the planned actions under the Kyoto Protocol appear to be ineffective, at best. Because of the enormous stakes associated with the protocol, it is incumbent upon scientists to take steps to better understand complex natural systems, how they operate, and their overall effects on climate change. Only then can anthropogenic overprints be properly evaluated and analyzed.

The editors and individual authors were not financially supported by any industrial or similar organizations. Most are academicians who design and conduct their individual research programs without regard for potential policy implications of their work. William E. Harrison and Lee C. Gerhard are employed by a state geological survey. Bernold M. Hanson, a past president of the American Association of Petroleum Geologists, is deceased. None of the contributing authors is currently a practicing petroleum geologist, although several have had broad and successful experiences in that profession.

We have made a conscientious effort to bring pertinent science to the question of climate change, regardless of whether the evidence challenges or supports the global-warming hypothesis. The reader will probably note that each individual author may bring professional opinions into his or her work, but in all cases, the scientific material has been subjected to a rigorous peer-review process. Although the publisher of this book is the American Association of Petroleum Geologists (AAPG), we have worked diligently to avoid an industrial or environmental bias to the volume. From the outset, we viewed our task to be to compile the best and most appropriate scientific investigations, and to let those studies speak for themselves.

AAPG is one of the largest professional geological societies in the world, and most of its members concentrate on science that supports their quest to supply global society with energy while vigorously maintaining the highest level of environmental stewardship. Over the last 18 months, we have been privileged to be associated with numerous earth scientists who have contributed indirectly to this volume by virtue of their interest and support. These colleagues often made us aware of relevant (but sometimes obscure) scientific investigations, provided encouragement and insight, and invariably served as a critical, but frank, first audience.

Much of the work set forth in this volume resulted from oral presentations at national meetings, efforts from members of the AAPG Ad Hoc Committee on Global Climate Issues, and activities of the AAPG Division of Environmental Geosciences. We appreciate the association's support and encouragement to continue this work and bring these studies together in a single collection of papers.

Finally, we recognize the leadership of our departed colleague, Bernold M. "Bruno" Hanson, whose inspiration and guidance always concentrated on integrity, openness, and high citizenship. He lived his beliefs of public and professional participation and responsibility. We dedicate this book to his memory, with the hope that he would have approved of what we have done.

Lee C. Gerhard
William E. Harrison

Acknowledgments

The editors wish to acknowledge the assistance of the many people who helped produce this book. First and foremost, this work arose from the efforts of the AAPG Ad Hoc Committee on Global Climate Issues, appointed by President Eddy David and continued by President Richard Bishop. The committee members who helped make this book possible are Roy D. Adams, John P. Bluemle, Elizabeth B. Campen, William L. Fisher, Richard G. Green, David A. L. Jenkins, Michael Johnson, Robert R. Jordan, Fred T. Mackenzie, Charles J. Mankin, Robert J. Menzie, Alfred H. Pekarek, Ronald W. Pritchett, Perry O. Roehl, Joseph M. Sabel, Eugene A. Shinn, Douglas Swift, and Donald E. Wilde.

At the Kansas Geological Survey, the efforts of Pat Acker, Marla Adkins-Heljeson, Debbie Douglas, and Jennifer Sims made it possible for the editors to expedite this volume. Rowena Mills, Ken Wolgemuth, and Anne Thomas of the Publications Department at AAPG headquarters all contributed to this effort, and we gratefully acknowledge their individual roles. Their friendly and professional interactions with editors and authors both facilitated and accelerated the publication process.

Each chapter was reviewed by multiple individuals (several of whom reviewed more than one paper), and we appreciate the candid remarks, insightful comments, and technical perspectives of each of them. The reviewers were Fred T. Mackenzie, Robert E. Stevenson, David Hodell, William F. Ruddiman, Henry F. Diaz, William W. Hay, Algimantas Grigelis, Neeraj Gupta, Brian Hitchon, John Hoganson, Stephen Porter, Joseph Hartman, Perry O. Roehl, William Wilbert, Sallie Baliunas, Wallace Broeker, Robert A. Berner, Grant R. Bigg, Richard H. Sams, Ronald K. Stoessell, Leon H. Allen Jr., Michael Krings, Eugene A. Shinn, Terry Edgar, Richard Von Herzen, Alan Beck, Andrew Fisher, Eric T. Sundquist, Tim Coburn, John Grace, Donald E. Meyers, Robert F. Dill, Peter Stuart, Ellen R. M. Druffel, Marek Kacewicz, David A. L. Jenkins, and Robert J. Menzie. We are grateful to them for their contributions to the final product.

Finally, we thank numerous friends and colleagues who offered continuous encouragement and support for this effort, from conception through the draft phases to final production. We recognize that many of these people help constitute the potential audience for this work, and thus their comments, advice, and counsel were especially appreciated.

Lee C. Gerhard
William E. Harrison

AAPG
wishes to thank the following
for their generous contributions
to

Geological Perspectives of Global Climate Change

Kansas Geological Survey
United States Geological Survey

Contributions are applied toward the production
costs of publication, thus directly reducing the book's
purchase price and making the volume
available to a larger readership.

Dedication

To Bernold M. "Bruno" Hanson, a pioneering and legendary oilman, past president of the American Association of Petroleum Geologists, and a friend and mentor to all. His vision led to the founding of the AAPG Division of Environmental Geosciences, and in so doing, he found another way to help us focus on our responsibilities as practicing earth scientists. For those who knew him, he lived larger than life. He represents our heritage, our aspirations, and our legacy.

About the Editors

Lee C. Gerhard is principal geologist of the Kansas Geological Survey at the University of Kansas. He received his B.S. in geology at Syracuse University and M.S. and Ph.D. at the University of Kansas. He has combined academic, government, and industry leadership and technical appointments, including petroleum exploration, management of research and exploration programs, oil and gas regulation, and reservoir geology. His research interests are in carbonate sedimentology, petroleum geology, and environmental public policy. Gerhard was state geologist of North Dakota and directed a marine laboratory in the U.S. Virgin Islands. Prior to returning to Kansas, he was the Getty Professor of Geological Engineering at the Colorado School of Mines, and he operated Gerhard and Associates, an independent petroleum-exploration company.

Gerhard is an honorary member of the American Association of Petroleum Geologists (AAPG), the Association of American State Geologists, and the Kansas Geological Society. He is a former president of the AAPG Division of Environmental Geosciences and an honorary member of that organization. He has published more than 150 papers and books on geology, petroleum exploration, natural resources, and environmental policy, and he was cochairman of the AAPG Ad Hoc Committee on Global Climate Issues.

William E. Harrison is deputy director of the Kansas Geological Survey. He received B.S. and M.S. degrees in geology from Lamar University and the University of Oklahoma, respectively. He joined Shell Oil Company as an exploration geologist in the Gulf Coast Division, and worked in the Houston and New Orleans Districts. Harrison left industry to attend Louisiana State University, where he received his Ph.D. in organic geochemistry. He worked in the Geologic Research Department of ARCO before going to the Oklahoma Geological Survey and the University of Oklahoma, where he held the Klabzuba Chair of Geology in the School of Geology and Geophysics.

Harrison rejoined ARCO as research director and was responsible for developing programs in petroleum geochemistry, quantitative basin modeling, reservoir characterization, and predrill porosity prediction. He joined the DOE Idaho National Engineering and Environmental Laboratory (INEEL) and held management positions with EG&G and Lockheed-Martin. He was responsible for a nationally recognized program in earth, life, and environmental sciences, and successfully transferred several INEEL technologies to the private sector. Harrison is president-elect of the Division of Environmental Geosciences of the American Association of Petroleum Geologists.

Bernold M. "Bruno" Hanson (1928–2000) was a mentor who touched the careers and lives of many of today's professional geologists. He served as president of the American Association of Petroleum Geologists and was a leader of alumni advisory committees for the advancement of education in geology. A native of North Dakota, Hanson was notably successful in the discovery of oil and gas. He was president of the Hanson Corporation, of Midland, Texas, for 40 years. Previously, he was employed by Magnolia (Mobil) Corporation and held several management positions with Exxon Corporation. He received his baccalaureate degree in geology from the University of North Dakota, and his master's and honorary doctorate from the University of Wyoming. Hanson published geological articles about Alaska, west Texas, and North Dakota oil fields and reservoirs.

Among his many activities in AAPG, Hanson was the founder and first president of the Division of Environmental Geosciences and cochairman of the Ad Hoc Committee on Global Climate Issues. He received the Sidney Powers Medal, AAPG's highest award, in 1996. He also received honors from the University of Wyoming, the University of North Dakota, the Boy Scouts of America, and the city of Midland, Texas. The Permian Chapter of Professional Secretaries International named him "Boss of the Year."

Table of Contents

Preface iii

Acknowledgments v

Introduction and Overview 1
  Lee C. Gerhard, William E. Harrison, and Bernold M. "Bruno" Hanson

Part I—Climate Drivers

Chapter 1
  Solar Forcing of Earth's Climate 19
    Alfred H. Pekarek

Chapter 2
  Distribution of Oceans and Continents:
  A Geological Constraint on Global Climate Variability 35
    Lee C. Gerhard and William E. Harrison

Chapter 3
  Recent Past and Future of the Global Carbon Cycle 51
    Fred T. Mackenzie, A. Lerman, and L. M. B. Ver

Chapter 4
  Are We Headed for a Thermohaline Catastrophe? 83
    Wallace S. Broecker

Part II—Methods of Estimating Ancient Temperature

Chapter 5
  Stable Isotopes and their Relationship to Temperature
  as Recorded in Low-Latitude Ice Cores 99
    Lonnie G. Thompson

Chapter 6
  Century-Scale Variation of Seafloor Temperatures Inferred from
  Offshore Borehole Geothermal Data 121
    Seiichi Nagihara and Kelin Wang

Chapter 7
  Sclerosponges: Potential High-Resolution Recorders of
  Marine Paleotemperatures 137
    Gary B. Hughes and Charles W. Thayer

Chapter 8
  Perspectives on Quaternary Beetles and Climate Change 153
    Allan C. Ashworth

Chapter 9
Using Fossil Leaves for the Reconstruction of
Cenozoic Paleoatmospheric CO_2 Concentrations 169
Wolfram M. Kürschner, Friederike Wagner, David L. Dilcher, and Henk Visscher

Part III—Natural Variability and Studies of Past Temperature Changes

Chapter 10
Rate and Magnitude of Past Global Climate Changes 193
John P. Bluemle, Joseph M. Sabel, and Wibjörn Karlén

Chapter 11
The Search for Patterns in Ice-Core Temperature Curves 213
John C. Davis and Geoffrey C. Bohling

Chapter 12
Sea-Level Change in the Baltic Sea:
Interrelation of Climatic and Geological Processes 231
Jan Harff, Alexander Frischbutter, Reinhard Lampe, and Michael Meyer

Chapter 13
Coral Reefs and Shoreline Dipsticks 251
E. A. Shinn

Part IV—Policy Drivers

Chapter 14
Microbial Lime-Mud Production and Its Relation to Climate Change 267
K. K. Yates and L. L. Robbins

Chapter 15
Geological Sequestration of Anthropogenic Carbon Dioxide:
Applicability and Current Issues 285
Stefan Bachu

Chapter 16
Near-Term Climate Prediction Using Ice-Core Data from Greenland 305
Sergey R. Kotov

Chapter 17
Carbon-Dioxide-Induced Global Warming:
A Skeptic's View of Potential Climate Change 317
Sherwood B. Idso

Chapter 18
Potential Impact and Effects of Climate Change 337
David A. L. Jenkins

Epilogue 361

Index 363

Gerhard, L. C., W. E. Harrison, and B. M. Hanson,
Introduction and overview, 2001, *in* L. C. Gerhard,
W. E. Harrison, and B. M. Hanson, eds., Geological
perspectives of global climate change, p. 1–15.

Introduction and Overview

Lee C. Gerhard
William E. Harrison
Kansas Geological Survey
Lawrence, Kansas, U.S.A.

Bernold M. "Bruno" Hanson
Independent Petroleum Producer (Deceased)

Global climate has varied since the most primitive atmosphere developed on earth billions of years ago. This variation in climate has occurred on all timescales and has been continuous. The sedimentary rock record reflects numerous sea-level changes, atmospheric compositional changes, and temperature changes, all of which attest to climatic variation. Such evidence, as well as direct historical observations, clearly shows that temperature swings occur in both directions. Past climates have varied from those that create continental glaciers to those that yield global greenhouse conditions. Many people do not comprehend that this means their living climate also varies—it gets warmer or cooler—but typically does not remain the same for extended periods of time. Human history shows us that in general, warmer conditions have been beneficial, and colder conditions have been less kind to society (Lamb, 1995). We currently are living in a not-yet-completed interglacial stage, and it is very likely that warmer conditions lie ahead for humanity, with or without any human interference. Interglacial stages appear to last for about 11,000 years, but with large individual variability. We have been in this interglacial for about 10,000 years.

Geologic processes are not in equilibrium. Geologic processes are one set of drivers for climate and therefore climate cannot be in equilibrium. This makes assessing any cumulative human impact on climate difficult. What is the range of natural global climatic variability? What percentage of this may be due to human-induced activities? We may perceive the activities of humans to have the greatest impact on global climate simply because we are the advanced life form on earth. Perhaps it is a human trait to assume that the consequences of mankind on global climate must be significantly more important than the impact of "natural processes." Geology is the one scientific discipline that routinely works backward through significant periods of time to evaluate earth's natural processes. Our discipline brings both data and interpretation to the debate of the past history of climate, greenhouse gases, and global temperature behavior. Geology also brings to the debate a collection of climate drivers including tectonism (large and small scale), volcanism,

1

topography, glaciation, denudation, evolution of biota, the hydrologic cycle, carbon sequestration, and interactions between geologic and astronomic events.

The chapters of this book are arranged to take the reader through examples of geologic climate drivers ("Climate Drivers"), documentation of a number of methodologies for establishing ancient temperature and atmospheric gas concentrations ("Methods of Estimating Ancient Temperature"), followed by climate history ("Natural Variability and Studies of Past Temperature Change"), and then examination of some of the geological, engineering, and political effects of climate change ("Policy Drivers"). Many of the chapters are written for the educated lay public rather than technically trained scientists, in an effort to better communicate the geologic science of climate to those who make policy. The authors have summarized the papers of this volume in this introductory chapter. Detailed references for our summaries are in the specific chapters.

One of the most difficult concepts to communicate to the lay public is the scale of temperature changes through time compared with climate change drivers. We have tried to address this with an ordering of climate drivers and timescales **(Figure 1)**, which suggests there is a direct relationship

(continued on p. 4)

Figure 1 (opposite page) Climate drivers can be categorized by the dual attributes of range of temperature change forced by the driver and the length of time that the driver cycles through. In this graph, the vertical axis is logarithmic (units are powers of 10), and the horizontal axis is arithmetic. This interpretation permits comparison of climate drivers with their potential effects, and separation of drivers of different magnitude. Note that human intervention, which may or may not exist, is in the same category as some other small natural drivers. Diamond shapes are interpreted and literature-documented ranges of values; dots indicate possible ranges of values.

First-order climate controls: Earth has a life-supporting climate because of its distance from the sun, solar luminosity, and the evolution of a greenhouse atmosphere of water vapor, methane, CO_2, and other gases that trap solar energy and make it usable. This atmosphere evolved over the last 4.5 billion years and continues to evolve. For instance, Berner (1994) suggested that the carbon-dioxide content has decreased over the past 600 million years from 18 times the current concentration (see also Moore et al., 1997, p. 27). The greenhouse effect itself makes the earth 20°–40°C warmer than it would otherwise be (Pekarek, this volume, Chapter 1; Moore et al., 1997, p. 10, 12).

Second-order climate controls: Distribution of oceans and continents on the surface of the earth controls ocean currents, which distribute heat. This fundamental concept (Gerhard and Harrison, this volume, Chapter 2) explains the 15°–20° climate variations over hundreds of million of years (Lang et al., 1999; Frakes, 1979, p. 203). Such variations are exemplified by the two major earth cycles between glacial "icehouse" and warm "greenhouse" states. The late Precambrian "icehouse" evolved into the Devonian "greenhouse," then the Carboniferous "icehouse," then the Cretaceous "greenhouse," which evolved to the present "icehouse" state. Redistribution of heat around the earth is determined by the presence of equatorial currents that keep and thrust warm water masses away from the poles. Blockage of such currents, which permits the formation of gyres that move warm waters to the poles, creates the setting that allows continental-scale glaciation.

Third-order climate controls: Solar insolation variability has emerged as a major climate driver, as are the orbital variations that change the distance between the earth and the sun (Hoyt and Schatten, 1997; Frakes, 1979, p. 9; Pekarek, this volume, Chapter 1; Fischer, 1982; Berger et al., 1984). In addition, large-scale changes in ocean circulation through changes in current structure can be significant climate drivers (Broecker, 1997, 1999). Large-scale ocean tidal cycles may drive climate, including the large-scale maximum and minimum associated with the Medieval Climate Optimum and the Little Ice Age, on an 1800-year cycle with a 5000-year modulation (Keeling and Whorf, 2000). These drivers may cause temperature changes of 5°–15°C over hundreds to hundreds of thousands of years.

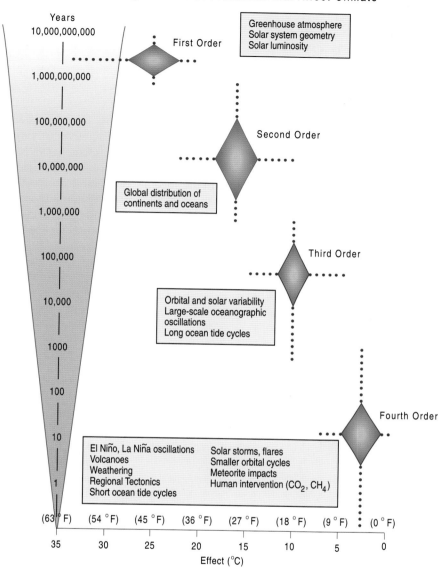

Relative Significance of Processes that Affect Climate

Fourth-order climate controls: There are many drivers that control small temperature changes (up to 5°C) over short periods of time (up to hundreds of years). Many are natural phenomena, including smaller-scale oceanographic oscillations (La Niña and El Niño), volcanic activity (such as the eruptions of Pinatubo and Krakatoa) (Moore et al., 1997, p. 47), solar storms and flares (Hoyt and Schatten, 1997, p. 168, 198, 199), small orbital changes (Frakes, 1979; NASA Web site), meteorite impacts, and human intervention (such as human-derived carbon dioxide [CO_2] and methane [CH_4] alterations to atmospheric composition). Tectonic and topographic uplift have small temperature effects and are regional rather than global, but may be of long duration (Ruddiman, 1997, p. 178, 502). Eighteen-, 90-, and 180-year cycles driven by ocean tides have been recognized by Keeling and Whorf (1997), and they drive climate by changing heat transfer rates between oceans and atmosphere.

The authors recognize that the error bars for these interpretations are broad, but the phenomena appear to fall into significant categories of effects and time.

between the range of absolute temperature change and the amount of time over which the drivers operate. This is a general statement, and simply an attempt to place in perspective the various climate drivers, the effect of the drivers on global climate, and the amount of time over which selected natural processes take place. A devolved opinion of the writers is that the less pronounced the effect, the less we understand its origin.

Several general statements are appropriate about natural systems. Climate can vary rapidly and over a range that can have a profound influence on human society. There are predictable geologic effects of climate change that occur in second through fourth order changes **(Figure 1)**, such as sea-level changes, rates of glacial movement, ecosystem migration, methane hydrate formation, and agricultural productivity. Finally, we are able to predict that the next ice age cannot take place unless additional global warming occurs. Such warming must be sufficiently large so that the ice of the Arctic Ocean is thawed, thus providing a moisture source for the immense snow and ice accumulations necessary to build continental-scale glaciers (Ewing and Donn, 1956).

CLIMATE DRIVERS

The sun is the primary source of energy for the climate of the earth. Earth's distance from the sun, a function of the geometry of the solar system, is the major factor controlling the base temperature of the earth. During earth's earliest history, solar irradiance was low, but internal radioactive decay likely produced enough heat to melt the earth's crust. Slowly, solar irradiance has grown, perhaps by as much as 25%. Eventually, in perhaps a billion years, the growing solar irradiance will have burned away the earth's atmosphere and made the planet uninhabitable (Hoyt and Schatten,1997; Pekarek, this volume, Chapter 1). According to current theory, based on solid geophysical measurements of density distribution in the earth, differentiation of the hot earth segregated a core, a mantle, a crust, and an atmosphere of light gases. These gases, which constitute the atmosphere that exists today, have varied somewhat in composition through geologic time, and have caused greenhouse conditions to develop. We know of no scientists who will argue that the greenhouse envelope of the earth's atmosphere, evolved over 4.5 billion years' time, is not the primary temperature control that permits life, as we know it, to exist. Without that envelope, it is likely that the earth's temperature would be 15°–30°C below its present level (Moore et al., 1996, p. 12; Pekarek, this volume, Chapter 1). About 80% to 95% of the total greenhouse gas budget is water vapor (including clouds) and the remainder consists of CO_2, methane, and other gases.

We consider these two effects, solar-system geometry and fundamental greenhouse conditions, to be first-order climate controls **(Figure 1)**.

Second-order climate control is by the distribution of continents and oceans upon the planet (Gerhard and Harrison, Chapter 2) **(Figure 1)**. The earth has undergone several cycles of icehouse and greenhouse climates, from at least the Vendian (late Precambrian) through the present. Glacial activity is reasonably interpreted at various locations as early as about 3 billion years ago (Crowell, 1999). Temperature variability between the colder and warmer climates is likely between 10° and 15°C (for instance, see Frakes, 1979, p. 170, figures 6–7). Gerhard and Harrison theorize that when continental landmasses are positioned so that equatorial oceanic circulation patterns exist, general global climate conditions are warmer. Conversely, when landmasses are positioned so as to impede or prevent equatorial circulation, "icehouse" conditions prevail. When warm waters are moved to polar regions, high rates of evaporation create continental glaciers and facilitate widespread global cooling. Conversely, strong and persistent equatorial currents preclude heat transfer to high latitudes, and warm conditions prevail. These relationships help to illustrate that thermal energy or heat is transferred around the earth much more effectively by oceanic circulation patterns than by atmospheric circulation.

Temperature variation under this scenario ranges up to 15°C, with major global impacts. Second-order control of global temperature is natural, driven by earth dynamics, and occurs over tens to hundreds of millions of years. For instance, the Cretaceous "greenhouse" condition changed to the modern "icehouse" condition over a period of 60 million years. There appears to be a relationship between the intensity of temperature variation and the length of time over which it occurs.

Third-order temperature drivers **(Figure 1)** include variability of solar energy caused by actual luminosity changes and by orbital "wobbles" of the planet earth (Pekarek, Chapter 1). Temperature changes caused by these wobbles are named "Milankovitch cycles" (the orbital wobbles were first popularized as climate drivers by Milankovitch; see Pekarek, Chapter 1, for references). Also among third-order drivers are temperature changes that occur by major reorganization of ocean currents (Broecker, this volume, Chapter 4).

Variations in solar energy reaching the earth's surface modify the climate. Several factors control the influx of solar energy, including variations in (1) the earth's albedo, (2) earth's orbit and rotation, and (3) solar energy output. Potential temperature changes driven by these variations can be as much as 10°C and the changes may take thousands of years (Pekarek, Chapter 1). Minor climate changes and those that mark the changes from glacial to interglacial may be the signature of these events (i.e., Frakes, 1979, figures 8–9).

In the short time since 1978, direct measurement of total solar irradiance (TSI) by satellites has shown cyclical variations in solar energy of 0.1% in conjunction with the 11-year sunspot cycle. Indirect evidence from the sun and other sunlike stars indicates that TSI has had significantly greater variation as the sun goes through its energy output cycles.

Correlations between climate and TSI variations are statistically solid. Small variations in TSI initiate indirect mechanisms on earth that yield climate changes greater than that predicted for the TSI change alone. At least three solar variables are known to affect earth's climate: (1) TSI, which directly affects temperatures; (2) solar ultraviolet radiation, which affects ozone production and upper atmospheric winds; and (3) the solar wind, which affects rainfall and cloud cover, at least partially through control of earth's electrical field. Each affects the earth's climate in different ways, producing indirect effects that amplify small changes in TSI. Individually, they do not cause the entire observed climatic changes. Collectively, they create changes. Because solar forcing of earth's climate is still an emerging science, some effects may not be fully understood. Many of these changes take place in a more or less regular cycle, as if the sun itself has periodic changes. Earth orbital changes are periodic. The most commonly observed solar cycle is the 11-year sunspot cycle. Statistically optimized simulations suggest that direct solar forcing can account for 71% of the observed temperature change at the earth's surface between 1880 and 1993, corresponding to a solar total irradiance change of 0.5%. The full effects of solar irradiance changes, including Milankovitch effects, must be interpreted from imprecise historical data, because direct measurements have been systematically available only since 1978.

The world ocean is vast, and owing to the specific heat of water, it contains vast amounts of thermal energy. Much of that energy is transferred around the earth through ocean currents. Some is transmitted to and from the atmosphere and controls local weather. Ocean currents are driven by wind, the Coriolis effect, and thermohaline circulation (which depends on density differences of waters entering the ocean). Fresh waters are lighter than saline waters, and warm waters are lighter than cold waters. The density of a water mass will determine whether it rises or sinks in the water column.

Broecker (this volume, Chapter 4) suggests that warm events after glacial events provide low-salinity, thus low-density, meltwaters into the ocean, which tend to float rather than sink in areas where sinking (downwelling) normally takes place. This action interferes with total normal oceanic circulation; for instance, the Gulf Stream may be diverted eastward by meltwaters from the North American arctic, depriving England and northern Europe of heat now moderating their climate. If massive events such as floods of fresh water backed up in proglacial lakes were to spill into the ocean, then very rapid climate changes could take place by changing the ocean circulation patterns. Broecker argues that at least two such events took place near the beginning of the present interglacial stage, with dramatic (up to 4°–5°C) rapid temperature swings in the Sargasso Sea.

Broecker postulates that this process would become effective around the time that atmospheric carbon-dioxide concentrations reach about 750 parts per million (ppm), or about twice ambient. He believes that such concentrations might be reached by 2100. If this is a valid cause and effect, then at some time in the future, assuming that the current interglacial stage continues to full melting of the Arctic Ocean, Broeker expects such dramatic climate changes to occur, regardless of whether human influence is shown to affect climate. If Broecker is correct, then the process to the next "flickering switch" climate change is toward the end of this century. If he is not correct in his assumption of human impact on climate, then the change could occur any time in the next thousand or more years. Fitzpatrick (1995) has discussed such changes without regard to thermohaline circulation. There are few other explanations for the dramatic temperature swings seen in the ice-core records.

A large number of smaller-scale geological climate drivers exist. Among these are volcanic eruptions, meteorite impacts, solar storms and flares, shorter solar cycles, small orbital changes, tectonism (mountain building and erosion), weathering of rocks, small ocean-circulation changes (i.e., La Niña and El Niño oscillations, North Atlantic Oscillation, etc.) and, although not geological, human interventions through increased greenhouse gas emissions. We consider these effects, which may alter climate up to perhaps 3° or 4°C, to be fourth-order climate drivers **(Figure 1)**. Some new concepts involving sudden release of massive amounts of methane through methane hydrate disassociation along continental edges have been advanced (Katz et al., 1999). Ruddiman (1997) has compiled an extensive collection of papers detailing tectonism. Well-known effects of the 1883 Krakatoa eruption that affected climate for two years or more and the recent Pinatubo eruption in the Philippines include the particulates blown into the upper atmosphere that caused cooling to take place.

The recent scientific literature is crowded with examples of short-term or moderate- to low-intensity, fourth-order climate-change demonstrations. The public media have documented El Niño and La Niña events. We will not further elaborate on details of fourth-order climate drivers except to present one discussion of the role of human activities' potential for climate effects (Mackenzie et al., Chapter 3).

Mackenzie et al. in Chapter 3 have modeled the cycles of carbon, nitrogen, phosphorus, and sulfur with the purpose of extracting human contribution to those cycles over the last 300 years. Their conclusions are that fossil-fuel emissions, land-use changes, agricultural fertilization of croplands, and organic sewage discharges to aquatic systems, all coupled with a slight temperature rise, have disturbed the carbon cycle. Land-use changes affect the uptake or release of carbon. Deforestation of Russia and the tropics may result in weakening of the terrestrial carbon sink in the future. Reduction of the thermohaline current transport of CO_2 into the deep world ocean could be a result of increased global temperature, and the increased temperature could then be further increased because of the reduction in sequestration of CO_2 in the deep ocean. In addition, such a change in the intensity of the thermohaline circulation could increase the ability of coastal marine waters to store atmospheric CO_2.

Mackenzie et al. have modeled the effects of the proposed Kyoto protocol on future carbon-dioxide (CO_2) concentrations. Despite a reduction in emissions of CO_2 in their model, concentrations of CO_2 would continue to build because of the continuous emissions of fossil-fuel and land-use CO_2 to the atmosphere, albeit at reduced rates for the former, complex interactions, and a sluggish response of the system to the reduction in emissions. This would be in part due to continued rise of CO_2 in the atmosphere owing to land-use changes and to the fact that CO_2 has an approximate 10-year residence time in the atmosphere. Mackenzie et al. suggest that the Kyoto approach might be useful in reducing the rate of increase of future atmospheric CO_2 concentrations, but only if the physical and biogeochemical mechanisms of redistribution of CO_2 among the atmosphere, land, and ocean do not significantly change from the present.

Other effects that may occur include reduced precipitation of carbonate minerals in biotic frameworks and increased transport of organic carbon, nitrogen, and phosphorus into coastal sediments. They suggest that human factors could impact mineral and organic carbon sedimentation, as already seen in lakes in industrialized countries.

METHODS OF ESTIMATING ANCIENT TEMPERATURE

The geologic record contains a large number of clues about former climate, with which we must compare any changes that we estimate might occur, either naturally or human induced. The methods are not infallible, but the vast majority of the proxies used to interpret past temperatures agree well enough to build a consensus about the more recent geologic past, and general acceptance of large-scale changes in the more distant past. The accuracy of interpretations of past climates declines as we go farther back into the past. Pleistocene climates are well documented, but those of 600 million years ago are less well documented, and those of billions of years ago are poorly documented. It is the nature of the geologic record that records become progressively destroyed by erosion and tectonic cycling as time passes. Parrish (1998) gives an excellent summary of paleothermometry methods.

One possible means to ensure that temperature proxies for past periods of time are robust and not subject to uncertainties related to erosion, tectonism, or sedimentation processes is to develop techniques based on materials that represent well-constrained and continuous time intervals. Reconstruction of paleoclimatic conditions based on the study of stable isotopes in ice core materials is one such proxy, and it has evolved into a well-recognized and powerful methodology. The review paper by Thompson (Chapter 5) illustrates how stable isotopic data are acquired and interpreted, and demonstrates how this technique has emerged as a climatic indicator. Ice-core materials from mid- to low-latitude glaciers represent a time span of approximately 25,000 years and provide valuable insight to regional variation in climate patterns.

Isotopic data are presented for several ice cores collected from three low-latitude but high-altitude locations of the Tibetian Plateau and from the Andes Mountains of South America. The results show that the Tibetian Plateau area experienced warm climates (like that of today) about 3000 years ago. In the Andes, such conditions existed about 5000 years ago.

Thompson suggests that tropical warming, as demonstrated by significant losses of ice mass from low-latitude glaciers, is enhanced by (1) evaporation of oceanic water and (b) the release of latent heat (due to condensation) at higher elevations. He also proposes that water vapor, the most important greenhouse gas on earth, is pumped into the atmosphere at low latitudes. Any increases in CO_2 levels may contribute to an increasing global inventory of water vapor and may strongly influence global changes.

The technique developed by Nagihara and Wang (Chapter 6) relies on data from boreholes drilled into the seafloor as a means to determine variation in bottom-water temperature (BWT) over the last few hundred years. The study is based on 18 temperature measurements collected from an ODP (Ocean Drilling Program) site in the Straits of Florida; the water depth at this location is 669 m. Temperature measurements were made at subseafloor depths between 26 and 348 m, and those from the seafloor down to about 100 m show a different temperature trend from those taken at greater depths. The authors use an inversion technique to detect temporal variation of BWT at the ODP site.

The variation of BWT is compared with Key West surface air temperatures collected between 1850 and the late 1990s and found to be in good agreement. The air-temperature data show higher values between about 1860 and 1885, a rapid cooling during the late 1880s, and a progressive increase for the first half of the twentieth century. The BWT reconstructed history at this location shows the same relationship. Nagihara and Wang use published data for locations off the coasts of Massachusetts and New Jersey to further demonstrate the methodology. The second example involves temperature data collected since 1925 at a water depth of 800 m. These measurements show systematic warming from the late 1920s through the 1970s, and the reconstructed BWT curve shows a similar history.

This technique requires that (1) borehole temperatures are free of drilling disturbances, (2) a sufficient number of measurements are made over small vertical depth intervals, and (3) effects of sedimentation, erosion, and pore fluid movement can be accurately determined. When such conditions can be satisfied, BWT-reconstructed history curves may become very powerful tools for assessing global climate change.

Coralline sponges or sclerosponges live in tropical marine waters but cannot compete with reef-building corals and primarily exist at depths to which light does not greatly penetrate (i.e., below the photic zone). These organisms precipitate calcium-carbonate ($CaCO_3$) growth rings from seawater to form their exoskeleton structure. Hughes and Thayer (Chapter 7) present evidence that Mg:Ca and Cl:Ca elemental ratios of the precipitated $CaCO_3$ material have potential application for evaluating past seawater temperatures and salinity conditions. In addition, these workers review the literature concerning precipitated $CaCO_3$ in fossil ostracodes, mussels, bivalves, etc., and paleotemperature conditions.

A sclerosponge began growing on a submerged marker plate that was attached to a submarine cave wall in July 1989, and it was collected for study nine years later. Chemical analysis of the sectioned specimen was made using energy dispersive spectroscopy (EDS), and elemental ratios were determined at discrete locations across individual growth bands. Temperature and salinity data are known from the specific location where the sponge was growing, and a correlation was developed between them and the measured elemental ratios. The preliminary result was the creation of a curve that provided submonthly temporal resolution and temperature variations on the order of tenths of a degree Centigrade.

This technique has potential as a means to evaluate marine paleotemperature and paleosalinity conditions. Because individual sclerosponges can live for centuries, it may be possible to reconstruct high-resolution seawater temperature profiles for the past 2000 years. Such a tool would be quite powerful in helping to understand and quantify past global climatic conditions.

In addition to techniques that may reveal paleoclimate information from marine settings, some lines of evidence are based exclusively on terrestrial organisms (Ashworth, Chapter 8). Insects, specifically fossil beetles, provide such evidence. The estimated 1 million to 7 million species of beetles account for approximately 20% of all species on earth. They are found in almost all habitats, ranging from high mountain settings to rain forests. Beetles occupied the interior part of Antarctica until the Neogene, when the growth of the polar ice sheets resulted in their extinction. The fossil record indicates that beetle diversity has not increased due to rapidly changing climate changes. Hundreds of species now in existence have been reported from Pleistocene and Holocene studies. This indicates that different species of beetles do not readily develop over extended periods of time and suggests a pattern of stasis or constancy.

A fossil beetle assemblage collected from northern Greenland demonstrates an unusual level of stasis through approximately 2 million years, even though the climate at that time was significantly warmer than that of today. The number of species remained relatively constant through several episodes of glaciation.

Beetles are highly mobile insects, and their most apparent response to variation in climate is migration from one location to another. This is well illustrated by the replacement (21,500 B.P.) of a forest fauna along the North American Laurentide ice sheet by a glacial fauna when temperatures were approximately 10°–12°C below current ones. When the ice sheet receded (approximately 12,500 B.P.), a forest fauna returned. Overall, beetles have survived global climatic changes due to their mobility, but reduction in areas suitable for habitat make future extinctions more likely.

The use of fossil biosensors to estimate paleoatmospheric levels of CO_2 is an area of active research, and Chapter 9 by Kurschner et al. is especially interesting due to early results as well as potential applications. The technique is based on the fact that all higher land plants rely on stomata to regulate the exchange of gas between the atmosphere and leaf tissue. Herbarium materials collected during the last 200 years and growth experiments conducted under preindustrial conditions all demonstrate an inverse relationship between stomata development and atmospheric levels of CO_2. Elevated levels of CO_2 produce fewer stomata than do decreased levels.

SI (Stomatal index) values for birch trees dated (by radiocarbon) at 10,070 years are in the 12 to 14 range and correspond to CO_2 levels of approximately 240 to 280 parts per million, volume (ppmv). Birch trees that are 9370 years old have SI values of 6 to 8 and indicate CO_2 concentrations of 330–360 ppmv. Thus, these findings indicate an 80–90 ppmv variation in naturally occurring CO_2 levels over a 700-year period. Additionally, SI data suggest a dramatic change of 65 ppmv CO_2 levels in less than a century.

Several lines of evidence indicate that early to middle Miocene time was one of the warmest periods of the entire Cenozoic. Kurschner et al. studied exceptionally well preserved fossil angiosperm materials of middle Eocene age and found that SI values were elevated. By extrapolation of SI data, the authors offer an estimate of 450–500 ppmv for the CO_2 level of the middle Eocene atmosphere. Based on these results and those obtained over the last few years, it appears that plant biosensor technology has good potential as a proxy for assessing paleoatmospheric CO_2 concentration levels.

NATURAL VARIABILITY AND RECORDS OF CHANGE

Interpretation of geological evidence for climate change results in development of curves that illustrate climate change through time and effects of climate change through time. This section offers four papers that document climate changes over the last few million years (Bluemle et al., Chapter 10), effects of climate change on shorelines and thus population in northern Europe (Harff et al., Chapter 12), sea-level changes that are a global response to climate variability (Shinn, Chapter 13), and, finally, some biogeochemical effects of climate variability (Yates and Robbins, Chapter 14).

In a previously published article, Bluemle et al. (1999; also this volume, Chapter 10) have recreated the Pleistocene climate variability in sufficient detail to draw conclusions about the impacts of fourth-order climate changes. Pleistocene glaciations are reflected in second- and third-order drivers (**Figure 1**), but have superimposed on them all of the variability of fourth-order drivers.

Bluemle et al. present several lessons. First, over the last 60 million years, in central Europe, temperature has dropped by more than 20°C (their **figure 1**). In relatively recent times, the Medieval Climate Optimum, or Medieval Warm Event (MWE) (A.D. ~1200–1350), was the time of building of castles, growing of vineyards in northern Europe, and the settlement of agricultural colonies of Vikings in Greenland. This warm period in human history was followed by the Little Ice Age (LIA), from the end of the MWE to about 1850, when temperatures plunged to isolate the Greenland colonies by sea ice and cause their demise by starvation. As Lamb (1995) pointed out, society prospered during the MWE, but suffered greatly from starvation, plague, and pestilence during the ensuing cold years.

The second lesson from the paper by Bluemle et al. involves the range of variation of temperature. The variability of temperature has been constant, especially for those processes considered in the third-order range. Temperature changes such as those discussed above have been regular, although not clearly cyclical, parts of the current interglacial stage. That temperature varies continuously and at all scales is apparent from any cursory or detailed examination of climate data (see also Davis and Bohling, Chapter 11).

Third, none of these temperature changes is human induced, suggesting that the fourth-order changes that might be induced by human activities may be transitory and of relatively low importance. However, there is a major point about climate to be made in context of the Pleistocene record. The graphs clearly show a range of variability, but it is important to note that cold periods are at least as frequent and perhaps longer lasting than warm periods. Bluemle et al. have documented many anthropological records that clearly show the effects of climate change on society through recorded history, as does Lamb (1995).

General circulation models (GCMs) are the basis for much discussion of the role of CO_2 in the atmosphere and as a climate driver. These models tend to examine relatively short time spans for their historical perspectives, and predict atmospheric and climate changes from that perspective. The advent of data from ice cores on Greenland has provided a time series of data that might be usable to extend the GCMs' historical perspective and to validate algorithms. Before using the data, a statistical validation of the data set can identify anomalies in the data or problems in data quality and continuity. Davis and Bohling (Chapter 11) have used 20-year averages to examine oxygen ^{18}O data from the Greenland Ice Sheet Project II (GISP2) "core." The results of their analysis demonstrate that the last 10,000 years exhibit a general cooling trend and that the current rate of increase in temperatures and warming trend are not unusual compared to the last 10,000 years. Further, past periods of consistently changing temperature have not persisted much longer than the current interval, so temperature trends may well reverse in the near future. The data exhibit distinct cyclic patterns, including a 560-year sequence of relatively abrupt change followed by a gradual reversal (it is possible that the present trend is the initial phase of such a pattern). Determination of the direction of global temperature change is a function of the time span used to make the determination (Davis and Bohling, their **figure 2**). On the 10,000-year interglacial scale, the earth is cooling from the high temperatures of the early interglacial. However, on a 16,000-year record, which initiates in the late Pleistocene glacial episode, the overall effect is global warming. Similarly, the last 2000 years show that the earth is cooling, over the last 600 years it is slightly warming, and over the last decade it is warming. The point of this comparison is to illustrate that picking the time constraint for a model determines the model's outcome, without regard to the complexity of the model.

These are important points. That the current rate of temperature increase is not unusual, despite the human-induced addition of CO_2, implies that it is not possible to detect a human imprint on earth temperatures. But the fact that the temperature has been declining slowly since the end of the Little Dryas, 10,000 years ago, implies that the agricultural base of human society may be threatened by continued cooling. Human population has dramatically increased since the beginning of the industrial revolution, corresponding to major advances in agricultural output and the advent of modern medicine. Technology has permitted population growth. If the climate becomes colder, will the consequent decline in agricultural productivity reduce global population, or will new technology derive more nutrition from existing crops? Davis and Bohling have made it clear that the historical record of climate is a serious backdrop to future agricultural policy.

Geological interpretations are four-dimensional, that is, they consider time as well as space. Many lay people wrongly assume that sea-level effects are unidirectional, whereas they are actually relative (that is, land elevation changes as well as actual sea level). Harff et al. in Chapter 12 demonstrate that the rise and fall of sea level in the Baltic Sea region is a function of climate-driven eustatic sea-level changes coupled with tectonism and glacio-isostatic rebound. For coasts in the north (Baltic shield), glacio-isostatic rebound is clearly dominant, with rates of uplift up to +9 mm/year. In the south, there is a small amount of subsidence added to a climate-driven eustatic sea-level rise that totals a relative sea-level rise of 2.5 mm/year. In comparison, Shinn (Chapter 13) demonstrates that the rate of sea-level rise in Florida is on the order of 10–20 cm per 100 years (1–2 mm per year).

This means the area south of the Baltic Sea, a densely populated area on a low-lying coastal plain, is faced with climatically controlled retreating shorelines. Harff et al. lay the groundwork for development of strategies for the protection of subsiding coastal areas. The pattern and high rate of sea-level change in the Baltic qualify the Baltic Sea to serve as a "model ocean" for studies that reveal the effects of natural geological processes, including climate change, on human populations along coastlines.

Parenthetically, measurement of sea level is very difficult because of the roughness of the water surface, continually varying tides, slow tectonic movement of the land surface, land subsidence by water withdrawal, and wind effects. Continual improvement in techniques for measurement of actual sea level are critical to predicting the effects of climate change on coastal regions.

One issue that receives popular attention and is of very personal interest to those who live on low-relief coral atolls and islands in the Pacific is the rate and magnitude of sea-level changes to be expected under various climate scenarios. The best way to evaluate such changes is to look at the record of the recent past, the Pleistocene. Shinn (Chapter 13) has summarized evidence to demonstrate that the last interglacial sea-level rise maximum was about 6 m above the present sea level, between about 135,000 and 115,000 years ago (Stage 5e, Shinn's, **figure 1**). This gives a reasonable maximum that can be expected for the current interglacial, that is, based on the assumption that much polar glacial ice will have melted by the end of the interglacial, a 6-m sea-level rise can be predicted. Shinn uses elevated coral reefs and wave-cut notches to correlate the highstand. Much more difficult, according to Shinn, is correlation of lowstands and identification of the total range of sea-level change between glacial (lowstand) and interglacial (highstand) events.

Using new diving technologies and accessing resource exploration seismic and drilling data as well as research drilling data, Shinn has documented lowstand beaches and coral reefs whose elevations are approximately 80 m below present sea level. However, his **figure 1** indicates that the last lowstand was approximately 130 m below standard. Thus a minimum range for sea-level change through a full glacial cycle is 86 m in tropical areas where glacial rebound is not significant.

Of interest is that the sea-level record (Shinn's **figure 1**) approximates the "sawtooth" effect seen in the temperature records, that is, sharp warming episodes that gradually cool. The causes of this sawtooth effect are still being debated, but for glacial and interglacial episodes it may be the result of polar ocean freezing and cutting off moisture to feed continental glaciers, with rapid warming and sea-level rise as a consequence, or thermohaline circulation changes as postulated by Broecker, Shinn, and others (Broecker, Chapter 4; Shinn, Chapter 13).

POLICY IMPLICATIONS

Whether human-induced CO_2 is a significant factor in climate change or not, many businesses are preparing for political intervention in industrial processes to reduce CO_2 emissions. Scientists must recognize that scientific conclusions or data may not be a driver for political action; rather, social and economic issues affecting the immediate well-being of voting citizens supersede more academic approaches. In many cases, scientists have found it difficult to communicate sophisticated and complex conclusions to the lay public and government. Frequently, immediate needs of government supersede what scientists may argue is prudent action, and public perception of scientific issues may deviate from the science. Because of all those factors, it is sometimes prudent to look at technical solutions to issues, regardless of their scientific merit. If political action takes place to reduce anthropogenic emission of CO_2, it is likely best to lead in introducing sound methods that are the most effective and the least expensive. Therefore, we present several papers that address technology of carbon sequestration and hopefully will help the political process appreciate the costs and effects of considered actions.

Sequestration is the common term used to describe methods for placing CO_2 where it cannot affect atmospheric concentrations. This book contains papers that describe two possible methods. The first, an original piece of scientific research into a generally difficult problem of the origin of lime mud in the ocean (Yates and Robbins, Chapter 14), documents that microbial activity in the ocean may be responsible for generation of lime mud. The chemistry of organic catalysis takes up CO_2 and thus provides a possible method to sequester CO_2 through sedimentation in the shallow ocean. The details of the original research are somewhat technical and may be difficult for the lay scientist to follow, but their summary provides the reader with an overview of the potential for a new geological sequestration process. Biologic catalysis can remove CO_2 from water while producing lime mud, either in sheaths around the microbe or on (or near) the surfaces of cells without sheaths. Both processes result in direct precipitation owing to the microchemical conditions surrounding the microbe. This process sheds light on the controversial question of production of large amounts of lime mud that cannot be accounted for by controlled biologic precipitation or by skeletal abrasion and disintegration.

The second paper (Bachu, Chapter 15) addresses use of subsurface (underground) sedimentary rocks as a long-term (or "permanent") host for anthropogenic CO_2 as one of several possible means to dispose of unwanted CO_2. The physical state of CO_2 changes phase, depending on temperature and pressure conditions. This permits the gas to be a liquid or hydrate solid in the deep ocean, but the technologies to handle large volumes of CO_2 in this manner are not yet developed, nor are the long-term effects of ocean disposal well understood. On the other hand, the petroleum industry is skilled at subsurface geological exploration and development, and can access potential geological disposal sites with economical technology used in the production of oil and gas. Defining the best potential conditions for geologic disposal of CO_2 means extensive geological analysis of buried rocks and careful interpretation of the geological history of confining structures. Bachu argues that CO_2 can be sequestered in geologic traps as a gas, a liquid, or in a supercritical state, similar to the natural trapping of hydrocarbons (oil and gas), if the proper conditions exist. Geologic trapping, hydrodynamic trapping, solubility trapping, mineral trapping, and cavity trapping are all geologic possibilities. Trapping in geologic media, whatever the method, may be the easiest and cheapest method to permanently dispose of anthropogenic CO_2. Bachu treats the subject in detail, identifying the criteria that should be used to find and effectively use geologic media for sequestration. Interestingly, some of the CO_2 can be used to produce more petroleum because of the miscibility of CO_2 with oils and coalbed methane, perhaps forming a closed loop of energy production that sequesters as much CO_2 as is emitted. A current research project of the Kansas Geological Survey and the U.S. Department of Energy is addressing this opportunity.

Using statistical methods, Kotov (Chapter 16) has examined the Greenland ice-core record for patterns that can be projected into the future. Using the geological past to understand the present is part of every geologist's training, as is the reverse. Statistical analysis of past climate variability from oxygen isotope and other ice-core data suggests that there are patterns within the apparently chaotic data. Testing these patterns suggests that there is some regularity to climate patterns, and it is this statistically identifiable regularity that forms the basis for future projection of natural variability.

Study of the past permits Kotov to predict the future. The present climate is not significantly different from much of the past, and projected future temperature variations fall comfortably within the range of the variance of the past. Kotov theorizes that at least 200 years of continued natural warming will be likely, followed by a period of natural cooling.

Corroboration of this prediction appears in a very recent related paper (Keeling and Whorf, 2000) about 1800-year tidal cycles that may be responsible for the twelfth-century climate maximum and fifteenth-century climate minimum (Medieval Climate Optimum and Little Ice Age, respectively). Based on their analysis of large-scale tidal cycles, these workers predict that continued natural warming is likely to "continue in spurts for several hundreds of years," before the next cooling episode starts. It is important to note that Keeling and Whorf expect that, "Even without further warming (from greenhouse gases), . . . this natural warming at its greatest intensity would be expected to exceed any that has occurred since the first millennium of the Christian era . . . independent of any anthropogenic influences."

It is usually significant when two completely different methodologies arrive at a similar conclusion independently of each other, and thus we offer these observations to assist readers in making up their own minds on this matter.

The paper by Idso (Chapter 17) is one of two reprinted in this volume by permission of the original publishers. It represents an unusual set of observations, measurements, and interpretations carried out while the author was studying meteorological processes over a two-decade period, wherein he and his colleagues quantified the climatic consequences of several naturally occurring atmospheric phenomena: variations in atmospheric dust content, cyclical patterns of solar radiation receipt, the greenhouse effect of water vapor over the desert Southwest of the United States, etc. This paper, originally published in *Climate Research* in 1998, represents one of the major studies to question the relationship between increasing CO_2 levels in the atmosphere and the concurrent increase in mean annual global surface air temperature, primarily on a meteorological basis.

Based on measurements resulting from eight types of "natural experiments," Idso deduced that raising the air's CO_2 concentration from 300 to 600 ppm should result in an increase in mean surface air temperature of no more than 0.4°C. This estimate is only about one-tenth to one-third of the temperature increase typically projected by numerical simulation results obtained from general circulation models used by climate modelers. After examining various data sets and finding little evidence that elevated CO_2 levels have affected the earth's surface temperature over the last hundred years, Idso offers the possibility that "the global warming of the past century may have been nothing more than a random climatic fluctuation."

Idso also provides a summary of relative temperature changes over the last millennium. He points out that there were two episodes, each of several hundred years' duration, during which temperatures may have been somewhat higher (the Little Climatic Optimum, or, Medieval Warm Event) and lower (the Little Ice Age) than they are today, and that CO_2 levels, deduced from ice cores, showed no changes over those periods.

Of particular interest in this paper are Idso's descriptions of mechanisms that might enhance the overall cooling properties of the earth if global temperatures were to increase slightly. These negative feedbacks include the likelihood that a 10% increase in earth's low cloud cover would completely cancel the warming predicted to result from a doubling of the air's CO_2 content, plus the fact that a warmer and CO_2-enriched world would produce clouds with an increased liquid water content and increased levels of cloud condensation nuclei (which allow clouds to last longer and cool the earth longer). And in what may be considered a bit of natural irony, some of these meteorological phenomena can be triggered by elevated levels of CO_2 alone, without an accompanying temperature increase. Finally, Idso indicates his skepticism about the ability of general circulation models of the atmosphere to correctly predict how opposing climatic forces will respond to increased levels of atmospheric CO_2, and he expresses serious reservations about the use of such models to develop national and international energy policies related to potential climate change.

The final paper in this volume is a philosophic view of the issue of climate change and of the effects of humankind on the earth in general. Jenkins (Chapter 18) addresses the myths of stability of natural systems and the reality of human technological prowess to adapt and prosper in naturally dynamic and unpredictable systems. While developing his thesis of natural variability and yet understanding the public perception of an unchanging world, Jenkins demonstrates the fallibility of assuming stasis. Particularly poignant are his arguments that although warming may be a bother, significant cooling, geologically predictable, could be disastrous for feeding the vastly increasing numbers of people on this planet. Warming may well increase the problems of sea-level rise and provision of fresh water. The natural rates of change may well be accelerated by human influence but likely will not move in different directions. Humans may easily adapt to changes in their environment by using technology and the inherent flexibility of intelligent beings. Once people accept that the world is not dynamically stable, on any timescale, then comprehensive adaptations can begin.

In conclusion, geologists know the earth is a single dynamic system, billions of years old, that is not in equilibrium. A flat, featureless, and uninteresting earth would be the result of equilibrium. Because change is constant, inevitable, and interesting, humankind must embrace change rather than fear it. Adaptation to the changes that continually occur on our planet requires flexibility, planning, and acceptance of the earth-system constraints. Political processes cannot change earth dynamics. Changes that do take place must be placed in context of their real effects. And finally, a major and recurring theme of this volume bears repeating once more. Climate drivers are variable in both time and intensity **(Figure 1)** and—regardless of the largely political belief that human consequences on global climate are pronounced—human influences are of comparatively low intensity and take place over short time spans. The nonequilibrium systems that control natural phenomena on earth very likely dwarf man's ability to affect climatic conditions on a global scale.

REFERENCES CITED

Berner, R., 1994, 3Geocarb IIA Revised model of atmospheric CO_2 over Phanerozoic time: American Journal of Science, v. 291, p. 56–91.

Berger, A. L., J. Imbrie, J. D. Hays, G. J. Kukla, and B. Salzman, eds., 1984, Milankovitch and Climate: Dordrecht, Netherlands, D. Reidal, 895 p.

Broecker, Wallace S., 1997, Thermohaline circulation, the Achilles heel of our climate system: Will man-made CO_2 upset the current balance?: Science, v. 278, p. 1582–1588.

Broeker, Wallace S., 1999, What If the Conveyor Were to Shut Down? Reflections on a Possible Outcome of the Great Global Experiment: GSA Today, v. 9, No. 1, p. 1–4.

Crowell, John C., 1999, Pre-Mesozoic ice ages: Their bearing on understanding the climate system: Geological Society of America, Memoir 192, 106 p.

Ewing, M., and W. L. Donn, 1956, A theory of ice ages: Science, v. 123, n. 3207, 15 June 1956, p. 1061–1066.

Fischer, A. G., 1982, Long-term climate oscillations recorded in stratigraphy: *in* Climate in Earth History: National Research Council, National Academy Press, p. 97–104.

Fitzpatrick, J., 1995, The Paradigm of rapid climate change; a Current Controversy: *in* L. M. H. Carter, ed., Energy and the Environment: Application of Geosciences to Decision-making: U.S. Geological Survey Circular 1108, p. 20–22.

Frakes, L. A., 1979, Climates through geologic time: Elsevier, Amsterdam, 310 p.

Hoyt, D. V., and K. H. Schatten, 1997, The Role of the Sun in Climate Change: Oxford University Press, New York, 279 p.

Katz, M. E, D. K. Pak, G. R. Dickens, and K. G. Miller, 1999, The Source and Fate of Massive Carbon Input During the Latest Paleocene Thermal Maximum: Science, v. 286, p. 1531–1533.

Keeling, C. D., and T. P. Whorf, 1997, Possible forcing of global temperature by the oceanic tides: Proceedings of the National Academy of Science, v. 94, p. 8321.

Keeling, Charles D., and T. P. Whorf, 2000, The 1,800 year oceanic tidal cycle: A possible cause of rapid climate change: Proceedings of the National Academy of Science, v. 97, p.3814–3819, 8328.

Lamb, H. H., 1995, Climate, History, and the Modern World: 2nd Ed., Routledge, NY, 433 p.

Lang, C., M. Leuenberger, J. Schwander, and S. Johnson, 1999, 16°C rapid temperature variation in central Greenland 70,000 years ago: Science, v. 286, p. 934–937.

Moore, Peter D., Bill Chaloner, and Philip Stott, 1996, Global environmental change: Blackwell Science, Oxford, England, 244 p.

NASA Web site: http://www.nasa.gov/today/index.htm.

Parrish, Judith Totman, ed., 1998, Interpreting Pre-Quaternary Climate from the Geologic Record, Columbia University Press, New York, 338 p.

Ruddiman, William F., ed., 1997, Tectonic uplift and climate change: Plenum Press, New York, New York, 535 p.

Scotese, Christopher R., 1997, Paleogeographic Atlas: PALEOMAP Progress Report 90-0497, PALEOMAP Project, Univ. Texas–Arlington, 21 p.

Part I | Climate Drivers

The climate around us changes constantly in response to major and minor physical controls, or drivers. There are many climate drivers other than greenhouse gases. Factors such as the variation in solar energy received by earth, major shifts in oceanic circulation patterns, and the distribution of land and water masses all exert major control on climate.

Pekarek, A. H., Solar forcing of earth's climate, 2001, *in* L. C. Gerhard, W. E. Harrison, and B. M. Hanson, eds., Geological perspectives of global climate change, p. 19–34.

1 | Solar Forcing of Earth's Climate

Alfred H. Pekarek
Department of Earth Sciences
Saint Cloud State University
Pierz, Minnesota, U.S.A.

ABSTRACT

The sun is the primary source of energy for the climate of the earth. Variations in solar energy reaching the earth's surface change the climate. Several factors control the influx of solar energy, including (1) variations in the earth's albedo, (2) variations in earth's orbit and rotation, and (3) variations in solar energy output.

In the short time since 1978, direct measurement of total solar irradiance (TSI) by satellites has shown cyclical variations in solar energy of 0.1% in conjunction with the 11-year sunspot cycle. Indirect evidence from the sun and other sunlike stars indicate that TSI has had significantly greater variation as the the sun goes through various cycles.

The correlations between climate and TSI variations are statistically solid. Small variations in TSI initiate indirect mechanisms on earth that yield climate changes greater than that predicted for the TSI change alone. At least three solar variables are known to affect earth's climate: (1) TSI, which directly affects temperatures; (2) solar unltraviolet radiation, which affects ozone production and upper atmospheric winds; and (3) solar wind, which affects rainfall and cloud cover, at least partially, through control of the earth's electrical field. Each affects the earth's climate in different ways, producing indirect effects that amplify small changes in TSI. Individually, they do not cause the entire observed climate changes. Collectively, they appear to be sufficient, especially because solar forcing of earth's climate is still an emerging science. Undoubtedly, other mechanisms of solar forcing are poorly understood, perhaps even unknown.

INTRODUCTION

The sun is the primary source of energy for the weather and climate of the earth. "The Earth's climate is a manifestation of how the radiation from the sun is absorbed, redistributed by the atmosphere and the oceans, and eventually re-radiated into space" (Svensmark and Friis-Christensen,

19

1997). Approximately 25% of the solar energy reaching the earth is reflected by the atmosphere (primarily clouds and water vapor), 25% is absorbed by the atmosphere (primarily clouds and water vapor), 5% is reflected by the earth's surface, and 45% is absorbed by the earth (Schneider, 1989, p. 15; Parsons, 1995, p. 189–192). The surface heat of the earth is transferred to the atmosphere by a variety of mechanisms including evaporation of water, rising hot-air thermals, and infrared radiation. Most of this energy is absorbed by clouds, water vapor, and other minor greenhouse gases. The atmosphere radiates some of the heat energy into space and returns some to the earth via the greenhouse effect, a natural phenomenon that increases the earth's surface temperature by approximately 15° C, making life on earth possible (Schneider, 1989, p. 13–17).

The energy budget of the earth must be in dynamic equilibrium with the influx matched by the outflow or a change in climate will take place. However, the climate system is buffered in many ways to maintain equilibrium with changing conditions. For example, increased energy input will raise the surface and atmospheric temperatures, resulting in additional infrared radiation to space. Conversely, decreased energy input reduces the surface and atmospheric temperatures, reducing the amount of infrared radiation. Surface temperatures rise and fall, respectively, to new equilibrium points.

Anything that changes the amount of solar energy reaching the earth's surface will affect the climate. Changes in the amount of solar energy received can result from several factors, including: (1) variations in the earth's albedo (reflectivity), (2) variations in the earth's orbit and rotation, and (3) variations in solar energy output. With constant solar energy, changes in the earth's albedo can alter the amount of solar energy available to the climate system of the earth. For instance, a higher albedo due to increased cloud cover or snow cover reduces the amount of energy received or retained, respectively, by the surface of the earth.

The effects of orbital and rotational changes on the earth's climate have been known for decades (Croll, 1867, 1875; Milankovitch, 1920, 1930, 1936, 1938, 1941, 1957). Because Milankovitch's work is published in German, French, and Serbo-Croatian, the reader is referred to Imbrie and Imbrie (1979, p. 77–122) for a lengthy discussion of his work and that of Croll. Basically, these authors show that cyclical variations in the earth's orbit, axial precession, and axial tilt provide a convincing causative mechanism for the glacial and interglacial climates of the Pleistocene. Hodell et al. (1991) show that climatic changes in the Caribbean and Africa over the past 10,500 years "can be largely explained by orbitally induced (Milankovitch) variations in seasonal insolation which modified the intensity of the annual cycle." However, they report that certain rapid climate shifts cannot be explained by the slow, orbitally induced changes in insolation. The unknown mechanism may be related to periodic changes in solar energy. However, a rapid warming would also follow the melting of large ice sheets (heat sinks).

Variations in solar energy output and its effects on earth's climate are the focus of this paper. Prior to discussing the main topic, it seems advisable to briefly review the inner workings of the sun to determine how the sun, the primary source of energy for earth's climate, generates and transmits that energy.

THE STANDARD SOLAR MODEL

The sun is approximately 4.65 billion years old. Its energy output has gradually increased over its lifetime and is now approximately 40% greater than at its birth (Gribbin, 1991). The energy output is expected to increase an additional 15% in the next 1.5 billion years. Lang (1995) provides a comprehensive discussion of the sun.

The interior of the sun is divided into four zones: the central core, the radiation zone, the convection zone, and the photosphere (the visible surface). Bahcall and Pinsonneault (1992) computed what is regarded as the standard solar model. According to their model, the core extends from the center to about one-quarter (25%) of the solar radius. In the center of the core, the density is 151 grams per cubic centimeter and the temperature is 15.6 million degrees Kelvin. At such high temperatures, atoms lose their electrons and become a plasma. Although containing only slightly more than 1.5% of the solar volume, the core contains almost half of the solar mass. Solar energy results from the thermonuclear fusion of hydrogen into helium in the core. The fusion of hydrogen into helium releases a photon of energy (an X ray) that immediately reacts with a charged particle, forcing the particle into an excited state. The excited particle returns to normal by emitting another photon in a random direction. The photon interaction is repeated billions of times before the energy contained in the original photon exits the core and reaches the surface of the overlying radiation zone, a journey that requires millions of years to complete (Hathaway, 1995; Parsons, 1995, p. 180).

The radiation zone extends outward from the core to 71% of the solar radius, where it borders the convection zone. Temperature decreases outward from 10 million degrees Kelvin at the base to 2 million degrees Kelvin at the outer edge. "Throughout this layer, the gas is hot enough to keep the nuclei of hydrogen, helium, and most of the heavier elements completely stripped of electrons" (Hathaway, 1995, p. 41). Average density in the radiation zone is about 10 grams per cubic centimeter (Parsons, 1995, p. 180).

At the base of the convection zone, the temperature is low enough to allow some heavy nuclei to capture electrons, making it more difficult for photons to pass through the gas because atoms absorb and scatter light more efficiently than bare nuclei and electrons do (Hathaway, 1995). The increased energy-absorbing ability of atoms traps some of the radiation as heat. The increased opacity and heat capacity make gas convection a more efficient way of carrying solar heat upward to the photosphere. Temperature decreases from 2 million degrees Kelvin at its base to 5780 degrees Kelvin at the top. With an average density of 0.005 grams per cubic centimeter (Parsons, 1995, p. 180), the convective zone comprises about 2% of the solar mass (Lang, 1995, p. 21).

The very thin photosphere (0.07% of the solar radius) is the visible surface of the sun from which energy is radiated as a black body at a temperature of 5780° K. Sunspots and other indications of solar activity visible from the earth occur in the photosphere. At the visible surface of the sun, density is about 10,000 times less than the air we breathe on earth (Lang, 1995, p. 95).

The sun is enclosed in an atmosphere, the chromosphere, in which gaseous emissions (solar flares and prominences) from the solar surface occur. Much of the radiation from the chromosphere is absorbed by the earth's atmosphere and can only be measured by satellite above the earth's atmosphere. Consequently, precise measurements of total solar energy output could not be made before the advent of satellites capable of measuring the total solar spectrum (the solar constant) from above earth's atmosphere.

The corona, or outermost part of the chromosphere, consists of a plasma at a few million degrees Kelvin (Lang, 1995). The corona expands into space as the solar wind of charged particles, primarily protons and electrons. The earth travels through the solar wind and is affected in various ways by changes in the solar wind. Some of the effects of the solar wind on earth's climate are discussed below.

SOLAR MAGNETISM

Gas motion in the convection zone controls the sun's magnetic fields (Hathaway, 1995). Nesme-Ribes et al. (1996), from which the following discussion is taken, discuss sunspots and the sunspot cycle, caused by systematic changes in the solar magnetic field. Movement of the charged particles of the plasma in the convection zone generates magnetic fields. Plasma conducts electricity very efficiently and tends to trap the magnetic field lines. The rotation of the sun tends to organize the random convection cells into an overall magnetic field. However, the nonuniform rotation of the sun has a period of roughly 25 days at the equator, 28 days at 45 degrees latitude, and progressively longer at higher latitudes. As a consequence, the north-south magnetic fields are progressively deformed in an east-west direction by the faster rotation at the equator. Eventually, the magnetic field floats to the surface and erupts as a pair of sunspots.

Sunspots, dark areas that periodically occur on the face of the sun, appear black because they are 2000° C cooler than the surrounding area (Lang, 1995). They are actually an orange-red versus the normal yellow of the sun. Sunspots form when strong magnetic fields suppress the flow of the surrounding gases, preventing them from carrying internal heat to the surface (Nesme-Ribes et al., 1996). The reduced radiation from the cool dark spots is more than offset by larger associated bright areas of strong magnetism called plages (Baliunas and Soon, 1996) and other bright areas called faculae (Lean et al., 1995). Sunspot activity is cyclical, varying from a minimum to a maximum number of spots with a period averaging 11 years and a range of at least 8 to 15 years (Baliunas and Soon, 1996). Solar magnetic polarity reverses with each cycle.

Variations in the first appearance and in the number of sunspots between the northern and southern hemispheres indicate that the solar magnetic field has a weak quadrupole component (Nesme-Ribes et al., 1996). Periodically, the strength of the quadrupole component increases, equaling that of the dipole field, and cancels the normal 11-year sunspot cycle, producing a Maunder minimum, a time of few sunspots, generally all in either the Northern or Southern Hemisphere. During the last Maunder minimum (1645–1715), very few sunspots occurred and all were in the Southern Hemisphere, moving much slower than normal (Eddy, 1976, 1977a, 1977b). This exceptionally cold period in European history is known as the Little Ice Age.

VARIATIONS IN SOLAR RADIATION

The sun radiates energy across the electromagnetic spectrum from very short wavelength gamma rays through X-ray, ultraviolet, visible light, infrared, and long wavelength radio waves. With the exception of visible light and radio waves, earth's atmosphere effectively prevents most of the remaining radiant energy from reaching earth's surface. As a result, total solar radiation (irradiance) output must be measured by satellite above the atmosphere. Fortunately, total solar irradiance (TSI) has been measured by satellite instruments with sufficient precision to detect intrinsic solar variability since 1978 (Willson, 1997). In addition, satellites in the last few decades have allowed us to examine solar radiation with precision in every spectral region.

In the short time since 1978, direct measurement of TSI by satellites has shown cyclical variations in solar energy of 0.1% in conjunction with the 11-year sunspot cycle (Willson, 1997) **(Figure 1)**. TSI, fairly steady during the minimum part of the cycle, can vary 0.2–0.3% over a period of several days during the active part of the cycle. The more rapid fluctuations with timescales of a few days are related to sunspots moving with the rotation of the sun, while variations over a few months result from faculae (Willson and Hudson, 1991). Faculae, bright regions of the photosphere seen in white light and visible only near the limb or edge of the solar disc, are brighter than the surrounding medium due to their higher temperature and greater density (Lang, 1995, p. 255).

Lacking direct observation of TSI before 1978, other methods are used to extend the useful record backward in time. For example, the number of sunspots, recorded more or less systematically since about 1610, has been shown to reflect TSI (Baliunas and Saar, 1992). These data show that sunspot cycles vary in length (7.3 to 17.1 years) and amplitude (number of spots). At solar maximum, the number of spots varies from a low daily average of 49 spots to a high of over 200 (Baliunas and Saar, 1992). Soon et al. (1996) use several proxies to develop profiles of solar

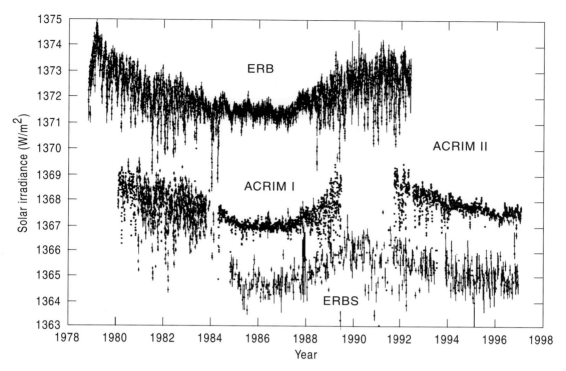

Figure 1 Satellite records of TSI during solar cycles 21 and 22. Acronyms are: ACRIM (Active Cavity Radiometer Irradiance Monitor), ERB (Nimbus 7 Earth Radiation Budget), and ERBS (Earth Radiation Budget Satellite) (after Willson, 1997).

irradiance prior to 1978: (1) the length of the sunspot cycle (Friis-Christensen and Lassen, 1991), (2) the mean sunspot number, and (3) a composite proxy that includes the two previous indicators plus the equatorial solar rotation rate, the fraction of penumbral spot coverage, and the rate of decay of the sunspot cycle. These profiles and variations in greenhouse gases were used to simulate global surface temperature changes during 1880–1993. Optimized simulations imply that solar forcing alone would account for 71% of the observed global mean temperature variance and suggests a solar total irradiance variation of 0.5% between 1880 and 1993 (**Figure 2**).

Baliunas and Soon (1996) describe another proxy, solar magnetic variability, which correlates closely with changes in TSI. A long-term record of the solar magnetic activity is deduced from the radioisotopes ^{14}C and ^{10}Be, which form as byproducts of energetic cosmic rays hitting the earth's upper atmosphere. The solar magnetic field, surrounding the earth as part of the solar wind, deflects some of these cosmic rays. The formation of ^{10}Be and ^{14}C is reduced by the stronger solar magnetic field in times of high TSI. The variable amounts of these two isotopes incorporated in tree rings and ice cores can be measured, yielding the history of solar magnetic variability, a proxy for TSI. That isotope record clearly shows by proxy a record of variations in TSI that correlates with the solar Maunder minimum of the Little Ice Age (low TSI) and the earlier Medieval Warm Period (high TSI) (Baliunas and Soon, 1995). Several similar periods of high and low TSI are recorded by these two radioisotopes during the last few thousand years (Baliunas and Soon, 1996; Beer et al., 1988).

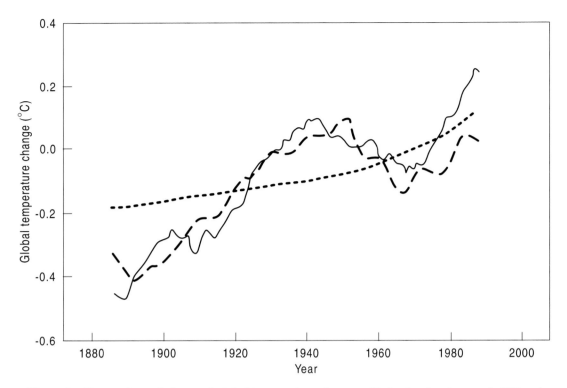

Figure 2 Comparison of observed global temperature change with forcing by changes in TSI and atmospheric carbon dioxide. Solid line is the 11-year running mean of the observed global surface temperature anomalies. Dotted line is carbon-dioxide forcing. Dot-dashed line is solar-radiative forcing (after Soon et al., 1996).

Stuiver et al. (1995) report that the oxygen isotope ratio in the Greenland Ice Sheet Project Two (GISP2) ice core correlates strongly with the sunspot cycle for at least the last two centuries. The amplitude of the oxygen isotope ratio, equal to a 2.6° C regional temperature change, is larger than expected as a direct result of total irradiance changes (Tinsley, 1997). Some physical mechanism is substantially amplifying the small changes in TSI. Part of the temperature change may be due to a poleward shift of about 3 degrees of latitude from solar maximum to solar minimum for storm tracks in the western north Atlantic (Tinsley, 1996).

Reid (1987), using total sunspot number as a proxy for TSI variations, shows a correlation between global sea surface temperatures (SST) and solar activity cycles for 120 years after 1860. Friis-Christensen and Lassen (1991), using solar cycle length as a proxy, report an even better correlation. Specifically, the global SST variation with a period of approximately 83 years has a good match with the 80–90-year solar Gleissberg cycle and a certain amount of similarity to the variable solar activity during the sunspot cycle. A simple ocean thermal model showed that the required solar change is less than 1%, an amount similar to the 0.6% observed in other sunlike stars during a single starspot cycle (Nesme-Ribes et al., 1996). These results suggest that the sun is probably capable of the variation required for the observed changes in SST.

Significantly, Willson (1997), using TSI data from four satellites, found an upward trend of 0.036% per decade in TSI between solar cycles 21 and 22 (1986 to 1996) **(Figure 3)**. However, Kerr

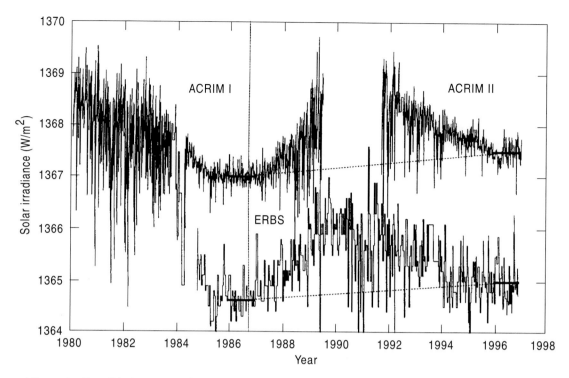

Figure 3 Possible increases observed in TSI data between the minima of solar cycles 21 and 22 (Willson, 1997).

(1997), providing additional analyses of the same data, suggests a much smaller increase. Interestingly, an increase in TSI should be expected as we continue to recover from the Little Ice Age on the way to the next Medieval Warm Period and as a result of the Gleissberg cycle. Additional TSI data from satellites during the next few solar cycles should be sufficiently precise to resolve the issue.

Ribes et al. (1987), who present evidence that the solar radius was larger by 1000 kilometers during the Maunder minimum, offer a possible mechanism for variations in solar activity. As the solar radius increases, TSI decreases, approximately 1% for the observed radius increase. This theory is supported by a long-term correlation of solar radius with the Gleissberg cycle and the 11-year sunspot cycle (Gilliland 1981).

Lean et al. (1995), using sunspot numbers to reconstruct TSI and ultraviolet irradiance back to 1610, calculate a total change in TSI since the Maunder minimum of 0.24%. The calculated record of TSI changes was used to investigate solar forcing of the Northern Hemisphere surface temperature changes since 1610. They found very close agreement through 1800 (preindustrial). Their modeling indicates that only half of the warming since 1860 is attributable to direct solar forcing and only a third since 1970. However, the calculated change in TSI of 0.24% by Lean et al. (1995) is only about half the 0.4% observed in sunlike stars (Nesme-Ribes et al., 1996). Further, Friis-Christensen and Lassen (1991) show that solar activity (sunspot) cycle length is a better proxy than sunspot number for calculating TSI. Finally, Lean et al. (1995) make no allowance for the effects of changes in solar radius (Ribes et al., 1987). Therefore, it seems probable that their estimated change in TSI of 0.24% since the Maunder minimum is too low by a factor of 2 or 3.

VARIABILITY IN SUNLIKE STARS

Our short direct measurement of variations in TSI and the longer data set obtained from proxy measurement can be augmented by data from sunlike stars with mass, age, and magnetic activity similar to the sun (Baliunas and Saar, 1992; Baliunas and Soon, 1995, 1996; Nesme-Ribes et al., 1996). Direct irradiance measurement of distant stars is difficult. Therefore, a proxy must be used. The solar magnetic field heats the outer atmospheric layers of the sun, causing it to radiate in certain spectral lines (Nesme-Ribes et al., 1996). For example, the intensity of the two violet emission lines of singly ionized calcium (K and H lines with wavelengths of 3968 and 3934 angstroms, respectively) closely follows the strength and extent of the magnetic field. Variations in these spectral lines can be used to study the magnetic activity in sunlike stars that are too far away for us to directly observe "starspots" (Baliunas and Soon, 1995, 1996).

Since 1966, the magnetic activity of approximately 100 stars similar to the sun has been monitored (Baliunas and Saar, 1992; Baliunas and Soon, 1995, 1996; Nesme-Ribes et al., 1996). These data show that the magnetic fields of young stars fluctuate over periods as short as two years, generally without a well-defined cycle. The cycle of magnetic activity gradually stabilizes and lengthens with age. Data from 20 to 30 stars similar in age and mass to the sun show that approximately 25% are in a magnetic calm, indicating a Maunder minimum phase. One star, HD3651, appears to be in transition between cyclical and Maunder minimum phases. These data suggest that sunlike stars may spend a significant part of their lives in a Maunder minimum phase.

For all the stars in cyclical mode, irradiance is strongest near the peak of the magnetic activity cycle. Total irradiance change over a cycle is proportional to the intensity of the star's activity (Baliunas and Soon, 1995, 1996). Although some stars vary as little as the sun—only 0.1% in the last solar cycle—other sunlike stars vary as much as 0.6% during a cycle (Nesme-Ribes et al., 1996). Consequently, it is very probable that the sun is capable of far greater variation than exhibited during our short period of direct observation. Stellar observations show that magnetic activity is lower during a Maunder minimum than during sunspot cycle minima (Baliunas and Saar, 1992). Lower magnetic activity results in lower TSI. The observed TSI variation of 0.1% during sunspot cycles since 1978 suggests that TSI during the last Maunder minimum (Little Ice Age) could have been as much as 1% less than present, sufficient to have caused the Little Ice Age (Baliunas and Saar, 1992).

The star-spot data indicate an average change in brightness of at least 0.4% between the cyclical and Maunder minimum phases, a solar energy decrease equal to one watt per square meter at the top of the earth's atmosphere (Nesme-Ribes et al., 1996). Such a reduction, occurring over several decades, is sufficient to lower the earth's average temperature by 1° to 2°C, the observed cooling during the last Maunder minimum (Little Ice Age).

SOLAR FORCING OF EARTH'S CLIMATE

Eddy (1976, 1977a, 1977b) was an early proponent of solar forcing of earth's climate. He found that earth's climate warmed and cooled as solar activity increased and decreased, respectively. The Maunder minimum (A.D. 1645–1715), the Sporer minimum (A.D. 1400–1510), and the Medieval Maximum (A.D. 1120–1280) are specifically identified. Eighteen similar features are recognized in tree-ring radiocarbon records for the last 7500 years. Slow changes in TSI of approximately 1% are proposed as the causative force producing the climate changes.

"In recent years increasing numbers of papers have been published and presented at meetings (at least 20 at the 1996 Fall AGU meeting) reporting correlations between climate and solar variations. The

Figure 4 Moving 11-year average of terrestrial Northern Hemisphere temperature as deviations from 1951–1970 mean (thinner line) versus solar magnetic activity as evidenced by solar magnetic cycle length (thicker line). The shorter the cycle, the more active and brighter the sun (Baliunas and Soon, 1996).

statistical confidence for the correlations is so high that there is now little reason to question that there is a physical link; the main puzzle now is the nature of the mechanisms involved" (Tinsley, 1997). There is probably a variety of ways, both direct and indirect, that variations in TSI affect the weather and climate of the earth. Small variations in TSI might initiate indirect mechanisms on earth that yield weather and climate changes far in excess of those predicted for the TSI change alone. Three solar variables currently favored as causes of change in terrestrial climate are: (1) TSI, which can directly affect temperatures; (2) solar ultraviolet radiation, which affects ozone production and upper atmospheric winds; and (3) the solar wind, which affects rainfall and cloud cover (Broad, 1997).

Much simulation and modeling have been done to evaluate the effect of TSI variability on the earth's climate (primarily historically observed temperature variation). Baliunas and Soon (1996) find that global temperatures correlate well with the sunspot cycle for the last 250 years **(Figure 4)**. Direct observations of TSI since 1978 show a variation of only 0.1% during a cycle whereas other sunlike stars vary as much as 0.6%, suggesting that the sun may be capable of greater variation. Soon et al. (1996) demonstrate that TSI variability can account for 71% of the observed temperature variation between 1880 and 1993 **(Figure 2)**. TSI variations, when amplified by indirect means, affect earth's climate significantly more than the direct effects of changes in TSI alone. Tinsley (1996, 1997) reports that a number of independent studies show correlations of tropospheric dynamics and temperatures on a daily basis with changes in the solar wind and energetic particle flux, rather than with solar photon changes. For example, coronal mass ejections decrease the lower-energy galactic cosmic-ray flux on earth and increase atmospheric magnetic storms.

Markson and Muir (1980) show that the solar wind controls the earth's electrical field on all timescales. They report that "a highly statistically significant relationship (with confidence limits greater than 99.9 percent) has been found between the solar wind and earth's fair-weather electric field intensity" (p. 988). The inverse correlations between solar wind velocity and both ionospheric potential and galactic cosmic radiation are inferred to show that solar variability modulates the earth's electric field by controlling ionizing radiation in the ionosphere.

The importance of atmospheric electrical processes on clouds is demonstrated by a 3-4% change in the low-latitude cloud cover, observed over a solar cycle from geosynchronous satellites, that correlates with changes in the galactic cosmic-ray flux (Svensmark and Friis-Christensen, 1997). An increase in the cosmic-ray flux decreases the downward air-earth current density (J) at low latitudes, lowering the rate of electrofreezing and resulting in an increase in cloud cover. J, the downward air-earth electrical current, flows between the air and the surface. Tinsley (1996) reports that solar wind changes and the resulting cosmic-ray flux changes produce daily and decadal changes of as much as 10% in J. Increases in J can increase the precipitation efficiency, amplifying the effect of the cosmic-ray energy influx by a factor of more than 100 (Tinsley, 1994). Thus it appears that relatively small solar changes can produce larger than expected weather and climatic changes on earth.

Tinsley (1997) discusses the electrically induced changes in the microphysics of clouds, specifically changes in ice nucleation rates, affecting both the latent heat release in storms and the albedo and opacity of clouds. Changing the albedo and opacity of clouds can directly affect the amount of solar energy reaching the surface of the earth, with consequent additional effects on the weather and climate (Tinsley, 1994). "The effect of clouds on the radiation balance of the earth (cloud forcing) is such that, on average, they cause the mean surface temperature to decrease" (Platt, 1989, p. 428). "The net effect of clouds is to provide a negative feedback on surface temperature, rather than the positive feedback found in earlier general circulation model studies without considering cloud optical depth feedbacks" (Roeckner et al., 1987, p. 138).

Clouds affect the climate in two ways: (1) They reflect a significant amount of the incoming solar radiation (short wave, SW), and (2) they absorb part of the long-wave (LW) infrared radiation of the earth (greenhouse effect). These two effects are partially offsetting. Ramanathan et al. (1989) obtained quantitative data on the radiative forcing of clouds on earth's climate from the Earth Radiation Budget Experiment (ERBE) satellites launched in 1984. Cloud reflection (SW) can double the earth's albedo compared to an absence of clouds, and is a negative force on the climate, whereas the greenhouse effect is positive, reducing the infrared (LW) emission of the earth. "The greenhouse effect of clouds may be larger than that resulting from a hundredfold increase in the CO_2 concentration of the atmosphere" (Ramanathan et al., 1989, p. 57). For the April 1985 period, global SW cloud forcing (−44.5 watts/square meter [W/m^2]) exceeded the LW greenhouse forcing (31.3 W/m^2). Thus clouds have a net cooling effect, varying by region and largest (−100 W/m^2) over the mid- and high-latitude oceans. The observed net cloud forcing is about four times larger than that expected from a doubling of the CO_2 content of the atmosphere (Ramanathan et al., 1989, p. 57). The SW and LW components individually are approximately 10 times as large as the effect of doubling CO_2 in the atmosphere. Thus anything increasing the cloud cover, such as a decrease in the air-earth electrical field produced by a cosmic-ray flux increase resulting from a weaker solar wind (Tinsley, 1996, 1997), may have an effect on earth's climate that is much greater than expected from the initial change in TSI. A variation in cloud cover of 3% during an average 11-year solar cycle produces an effect of approximately 0.8 to 1.7 W/m² (Svensmark and Friis-Christensen, 1997). Significantly, this amount is similar to the total radiative forcing of 1.56 W/m² estimated for the increase in atmospheric CO_2 since 1750 (Lakeman, 1995).

The solar wind also affects the geomagnetic field. Using the statistical relation between geomagnetic activity and satellite measurements of the solar wind velocity, Feynman and Crooker (1978) calculated that solar wind velocities were low at the beginning of the twentieth century and had a steady increase from 1900 to 1950. The increase in solar wind velocity, when amplified by the indirect climate forcing of the earth's electric field and cloud dynamics, may explain the temperature increase recorded from 1900 to 1950. The cause of the changing solar wind may be related to variations in solar radius. Gilliland (1981) estimates that the solar radius was at a maximum in 1911 (with an error of four years). Following the Gleissberg cycle, the radius should have decreased through 1950 with increasing TSI, possibly accounting for some of the temperature increase observed in the first half of the twentieth century. The cooling that followed into the 1970s may have resulted from a decrease in solar radius.

Labitzke (1987) reports a correlation among the solar sunspot cycle, the winter stratospheric temperature in the Arctic, and the quasi-biennial oscillation (QBO). Above the equatorial zone, stratospheric winds oscillate between westerly and easterly flows with a period of 24 to 36 months, a phenomenon called the QBO. The north polar stratospheric vortex is affected by the QBO. The arctic middle stratosphere tends to be colder and the vortex stronger when the QBO is westerly than when it is easterly (Holton and Tan, 1980, 1982; Labitzke, 1982). When the QBO is westerly, the polar stratospheric temperature follows the sunspot cycle with temperatures rising and falling with sunspot numbers (Labitzke, 1987). No midwinter warmings occur in the westerly QBO phase when sunspot numbers are less than 100. There is no such correlation in the easterly QBO phase, when temperatures are generally higher and midwinter warmings occur independently of the solar cycle.

Tinsley (1994) reports a similar decadal cycle of latitude shifts of cyclone tracks in the eastern north Atlantic (the Brown effect) that appears to have been phase-locked to the solar cycle since 1925. The Brown effect is thought to be part of a larger coupled oscillation in the north Atlantic region that is reflected in surface pressure, air temperature, sea surface temperature, 700-mbar wind direction, sea ice extent, and cyclone frequency. The eastern north Atlantic oscillation is evident in records of sea level pressures from 1874 to 1974 (Kelly, 1977). The records show patterns with cycles of 11 years (sunspot cycle) and 2.2 years (QBO).

White et al. (1997) found significant correlations between variations in TSI and global sea surface temperatures (SST) for the past century. They conclude that changes in TSI are the driving force causing changes in SST. The ocean's temperature sensitivity is observed to be 0.08–0.14° K(W/m²)⁻¹ on decadal and interdecadal scales. Long-wave back radiation equilibrates upper ocean layer temperatures to changing TSI in only 1.5–3 years. Applying the ocean's climate sensitivity to the apparent trend in TSI over the past century yields an increase in global average SST of 0.2°–0.3° K in response to the observed rise in TSI of approximately 2.0 W/m² at the top of the atmosphere. "This increase is of the same order as that observed (i.e., 0.4° K), suggesting that global warming occurring over the past century was significantly influenced by the corresponding increase in solar irradiance" (White et al., 1997, p. 3265).

The entire solar spectrum must be analyzed to determine the effects of variations in the solar flux on earth's climate. It appears that changes in flux are skewed to the shorter wavelengths. For example, a 100-fold change has been measured in the soft X-ray range over the course of a sunspot cycle (Astronomy Mag., July 1996, p. 32). At least some of the shorter wavelengths (ultraviolet) are thought to play a significant role in earth's climate (Chamberlain, 1977; Hood et al., 1993). Variations in solar ultraviolet radiation affect the ozone content of the earth's upper atmosphere. Haigh

(1996) reports that wavelengths shorter than 400 nm, comprising about 9% of TSI, account for approximately 32% of the change in TSI over a solar cycle. Ultraviolet radiation produces ozone in the stratosphere. Ozone also absorbs ultraviolet radiation. Therefore, increases in ultraviolet radiation warm the stratosphere in two ways: by increasing the ozone content which in turn absorbs more ultraviolet radiation. Haigh (1996) reports model results suggesting that increases in stratospheric temperature in response to increased TSI result in stronger summer easterly winds, which penetrate into the tropical upper troposphere and force tropospheric circulation patterns poleward. Thus, solar-induced increases in ozone content can affect lower stratospheric temperatures and climatic changes. Haigh's model does not include the effects of QBO, which may play an important role in transmitting the solar effects from the stratosphere to the troposphere. The results of the model imply that changes in the ozone content of the stratosphere by any means can have an impact on tropospheric climate.

In summary, three solar variables are known to affect earth's climate: TSI, ultraviolet radiation, and solar wind. Each affects the earth's climate in different ways and can produce indirect effects that amplify small changes in TSI. Individually, they do not appear to cause the entire observed climatic changes. However, when taken together, they may be sufficient, especially because solar forcing of earth's climate is still an emerging science. Undoubtedly, there are still mechanisms of solar forcing that are poorly understood, perhaps even unknown.

It is now obvious that the effects of TSI changes alone do not represent total solar influence on earth's climate. Models and studies relying on solar forcing from TSI changes only (e.g., Schlesinger and Ramankutty, 1992; Kelly and Wigley, 1990, 1992) invariably find solar forcing insufficient to have produced the observed climate changes, especially in the twentieth century, and are erroneously used to support the dominance of greenhouse forcing.

On the other hand, Jaworowski et al. (1992) indicate that atmospheric CO_2 content is rising as a result of increasing sea surface layer temperatures. As the water warms, less carbon dioxide can be held in solution. Offering significant support, Reid (1987) found that TSI variations of approximately 0.6% could have produced the observed sea surface temperature changes since 1860. Therefore, it is possible that rising CO_2 content in the atmosphere and the resultant greenhouse effect may in fact be another indirect effect of an increase in TSI.

THE HOLOCENE CLIMATE RECORD

Karlen and Kuylenstierna (1996) present a well-dated chronology of Holocene climatic changes obtained in a climate-sensitive area of Scandinavia, enabling an effective test of the solar-forcing hypothesis. They compare a chronology of climate changes derived from a large number of ^{14}C dates of pinewood samples, proglacial lacustrine sediments, and alpine glacial moraines with an index of solar activity obtained from the deviation between dendrochronological age and the ^{14}C age of tree rings using data from Wigley and Kelly (1990). Karlen and Kuylenstierna (1996) conclude that "a correlation exists between these records, which suggests that solar forcing affected the climate during the Holocene."

The Scandinavian climate apparently stabilized in a long-term optimum warm period 5100–7800 years BP. The trend towards a colder climate in Scandinavia since that time, partly a result of glacial rebound elevating the area and affecting the tree line, correlates with similar trends elsewhere, even where glacial rebound has not occurred, such as in the Caribbean (Hodell et al., 1991). Karlen and Kuylenstierna (1996) conclude that the trend toward cooler summers in the Northern Hemisphere results from orbital forcing (Milankovitch variations), causing a decrease in TSI for July by about 35 W/m^2 (Rind et al., 1989) at 60°N in the past 10,000 years. Superimposed on

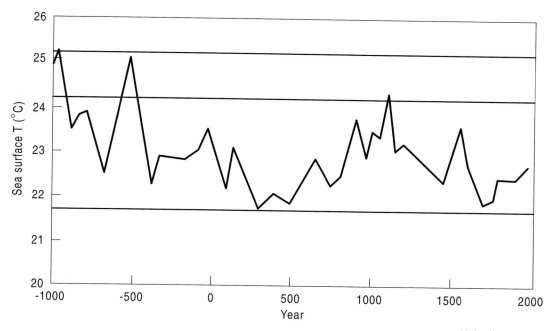

Figure 5 Sea surface temperatures in the Sargasso Sea, as determined by oxygen isotope ratios in dated planktonic foraminifera tests (Keigwin, 1996).

the long-term trend to a cooler climate are shorter fluctuations of relatively warmer and cooler events on a time scale of centuries to a few millennia. Seventeen of nineteen episodes of low solar activity coincide with times of inferred cold climate in Scandinavia. Several of these cold events in the Holocene were about as severe as the Little Ice Age of the seventeenth and eighteenth centuries. The coincidence of periods of cold climate and $\delta^{14}C$ anomalies indicates that variations in TSI are an important factor for climate change. In contrast, Nesje and Johannessen (1992) provide analyses suggesting that volcanic aerosols also played a significant role in Holocene climate changes.

Significantly, Karlen and Kuylenstierna (1996) maintain that data from an Antarctic ice core (Neftel et al., 1988) and from peat studies (White et al., 1994) show only small variations in atmospheric CO_2 concentration during the Holocene, with the exception of the last 100–200 years. They conclude that the CO_2 variations "are unlikely to account for changes in Holocene climate."

Using data obtained from sediment cores from the Sargasso Sea, Keigwin (1996) determined the climate for that part of the Atlantic for the last 3000 years **(Figure 5)**. Oxygen isotope ratios in the tests of planktonic foraminifera (*Globigerinoides ruber*) were used to determine sea surface temperatures at different ages as determined by radiocarbon dating. The results are in close agreement with the historically observed climate in Europe. The Little Ice Age (300–400 years B.P. [before the present) and the Medieval Warm Period (800–1000 years B.P.) are strongly reflected in these data. Significantly, a second cool period, another "Little Ice Age," occurred 1500–1700 years B.P., suggesting a solar Maunder minimum cycle of approximately 1300 years. These data suggest that earth's climate should be gradually warming during the next century or two. Similar warming over the next decade is expected from the Gleissberg cycle of 80–90 years that produced the warm, dry climate of the 1930s.

CONCLUSIONS

The data supporting solar forcing of earth's weather and climate are voluminous and growing. Short-term variations in TSI have been observed, measured, and documented. Longer-term variations are known by proxy measurements for the sun and supported by data from dozens of sunlike stars. It seems probable that the sun is capable of far greater variation than the 0.1% exhibited in the short-term data since 1978.

The observed climate effects are thought to be greater than can result directly from changes in observed and expected changes in TSI alone. However, it is becoming increasingly evident that there are mechanisms within the earth's climate system (e.g., the air-earth electric field, cloud forcing, and others) that amplify the effect of small changes in TSI, producing the greater climatic changes observed. It now seems probable from the growing volume of evidence that the sun is capable of significantly greater variability in TSI than found in our short period of direct observation (Willson, 1997). The consequent direct and indirect climatic effects are obviously sufficient to have produced the Holocene climatic record.

The cyclical nature of the variations in TSI strongly suggests that much, perhaps all, of the currently perceived global warming may result from increasing TSI. The increase in atmospheric carbon dioxide may result from rising sea surface temperatures, rather than be the cause of the increase in sea surface temperatures (Jaworowski et al., 1992). This is supported by White et al. (1994), who find three large increases in atmospheric CO_2 in the past 14,000 years: at 12,800 years ago during a warming period in the North Atlantic region, 10,000 years ago at the end of the Younger Dryas cold period, and 4400 years ago at the beginning of the modern climate.

Observational data prove what is intuitively obvious, that variations in TSI significantly affect earth's climate. More research is required to detail the direct and indirect solar forcing of earth's climate to determine the extent of anthropogenic forcing. In any case, much more data are required to eliminate totally natural causes for the observed global warming. Drastic measures are certainly not advisable to solve a problem that may not exist.

REFERENCES CITED

Bahcall, J., and M. H. Pinsonneault, 1992, Standard solar models with and without helium diffusion and the solar neutrino problem: Reviews of Modern Physics, v. 64, p. 885–926.

Baliunas, S., and S. Saar, 1992, Unfolding mysteries of stellar cycles: Astronomy, May, p. 42–47.

Baliunas, S., and W. Soon, 1995, Are variations in the length of the activity cycle related to changes in brightness in solar-type stars?: Astrophysical Journal, v. 450, p. 896–901.

Baliunas, S., and W. Soon, 1996, The sun-climate connection: Sky and Telescope, December, p. 38–41.

Beer, J., U. Siegenthaler, G. Bonani, R. C. Finkel, H. Oeschger, M. Suter, and W. Wolfli, 1988, Information on past solar activity and geomagnetism from [10]Be in the Camp Century ice core: Nature, v. 331, p. 675–679.

Broad, W. J., 1997, Another possible climate culprit: The sun: The New York Times, September 23, p. C1.

Chamberlain, J. W., 1977, A mechanism for inducing climatic variations through the stratosphere: Screening cosmic rays by solar and terrestrial magnetic fields: Journal of Atmospheric Sciences, v. 34, p. 737–743.

Croll, J., 1867, On the change in the obliquity of the ecliptic, its influence on the climate of the polar regions and the level of the sea: Philosophical Magazine, v. 33, p. 426–445.

Croll, J., 1875, Climate and time: New York, Appleton and Co.

Eddy, J. A. 1976, The Maunder minimum: Science, v. 192, p. 1189–1202.

Eddy, J. A., 1977a, The case of the missing sunspots: Scientific American, v. 236 (May), p. 80–88.

Eddy, J. A., 1977b, Climate and the changing sun: Climate Change, v. 1, p. 173–190.

Feynman, J., and N. U. Crooker, 1978, The solar wind at the turn of the century, v. 275, p. 626–627.

Friis-Christensen, E., and K. Lassen, 1991, Length of the solar cycle: An indicator of solar activity closely associated with climate: Science, v. 254, p. 698–700.

Gilliland, R. L., 1981, Solar radius variations over the last 265 years: Astrophysical Journal, v. 248, p. 1144–1155.

Gribbin, J., 1991, Blinded by the light, the secret life of the sun: New York, Harmony Books.

Haigh, J. D., 1996, The impact of solar variability on climate: Science, v. 272, p. 981–984.

Hathaway, D. H., 1995, Journey to the heart of the sun: Scientific American, v. 274, No. 1, p. 38–43.

Hodell, D. A., J. H. Curtis, G. A. Jones, A. Higuera-Gundy, M. Brenner, M. W. Binford, and K. Dorsey, 1991, Reconstruction of Caribbean climate change over the past 10,500 years: Nature, v. 352, p. 790–793.

Holton, J. R., and H.-C. Tan, 1980, The influence of the equatorial quasi-biennial oscillation on the global circulation at 50 mb: Journal of Atmospheric Science, v. 37, p. 2200–2208.

Holton, J. R., and H.-C. Tan, 1982, The quasi-biennial oscillation in the northern hemisphere lower stratosphere: Journal of the Meteorological Society of Japan, v. 60, p. 140–147.

Hood, L. L., J. L. Jirikowic, and J. P. McCormack, 1993, Quasi-decadal variability of the stratosphere: Influence of long-term solar ultraviolet variations: Journal of Atmospheric Sciences, v. 50, p. 3941–3958.

Imbrie, J., and K. P. Imbrie, 1979, Ice Ages: Hillside, Enslow Publishers, 224 p.

Jaworowski, Z. T. V. Segalstad, and V. Hisdsal, 1992, Atmospheric CO_2 and global warming: A critical review, Norwegian Polar Research Institute, Meddelelser NR. 119, 74 p.

Karlen W., and J. Kuylenstierna, 1996, On solar forcing of Holocene climate: evidence from Scandinavia: The Holocene, v. 6, p. 359–365.

Keigwin, L. D., 1996, The Little Ice Age and Medieval Warm Period in the Sargasso Sea: Science, v. 274, p. 1504–1508.

Kelly, P. M., 1977, Solar influences on north Atlantic mean sea level pressure: Nature, v. 269, p. 320.

Kelly, P. M., and T. M. L. Wigley, 1990, The influence of solar forcing trends on global mean temperature since 1861: Nature, v. 347, p. 460–462.

Kelly, P. M., and T. M. L. Wigley, 1992, Solar cycle length, greenhouse forcing and global climate: Nature, v. 360, p. 328–330.

Kerr, R. A., 1997, Did satellites spot a brightening sun?: Science, v. 277, p. 1923–1924.

Labitzke, K., 1982, On the interannual variability of the middle stratosphere during northern winters: Journal of the Meteorological Society of Japan, v. 60, p. 124–138.

Labitzke, K., 1987, Sunspots, the QBO, and the stratospheric temperature in the north polar region: Geophysical Research Letters, v. 14, p. 535–537.

Lakeman, J. A., 1995, Climate change 1995, the science of climate change: Intergovernmental Panel on Climate Change, Cambridge University Press, 572 p.

Lang, K. R., 1995, Sun, earth and sky: Berlin, Springer-Verlag, 282 p.

Lean, J., J. Beer, and R. Bradley, 1995, Reconstruction of solar irradiance since 1610: implications for climate changes: Geophysical Research Letters: v. 22, p. 3195–3198.

Markson, R., and M. Muir, 1980, Solar wind control of the earth's electric field: Science, v. 208, p. 979–990.

Milankovitch, M., 1920, Theorie mathematique des phenomenes thermiques produits per la radiation solaire: Paris, Gauthier-Villars.

Milankovitch, M., 1930, Mathematische Klimalehre und astronomische Theorie der Klimaschwankungen, in W. Koppen and R. Geiger, eds., Handbuch der Klimatologie: Berlin, Gebruder Borntraeger, p. 1–176.

Milankovitch, M., 1936, Durch ferne Welten und Zeiten: Leipzig, Koehler und Amalang.

Milankovitch, M., 1938, Astronomische Mittel zur Erforschung der erdgeschichtlichen Klimate, in B. Gutenberg, ed., Handbuch der Geophysik: v. 9, p. 593–698.

Milankovitch, M., 1941, Kanon der Erdbestrahlung und seine Andwedung auf das Eiszeitenproblem: Belgrade, Royal Serbian Academy, Special Publication, v. 133, p. 1–633.

Milankovitch, M., 1957, Astronomische Theorie der Klimaschwankungen ihr Werdegang und Widerhall: Serbian Academy of Science, Monograph 280, p. 1–58.

Neftel, A., H. Oeschger, T. Staffelbach, and B. Stauffer, 1988, CO_2 record in the Byrd ice core 50,000–5,000 BP: Nature, v. 331, p. 609–611.

Nesje, A., and T. Johannessen, 1992, What were the primary forcing mechanisms of high-frequency Holocene climate and glacier variations?: The Holocene, v. 2, p. 79–84.

Nesme-Ribes, E., S. L. Baliunas, and D. Sokoloff, 1996, The stellar dynamo: Scientific American, v. 275, No. 2, p. 46–52.

Parsons, M. L., 1995, Global warming, the truth behind the myth: New York, Plenum Press, 271 p.

Platt, C. M. R., 1989, The role of cloud microphysics in high-cloud feedback effects on climate change: Nature, v. 341, p. 428–429.

Ramanathan, V., R. D. Cess, E. F. Harrison, P. Minnis, B. R. Barkstrom, E. Ahmad, and D. Hartmann, 1989, Cloud-radiative forcing and climate: Results from the Earth Radiation Budget Experiment: Science, v. 243, p. 57–63.

Reid, G. C., 1987, Influence of solar variability on global sea surface temperatures: Nature, v. 329, p. 142–143.

Ribes, E., J. C. Ribes, and R. Barthalot, 1987, Evidence for a larger sun with a slower rotation during the seventeenth century: Nature, v. 326, p. 52–55.

Rind, D., D. Peteet, and G. Kukla, 1989, Can Milankovitch orbital variations initiate the growth of ice sheets in a general circulation model?: Journal of Geophysical Research, v. 94, p. 12,851–12,871.

Roeckner, E., U. Schlese, J. Biercamp, and P. Loewe, 1987, Cloud optical depth feedbacks and climate modelling: Nature, v. 329, p. 138–140.

Schlesinger, M. E., and N. Ramankutty, 1992, Implications for global warming of intercycle solar irradiance variations: Nature, v. 360, p. 330–333.

Schneider, S. H., 1989, Global warming: San Francisco, Sierra Club Books, 299 p.

Soon, W. H., E. S. Posmentier, and S. L. Baliunas, 1996, Inference of solar irradiance variability from terrestrial temperature changes, 1880–1993: An astrophysical application of the sun-climate connection: Astrophysical Journal, v. 472, p. 891–902.

Stuiver, M, P. M. Grootes, and T. F. Braziunas, 1995, The GISP $\delta^{18}O$ climate record of the past 16,500 years and the role of the sun, oceans, and volcanoes: Quaternary Research, v. 44, p. 341–354.

Svensmark, H., and E. Friis-Christensen, 1997, variation of cosmic ray flux and global cloud coverage—a missing link in solar-climate relationships: Journal of Atmospheric and Solar-Terrestrial Physics, v. 59, p. 1225–1232.

Tinsley, B. A., 1994, Solar wind mechanism suggested for weather and climate changes: Eos, v. 75, p. 369–376.

Tinsley, B. A., 1996, Correlations of atmospheric dynamics with solar wind induced changes of air-Earth current density into cloud tops: Journal of Geophysical Research, v. 101, p. 29, 701–29, 714.

Tinsley, B. A., 1997, Do effects of global atmospheric electricity on clouds cause climate changes: Eos, v. 78, p. 341–349.

White, J. W. C., P. Ciais, R. A. Figge, R. Kenny, and V. Markgraf, 1994, A high resolution record of atmospheric CO_2 content from carbon isotopes in peat: Nature, v. 367, p. 153–156.

White, W. B., J. Lean, D. R. Cayan, and M. D. Dettinger, 1997, Response of global upper ocean temperature to changing solar irradiance: Journal of Geophysical Research, v. 102, No. C2, p. 3255–3266.

Wigley, T. M. L., and P. M. Kelly, 1990, Holocene climatic change, ^{14}C wiggles and variations in solar irradiance: Philosophical Transactions of the Royal Society of London, v. A330, p. 547–560.

Willson, R. C., 1997, Total solar irradiance trend during solar cycles 21 and 22: Science, v. 277, p. 1963–1965.

Willson, R. C., and H. S. Hudson, 1991, The sun's luminosity over a complete solar cycle: Nature, v. 351, p. 42–44.

Gerhard, L. C., and W. E. Harrison, Distribution of
oceans and continents: A geological constraint on
global climate variability, 2001, *in* L. C. Gerhard,
W. E. Harrison, and B. M. Hanson, eds., Geological
perspectives of global climate change, p. 35–49.

2 | Distribution of Oceans and Continents: A Geological Constraint on Global Climate Variability

Lee C. Gerhard
William E. Harrison
Kansas Geological Survey
Lawrence, Kansas, U.S.A.

ABSTRACT

Major erathemic climate changes may result from redistribution of oceans and continents through
time. When plate tectonic reconstructions portray the presence of near-equatorial currents, greenhouse events are common, but when landmasses exist at the equator, such oceanic circulation patterns are not developed and the transfer of heat to the polar regions stimulates large-scale
glaciation. This pattern seems to have operated from the Vendian to the present. Observed tectonic
changes between Pleistocene icehouse and Cretaceous greenhouse events provide the basis for development of this hypothesis. As illustrated on Figure 1 in the introduction to this volume, the proposed relationship is a second-order driver and is relatively more important than those phenomena
that occur over short periods of time and result in smaller temperature changes.

INTRODUCTION

Understanding natural controls on earth's climate is crucial to assessing the potential for human-
induced climate change. Until we better understand the range of influence of major and naturally
occurring processes, it will not be possible to reasonably assess man's role in potentially influencing global climate patterns. Much attention has been given to general circulation models'

forecasts of anthropogenic carbon-dioxide–driven global warming for the future. In addition, popular suppositions of human influence on natural climate systems have been considered for short timescales, but there has been little focus on overall earth dynamics as controls on climate processes over geologic timescales. As several papers in this volume demonstrate, current temperature changes are very minor when compared with those that have occurred in the past. Further, anthropogenic impacts on global climate change may well be relatively insignificant when compared with other drivers.

Within the incredibly large energy flux of climate, the roles of oceans, solar isolation, orbital variations, and internal heat engines, combined with earth tectonism, have not been fully and systematically evaluated. The earth's climate is constantly varying and has exhibited many rapid and large-amplitude changes over both human-scale time and geologic time. The highly complex interactions among the varying elements of heat exchange have hampered efforts to effectively model the variability of climate. Much has been written in recent years about the variation of climate, especially recent decadal variability. Ice-core data and historical temperature data have been used to place modern climate variations into a Holocene or Pleistocene context (Fitzpatrick, 1995; Bluemle et al., 1999; Lamb, 1995). Extremely rapid climate changes have been interpreted from ice-core data (Fitzpatrick, 1995), suggesting that some significant climate changes may take place in less than a decade. However, in order to gain full understanding of these short-term events and their causes, it is necessary to understand larger-scale events in earth history and to decipher the origin of very large-scale alternation of earth climate as it has varied between greenhouse and icehouse events. Some of the major changes in earth climate have occurred as erathemic changes (i.e., they have occurred slowly over hundreds of millions of years) and are interpreted in this paper to be the result of global tectonic activities.

Heat is added to the earth by solar luminosity and internal radioactive decay; heat is transferred from the core and mantle of the earth to the near surface by convection. Oceans are the single greatest influence on the distribution of heat over the surface of the earth, by virtue of their volume, ferality, and the specific heat of water. The world's oceans serve as a means to absorb, transport, and release large quantities of thermal energy, and this flux exerts a major control on global climate. The Atlantic Ocean serves as an example of the phenomenon.

Climate effects of oceans on landmasses are well known, and have been subject to some conjecture about future changes in oceanic circulation (Broecker, 1999). The oceans are effective in absorbing and transporting extremely large quantities of thermal energy (heat) and thereby exerting a major control on global climatic patterns. This relation is well known from physical oceanography (Bigg, 1996; Tomczak, 1999) but is sufficiently important to warrant an overview here.

The world's thermohaline circulation system is driven by density contrasts that result from temperature and salinity differences. This system is characterized by major downwelling of cold, dense water, primarily generated in the Arctic and Antarctic regions, that drives the deepwater circulation component of the thermohaline system **(Figure 1)**. The area where the Norwegian and Greenland Seas come together is where the warm surface waters of the Gulf Stream cool rapidly, become dense, and sink to the bottom of the North Atlantic Ocean. This cold, dense mass of water is then transported to the Weddell Sea area, near Antarctica, in the extreme Southern Hemisphere. Because of ice formation in the southern polar region, the waters of the Weddell Sea area are highly saline and thus denser than normal ocean water. The highly saline water of the Weddell Sea area joins with the cold, dense water that originated in the Norwegian and Greenland Seas, and the combined flow moves northward toward the equator. Progressive upwelling of this highly saline, cold, dense water mass occurs very slowly in the Pacific and Indian Oceans and establishes the final connection for the deep-to-shallow water circulation pattern. The complete cycle may take up to 1000 years.

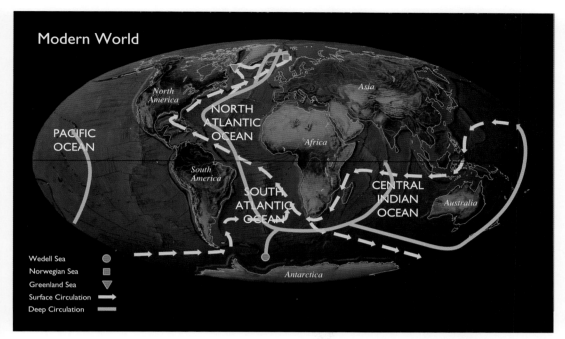

Figure 1   Highly generalized thermohaline circulation in the world ocean that transfers thermal energy (heat) and exerts a major influence on climate (modified from Bigg, 1996, and Scotese, 1997).

Relatively modest changes in thermal energy flux resulting from thermohaline circulation patterns may be responsible for 20- to 30-year climatic trends. For example, Gray et al. (1997) suggest that a reduction in heat transfer in the North Atlantic led to an overall cooling of ocean surface temperatures. This reduction in heat transfer is interpreted to have resulted in atmospheric circulation patterns that produced a 24-year pattern of reduced precipitation and lower hurricane activity.

The pole-to-pole deepwater component of today's thermohaline circulation pattern is possible because of the current locations of continental landmasses. If the positions of the landmasses change, as is known to have occurred in the geologic past, it is logical that thermohaline-controlled heat flux will be different. This heat flux will result in changes in weather patterns and climatic conditions. Oceanographers have offered generalizations regarding the relation between the locations of landmasses and climate, but few details have been presented to date. For example, Bigg (1996) suggests a tectonic control on basic climatic conditions and recognizes that maritime conditions have influenced the climate of the earth for the past 570 million years. He also indicates that increased snowfall on polar landmasses may be better protected from melting during the sunless winters common to such regions, and the overall process will eventually help form glacial ice sheets.

Our hypothesis that the distribution of continents on the earth's surface controls climate is based on the most likely thermohaline circulation patterns that will exist if major landmasses are located in equatorial positions. If this occurs, these landmasses will effectively prevent a dominantly equatorial oceanic circulation system from developing, which in turn will contribute to the transfer of thermal energy to polar regions and icehouse conditions. Plate tectonic reconstructions provide a reasonably rigorous means to assess the locations of landmasses through geologic time and thus a means to project the thermohaline circulation system. Climate effects on landmasses are well known and have been subject to some conjecture about future changes in oceanic circulation (Broecker, 1997, 1999). Unlike the rapid changes forecast by Broecker (1997, 1999) from changes in thermohaline circulation in the Atlantic Ocean, we envision the extensive climate changes caused by redistribution of continents and consequent ocean current changes to take place over extensive periods of time, perhaps punctuated by events of rapid redeployment of thermohaline circulation. For instance, the change from the Cretaceous greenhouse state to the current icehouse state took place during more than 60 million years and was punctuated several times by temperature inversions. The movement of continents is slow—on the order of a few centimeters per year—and the overall changes in ocean circulation caused by these movements must reflect that rate of movement.

One source of heat energy, internal heat from radioactive decay, occurs throughout the earth but is focused at spreading boundaries where higher-temperature mantle materials come in contact with seawater and where volcanism releases heat to the lithosphere or atmosphere. Radiogenic heat is the significant energy contributor to tectonic activity, whether by plume or by spreading center ridge activity (see Kerr, 1999, for a modern example), but it is not considered a significant direct climate driver (Figure 1, Gerhard et al., introduction to this volume).

Solar variability is a significant control on decadal to centennial scales, with sunspot minima being associated with colder climates and sunspot maxima with warmer ones. Long-term reduction in solar luminosity is apparent in the geologic context of climate control, but not in human time spans. Orbital variability (Milankovitch cycles) also has significant effects on short-term and intermediate-term climate cyclicity (Hoyt and Schatten, 1997).

Greenhouse gases are important to the stabilization of earth climate within the limits necessary to sustain life as we know it. If greenhouse gases did not exist to trap solar heat at the surface of our planet, it would be significantly colder. Water vapor is volumetrically the most important of the greenhouse gases, with carbon dioxide and methane distant, but theoretically significant, constituents. Methane, for example, is about 21 times more effective at trapping heat in the atmosphere than carbon dioxide.

There is ample evidence that the earth has undergone alternating icehouse and greenhouse events over at least the last 600 million years (see Crowell, 1999, for a comprehensive modern bibliography). Icehouse events in the late Proterozoic (Crowell, 1999, for example) are the earliest such events for which the extent of glaciation is reasonably well documented. Fischer (1982), in an examination of long-term climate oscillations, identified 100- and 500-year orbital and solar cycles, suggested oceanic processes and tectonic activities as possible controls, and speculated about mantle convection as a cyclical climate modifier. Other workers have recently discussed the possibility of ocean-current control of climate (Broecker, 1997, 1999; Ganapolski et al., 1998; Gerhard, 1999; Haug and Tiedemann, 1998; Crowell, 1999, p. 75–78; Huber, 1998). Veevers (1990) speculated that plate tectonics control carbon-dioxide levels in the atmosphere and thus indirectly control long-term climate changes. Fischer (1984) also suggested that carbon dioxide is the major controlling influence on climate through volcanism and sea-level changes and biotic activity.

We posit that the primary driving force behind erathemic-scale climate cycling is tectonic, specifically by controlling distribution of landmasses on the earth's surface, which in turn controls the geometry of ocean currents and thus the transfer of heat around the earth. When equatorial ocean currents exist, the earth tends to be in a greenhouse state. In contrast, when continents exist in positions that impede or block significant equatorial currents, the earth tends to be in the icehouse condition. Transitions between the two states are slow but may be punctuated by rapid shifts.

DISCUSSION

In discussions relating to climate change, a recurring fundamental question challenged us: What occurred in post-Cretaceous time to drive climate from the greenhouse condition to an icehouse state? Significant tectonic and geological changes (identified in the next paragraph) took place between Cretaceous and Holocene time. These changes resulted in the event(s) that caused the global climatic conditions to shift from greenhouse (Cretaceous) to icehouse (Pleistocene-Holocene). Possible temperature shifts of up to 16°C during the last 70,000 years suggest that similar or greater temperature variation has occurred over the last 70 million years (Lang et al., 1999). Although we do not speculate in this paper about possible causes of prominent but shorter-term climate changes, such dramatic shifts must be driven by major changes in the amount or distribution of heat. The shift was slow in completion, because glaciation is first reported from the late Miocene or early Pliocene and global temperatures trended toward cooling from the Cretaceous to the present (Bluemle et al., 1999). There have been other significant shifts during this 50-million-year span (i.e., the Eocene warm event described by Severinghaus and Brook, 1999).

Two obvious and prominent tectonic or geologic events occurred between the Cretaceous and the Pleistocene. The first was the construction of the Isthmus of Panama, and the second was the closure of the Tethyan seaway separating Africa and Europe and India and Siberia **(Figures 2, 4)**.

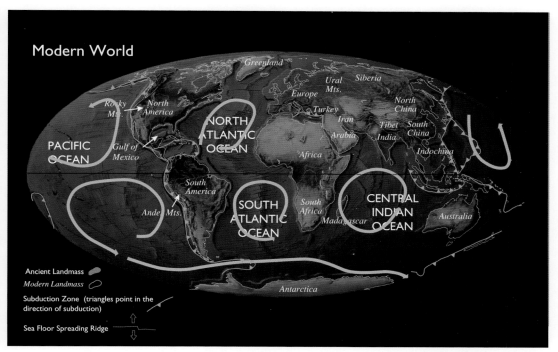

Reprinted by permission of C. R. Scotese

Figure 2 Holocene ocean circulation, the modern world. The heavy light blue lines and arrowheads represent the modern ocean gyres that help transfer thermal energy and moisture and thus exert influence on global climate (modified from Scotese, 1997).

Emergence of continental shelves and central cratons from areas which were submerged during Cretaceous time and perhaps a major bolide impact at about the end of the Cretaceous are tangential but potentially significant events. Tectonic change is most apparent, and may be typically identified, on the basis of climate, biotic, and topographic alterations.

For those who subscribe to the current and popular paradigm that carbon-dioxide variability drives global temperatures, it is noteworthy that extensive volcanism during Late Cretaceous time (and later) is somewhat coincident with the warm temperatures associated with that time interval (Tarduno et al., 1998). The carbon-dioxide-driver theme appears frequently in the current literature, but it should be carefully examined before adoption as the major climate driver through time (Fischer et al., 1999; Pagnani et al., 1999; Huber, 1998). Crowell (1999) offers the following pertinent summary statement: "Tectonobiogeochemical processes operating irregularly through geologic time and moderated by orbital variations and interrupted occasionally by bolide impacts are the basic causes of climate change." This overview understanding is especially important to any discussion of major climate controls.

Ocean currents now form large gyres in each of the world oceans **(Figures 1, 2)**. These gyres move equatorial waters to the poles, as previously described. Although glaciation requires cold temperatures that permit ice and snow to remain over summer seasons, it also requires massive moisture in order to provide the snow and ice necessary for maintenance of such conditions. Flow of atmospheric moisture to polar regions is enhanced with the ocean gyres, regardless of salinity gradients (Broecker, 1997). This oceanic circulation system is the conveyer belt that brings enough warm, moist air to the polar region to generate Northern Hemisphere snowfall. This conveyor is particularly effective when the Arctic Ocean is thawed (Ewing and Donn, 1956). Ewing and Donn suggest that the next thawing of the Arctic Ocean could precipitate the next glacial advance. Flow of warmer North Atlantic waters over the Greenland-Icelandic sill to replenish evaporating Arctic waters could help provide the moisture needed to initiate glaciation (see also Driscoll and Haug, 1998). A continental landmass at one polar position might be a necessary condition for establishment of continental glaciation and icehouse conditions (see Crowell, 1999, p. 75, for a brief discussion).

Prior to Laramide and Alpine orogenesis, Cretaceous seas extended over much of the global landmass. Continental masses were topographically low with respect to sea level. Although major Laramide and Alpine tectonic activity had not yet been initiated, widespread volcanism is evident from North American Late Cretaceous deposits. The climate during most of Cretaceous time appears to have been quite moderate and was obviously capable of supporting large vertebrates and luxuriant floras. Estimates are that Cretaceous global temperatures were significantly higher than current temperatures (Frakes, 1979, p. 180).

All plate reconstructions that show past continental and oceanic positions on the earth's surface suffer from a common relationship in geologic science. The degree of uncertainty generally increases as one goes increasingly farther back in geologic time. This leads to increased confidence for geologically young reconstructions and somewhat less confidence for geologically older ones. This is not to imply that the major plate reconstructions are seriously flawed. Overall, there are compelling arguments and enough agreement from various lines of geologic evidence to demonstrate that the major elements of plate reconstruction for a given period of geologic time are good. Typically, specific details are less obvious. These inaccuracies are primarily a consequence of increasingly limited data preservation and physical accessibility, and they primarily affect details of specific plate reconstructions. The reconstructions used in this study are mainly those of Scotese (1997), but reference is also made to the Cretaceous reconstructions of Hay et al. (1999). These reconstructions provide evidence that continental landmasses existed in at least one polar position

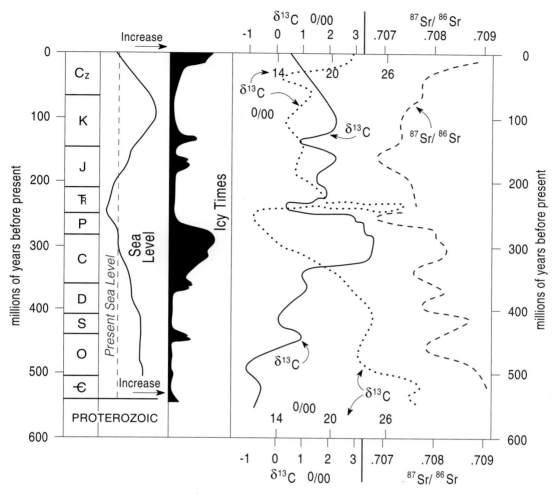

Figure 3 Generalized greenhouse and icehouse curve (modified from Crowell, 1999).

throughout most of Phanerozoic time. This relationship is not consistent with the concept that the presence of such landmasses contributes significantly to glacial episodes.

Glacial advances (icehouse events) are prominent during the Pleistocene, Late Carboniferous, and late Precambrian (Vendian) (see summary by Crowell, 1999) **(Figure 3)**. Corresponding greenhouse events occur in the Devonian and Cretaceous. Additional icehouse events of smaller magnitude are present in the Late Jurassic (Early Cretaceous) and Early Silurian (Crowell, 1999, figure 25). Tectonic reconstruction of the Cretaceous **(Figure 4)** shows that the most significant change in the earth's surface configuration from the modern world is the presence of a global circulation system that allows free and open oceanic circulation between North and South America and between Africa and Eurasia and India and Asia. This sinuous deep-sea channel provides opportunity for organization of strong ocean currents that circumnavigate the earth. Gyre current structure

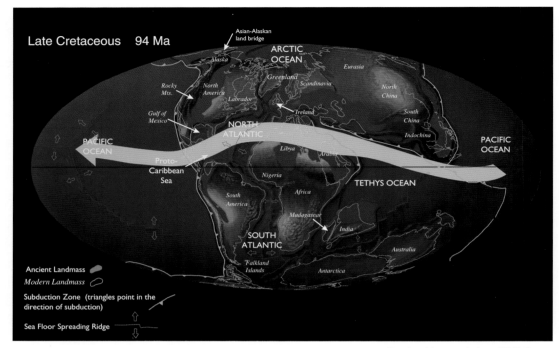

Reprinted by permission of C. R. Scotese

Figure 4 Paleoreconstruction of continental landmasses and oceanic bodies at 94 Ma (million years ago, in the Late Cretaceous). This geometry promotes near-equatorial ocean circulation, as shown by the broad blue arrow showing projected water-mass movement (modified from Scotese, 1997). Note the open connections between Africa and Eurasia and between North and South America.

organized by barriers to circumequatorial current, and thus equatorial heat transfer, dominates the modern and Pleistocene earth, transferring heat to the polar oceans and landmasses. This also helps provide the moisture necessary to sustain massive and prolonged glaciation.

Fundamental to the hypothesis is the transfer of heat to the poles during times of icehouse conditions, and reduced transfer during greenhouse times. All other times are somewhat intermediary, with mixed circulation patterns. For instance, Crowell (1999, p. 76) states, "The shapes of ocean bottoms and the topography of continents are significant in distributing heat in the system. They control the regional climate . . . and the poleward flow of surface waters and the equatorward flow of bottom waters. This oceanic conveyer-belt system plays a dominating role in planetwide heat distribution. . . ." The Crowell paper assumes that the hypothesized heat transfer is the dominant control for development of icehouse and greenhouse conditions but does not examine the geometry of continents and oceans through time.

Pleistocene and Cretaceous tectonic distribution of continents and oceans is easily visualized and substantiated with significant databases and observations (Barreta and Johnson, 1999; Crowell, 1999; Frakes, 1979). Older episodes of icehouse or greenhouse conditions are less well documented, but the relationship of continental and oceanic geometry to hypothetical climate events seems clear. The number and details of tectonic reconstructions do not permit as detailed a

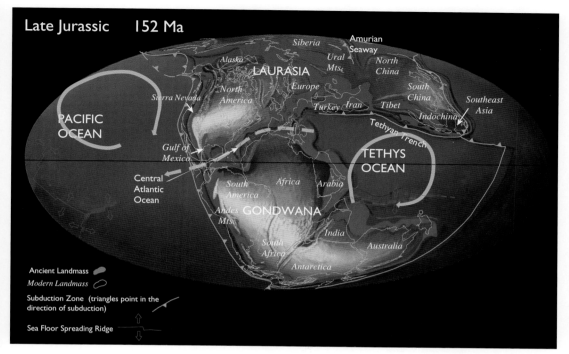

Figure 5 Paleoreconstruction of continental landmasses and oceanic bodies at 152 Ma (Late Jurassic). This geometry suggests that near-equatorial ocean circulation is minimal, as shown by the blue arrow showing projected water-mass movement (modified from Scotese, 1997). There is the possibility of continental glaciation, but glaciation is not likely to be extensive.

treatment of the interrelations of heat transfer, climate, and earth geometry as desired, but a major generalization can be derived. All episodes of major glaciation have occurred when landmasses occupy equatorial positions and thus prevent development of major equatorial current systems. Alternatively, whenever landmasses were positioned so that there was great potential for equatorial currents, little glaciation occurred.

There is apparent blockage of equatorial currents from mid-Carboniferous through the Early Jurassic (however, during interglacial episodes of the Carboniferous, perhaps indirect connections existed through epicontinental seas, suggested by cyclothermic deposits). Equatorial oceanic communication is reestablished during the Late Jurassic **(Figure 5)**. Mid- to Late Jurassic and possibly Early Cretaceous glaciation (Crowell, 1999) represents the last of the ice resulting from the equatorial Pangean continental assembly **(Figure 6)**, which occupied an equatorial position from Middle to Late Carboniferous time. This glaciation is apparently a relatively minor event when compared with Pleistocene and late Paleozoic icehouse events.

In the Early Carboniferous, oceanic communication through the Rheic Ocean **(Figure 7)** (Scotese, 1997) suggests relatively ice-free and favorable conditions for deposition of major thicknesses of Mississippian carbonate strata. This sinuous connection appears to have been established by the Middle Silurian, with a major greenhouse event during the Devonian **(Figures 8, 9)**. An

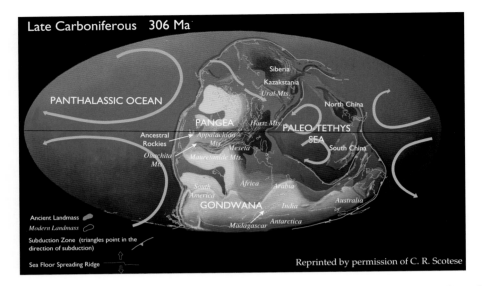

Figure 6 Paleoreconstruction of continental landmasses and oceanic bodies at 306 Ma (Late Carboniferous). This geometry promotes circulation from equator to poles, as depicted by the blue arrows showing projected water-mass movement (modified from Scotese, 1997). Note the closed circulation at the equator owing to landmasses extending far north and south of the equator. We ascribe a major glacial episode to this lack of equatorial circulation.

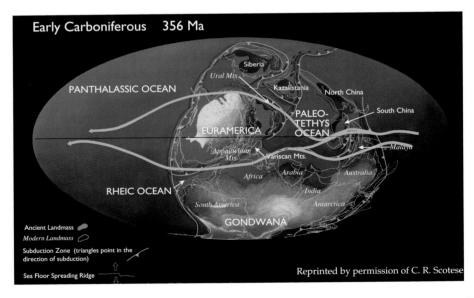

Figure 7 Paleoreconstruction of continental landmasses and oceanic bodies at 356 Ma (Early Carboniferous). This geometry restricts near-equatorial ocean circulation, as shown by the blue arrows showing projected water-mass movement (modified from Scotese, 1997). This model suggests declining circulation and the beginning of continental glaciation, fully developed in the Late Carboniferous (see **Figure 6** for comparison).

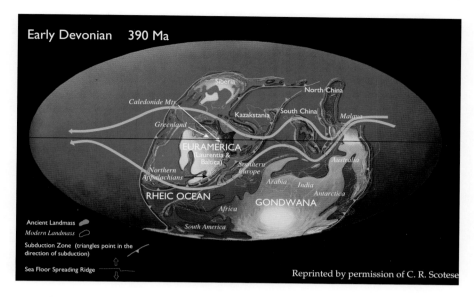

Figure 8 Paleoreconstruction of continental landmasses and oceanic bodies at 390 Ma (Early Devonian). This geometry suggests equatorial ocean circulation, as shown by the blue arrows showing projected water-mass movement (modified from Scotese, 1997). Note the open connections between Euramerica and Gondwana through the Rheic Ocean.

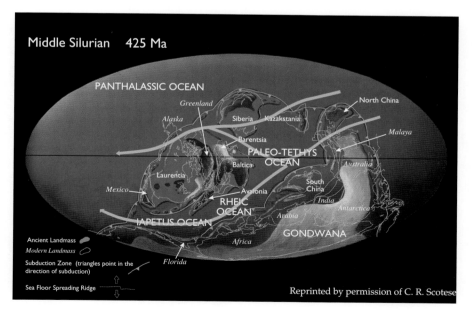

Figure 9 Paleoreconstruction of continental landmasses and oceanic bodies at 425 Ma (Middle Silurian). This geometry suggests near-equatorial ocean circulation, as shown by the blue arrows showing projected water-mass movement (modified from Scotese, 1997). Note the open connections between continents. This is interpreted to be the beginning of a greenhouse state.

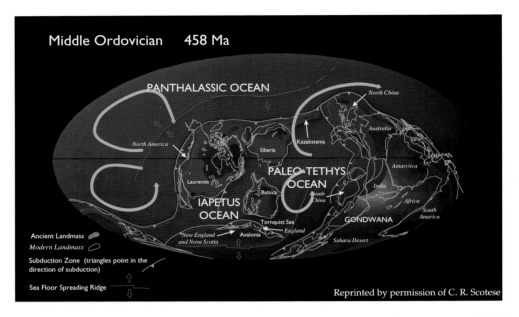

Figure 10 Paleoreconstruction of continental landmasses and oceanic bodies at 458 Ma (Middle Ordovician). This geometry restricts near-equatorial ocean circulation, as shown by the blue arrows showing projected water-mass movement (modified from Scotese, 1997). Note the closed connections between continents, suggesting that the world may have been entering or leaving an icehouse state.

equatorial landmass developed in the Middle Ordovician **(Figure 10)** and may have been responsible for a suggested minor Late Ordovician glacial episode (Crowell, 1999).

Of major interest to current students of the late Proterozoic, there appears to have been a full blockage of equatorial circulation during the Vendian (late Proterozoic) **(Figure 11)**. Hoffman et al. (1998) have argued that the earth was in a "snowball" state during the Vendian. They suggest that carbon fluxes controlled climate changes and that the earth was essentially entirely frozen, with glaciers and sea ice extending through all latitudes, accentuated by a greatly strengthened albedo from the sea-ice cover.

We find, without regard to the merits of such arguments, that the position of the continental cratonic plates of the time are congruent with our theory that equatorial current blockage is present and may well be the trigger for glaciation. Although the degree of certainty of the reconstructions decreases with time, we find that there is little disagreement about the general configurations of cratons, continents, and oceans. Observations of the congruency of these geometries with greenhouse and icehouse events suggest that future climate models may benefit from consideration of a tectonic "basement factor" that can account for all large-scale changes. These are undoubtedly readily modified by other, less-pronounced processes.

SUMMARY

Tectonic geometry (i.e., the distribution of continental and oceanic masses) of the earth through time appears to correlate well with those periods in earth's history when major climate changes took

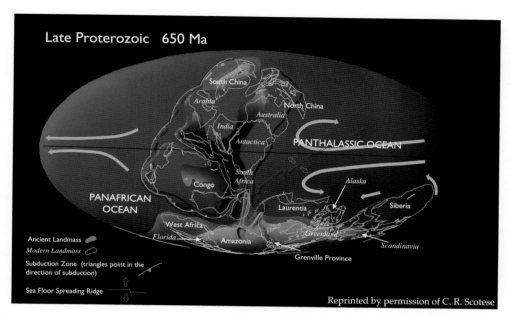

Late Proterozoic 650 Ma

PANTHALASSIC OCEAN

PANAFRICAN OCEAN

South China
Arabia
North China
Australia
India
Antactica
South Africa
Congo
Alaska
Laurentia
Siberia
West Africa
Florida
Amazonia
Greenland
Scandinavia
Grenville Province

Ancient Landmass
Modern Landmass
Subduction Zone (triangles point in the direction of subduction)
Sea Floor Spreading Ridge

Reprinted by permission of C. R. Scotese

Figure 11 Paleoreconstruction of continental landmasses and oceanic bodies at 650 Ma (Vendian). This geometry blocks near-equatorial ocean circulation, as shown by the blue arrows showing projected water-mass movement (modified from Scotese, 1997). This is the setting for the late Precambrian continental icehouse.

place. The locations of landmasses conducive to development of equatorial currents are consistent with the timing of greenhouse events, whereas physical blockage of equatorial current systems appears to coincide with icehouse events. We have theorized that the reason for the correlation lies in the distribution of heat by ocean currents (gyres). When equatorial currents are present, gyres are slower and less efficient, so that less warm tropical surface water reaches polar regions. Major circumequatorial current systems also decrease evaporation rates, which results in less snowfall. When equatorial currents are blocked, then the major circulation in the oceans is within gyres, which transfer heat rapidly and thus increase the likelihood that snow and ice can accumulate in polar regions. The overall effect of these processes enhances conditions that lead to continental glaciation.

Perhaps tectonic events are the major triggers for massive climate changes. Other forcing events modify the tectonism so that orbital forcing, solar variability, and tectonic/orographic events also modify the amplitude and frequency of significant climate shifts. Massive volcanism and bolide events may provide sufficient impetus to modify climate for significant periods of time and provide additional and smaller superposition of irregularities on the basic tectonic curve. The current emphasis on the possible role played by anthropogenic processes as climate-altering mechanisms warrants special consideration, given the geologic evidence of this volume. It is not necessary to invoke man-induced changes such as major increases in carbon-dioxide levels to understand climate changes, although we do not preclude such events. Preliminary studies indicate that carbon-dioxide increases are preceded by temperature increases (Fischer et al., 1999). This relationship suggests that changes in the concentration levels of atmospheric gases are not the universal driving force for climate change.

ACKNOWLEDGMENTS

We appreciate permission from Chris Scotese to use his continental reconstructions, which are a fundamental base for this interpretation of major earth-process control of global climate. The reviewers of this paper, especially Terry Edgar, made constructive and useful suggestions for improvement of the manuscript; we appreciate their efforts and help.

REFERENCES CITED

Barreta, E., and C. C. Johnson, 1999, eds, Evolution of the Cretaceous ocean-climate system, Geological Society of America Special Paper 332, 445 p.

Bigg, Grant R., 1996, The Oceans and Climate: Cambridge University Press, U.K., 266 p.

Bluemle, J. P., J. M. Sabel, and W. Karlen, 1999, Rate and magnitude of past global climate changes: Environmental Geosciences, v. 6, n. 2, p. 63–75.

Broecker, Wallace S., 1997, Thermohaline circulation, the Achilles heel of our climate system: Will man-made CO_2 upset the balance?: Science, v. 278, p. 1582–1588.

Broecker, Wallace S., 1999, What if the conveyor were to shut down? Reflections on a possible outcome of the great global experiment; GSA Today, v. 9, n. 1, p. 1–4.

Crowell, John C., 1999, Pre-Mesozoic ice ages: Their bearing on understanding the climate system: Geological Society of America Memoir 192, 106 p.

Driscoll, N. W., and G. H. Haug, 1998, A short circuit in thermohaline circulation: A cause for northern hemisphere glaciation?: Science, v. 282, p. 436–438.

Ewing, M., and W. L. Donn, 1956, A theory of ice ages: Science, v. 123, p. 1061–1066.

Fischer, A. G., 1982, Long-term climate oscillations recorded in stratigraphy, in Climate in Earth History: National Research Council, National Academy Press, p. 97–104.

Fischer, A. G., 1984, The two Phanerozoic supercycles, in W. A. Berggren and J. A. van Couvering, eds., Catastrophes and Earth History: Princeton University Press, Princeton, New Jersey, p. 129–150.

Fischer, H., M. Wahlen, J. Smith, D. Mastoianni, and B. Deck, 1999, Ice core records of atmospheric CO_2 around the last three glacial terminations: Science, v. 283, p. 1712–1714.

Fitzpatrick, J., 1995, The paradigm of rapid climate change; a current controversy, in L. M. H. Carter, ed., Energy and the Environment; Application of Geosciences to Decision-Making: U.S. Geological Survey Circular 1108, p. 20–22.

Frakes, L. A., 1979, Climates throughout Geologic Time: Elsevier, Amsterdam, 310 p.

Ganapolski, A., S. Rahmsdorf, V. Petouchov, and M. Claussen, 1998, Simulation of modern and glacial climates with a coupled global model of intermediate complexity: Nature, v. 391, p. 351–362.

Gerhard, Lee C., 1999, Geologic constraints on global climate variability (abst.): Environmental Geosciences, v. 6, n. 3, p. 152.

Gray, William M., John D. Sheaffer, and Christopher W. Landsea, 1997, Climate trends associated with multidecadal variability of Atlantic hurricane activity, in Henry F. Diaz and Roger S. Pulwarty, eds., Hurricanes—Climate and Socioeconomic Impacts, Springer Verlag, Berlin, p. 15–53.

Haug, G., and R. Tiedemann, 1998, Effect of the formation of the Isthmus of Panama on Atlantic Ocean thermohaline circulation: Nature, v. 393, p. 673–676.

Hay, W. W., R. M. DeConto, C. N. Wold, K. M. Wilson, S. Voigt, M. Schultz, A. R. Wold, W. Dullo, A. B. Ronov, A. N. Balukhovsky, and E. Soding, 1999, Alternative global Cretaceous paleogeography, in Evolution of the Cretaceous Ocean-Climate System, Barreta, E., and C. C. Johnson, eds., Geological Society of America Special Paper 332, 445 p.

Hoffman, P. F., A. J. Kaufman, G. P. Halverson, and D. P. Schrag, 1998, A Neoproterozoic snowball Earth: Science, v. 281, p. 1342–1346.

Hoyt, D. V., and K. H. Schatten, 1997, The Role of the Sun in Climate Change: Oxford University Press, New York, 279 p.

Huber, Brian T., 1998, Tropical paradise at the Cretaceous poles?: Science, v. 282, p. 2199–2200.

Kerr, Richard A., 1999, The Great African Plume emerges as a tectonic player: Science, v. 285, p. 187–188.

Lamb, H. H., 1995, Climate, History, and the Modern World, 2nd ed.: Routledge, New York, 433 p.

Lang, C., M. Leuenberger, J. Schwander, and S. Johnson, 1999, 16°C rapid temperature variation in central Greenland 70,000 years ago: Science, v. 286, p. 934–937.

Ruddiman, William F., Ed., 1997, Tectonic Uplift and Climate Change: Plenum Press, New York, 535 p.

Scotese, Christopher R., 1997, Paleogeographic Atlas: PALEOMAP Progress Report 90-0497, PALEOMAP Project, Univ. Texas–Arlington, 21 p.

Severinghaus, Jeffrey P., and Edward J. Brook, 1999, Abrupt climate change at the end of the last glacial period inferred from trapped air in polar ice: Science, v. 286, p. 930–934.

Tomczak, Matthias, 1999, Introduction to Physical Oceanography, internet notes for course offered by Flinders University of South Australia; Lecture 8, The Ocean and Climate: http://gaea.es.flinders.edu.au.

Tarduno, J. A., D. B. Brinkman, P. R. Renne, R. D. Cottrell, H. Scher, and P. Castillo, 1998, Evidence for extreme climatic warmth from Late Cretaceous Arctic vertebrates: Science, v. 282, p. 2241–2244.

Toon, O. B., and S. Olson, 1985, The Warm Earth: Science, v. 85, p. 50–57.

Veevers, J. J., 1990, Tectonic-climatic supercycle in the billion-year plate-tectonic eon: Permian Pangean icehouse alternates with Cretaceous dispersed-continents greenhouse: Sedimentary Geology, v. 568, p. 1–16.

Mackenzie, F. T., A. Lerman, and L. M. B. Ver, Recent
past and future of the global carbon cycle, 2001, *in*
L. C. Gerhard, W. E. Harrison, and B. M. Hanson,
eds., Geological perspectives of global climate
change, p. 51–82.

3 | Recent Past and Future of the Global Carbon Cycle

Fred T. Mackenzie

Department of Oceanography, SOEST
University of Hawaii
Honolulu, Hawaii, U.S.A.

A. Lerman

Department of Geological Sciences
Northwestern University
Evanston, Illinois, U.S.A.

L. M. B. Ver

Department of Oceanography, SOEST
University of Hawaii
Honolulu, Hawaii, U.S.A.

ABSTRACT

The global carbon cycle has been affected by four major perturbations owing to human activities on land, and global temperature change since the year 1700. The cycle has been, and continues to be, forced by global emissions from fossil-fuel burning, land-use change, agricultural fertilization of croplands, organic sewage discharges, and a slight temperature rise. The atmospheric carbon-dioxide (CO_2) change in the past 300 years, as computed from our model analysis, agrees well with the observed increase. The anthropogenic perturbations on land have resulted in an increased delivery of carbon to the coastal ocean and changes in its trophic status towards increased net heterotrophy (remineralization of organic carbon exceeding its *in situ* production). Future increases in emissions of carbon from land, based on the projections of the Intergovernmental Panel on Climate

51

Change (IPCC) and the Kyoto Protocol to the United Nations Framework Convention on Climate Change (UNFCCC), suggest increases in atmospheric CO_2 to 495 and 435 parts per million by volume (ppmv), respectively, by the mid-twenty-first century. The net release or uptake of carbon on land involving phytomass and soil organic carbon depends strongly on patterns of land use. In particular, significant, continuous deforestation of tropical and Russian forests, along with continuous increase in global mean temperature, could lead to a weakening of the hypothesized terrestrial sink for the 1990s during the twenty-first century. Furthermore, an increase in atmospheric CO_2 leads to a lower supersaturation state of coastal ocean water with respect to calcite and aragonite, which may result in lower rates of carbonate storage in shallow oceanic areas. Weakening of the oceanic thermohaline circulation ("the conveyor belt") may result in a greater transport of atmospheric CO_2 to coastal ocean waters, at the expense of its reduced transport to the surface open ocean.

GLOBAL CARBON CYCLE

The behavior of the global carbon (C) cycle during recent earth history and the partitioning of atmospheric CO_2 among its various sinks, as well as projections for the near future, are of fundamental concern today to scientists and policy makers. Human activities in the past 300 years have become an increasingly important geological factor in the biogeochemical cycling of carbon and the life-essential elements nitrogen (N), phosphorus (P), and sulfur (S) associated with it. In particular, four major environmental perturbations due to human activities have come to play particularly prominent roles in the carbon cycle and its coupling to the other bioessential elements. These include: (1) C, N, and S emissions to the atmosphere from fossil fuel combustion; (2) changes in land-use activities, which include deforestation, reforestation, logging, and shifting cultivation (these changes have resulted in gaseous C emissions to the atmosphere, and increases in the dissolved and particulate loads of world rivers, including organic matter, N, and P); (3) application of nitrogen- and phosphorus-bearing fertilizers to croplands and the subsequent mobilization of N to the atmosphere and N and P to aquatic systems; and (4) discharges to aquatic systems of sewage containing reactive C, N, and P. In addition, the global mean surface temperature of the planet has increased during the past 300 years by approximately 1°C. This increase, at least in part, could be a result of human activities that lead to the emissions of greenhouse gases to the atmosphere, their accumulation in this reservoir, and consequent global warming.

The past and projected emissions of CO_2 from fossil fuel burning and land-use activities **(Figure 1)** have been, so far, the two main human forcings on the global carbon cycle. Future projections for fossil-fuel emissions according to two different scenarios, "Business as Usual" (BAU) and the 1997 Kyoto Protocol, differ in magnitude by a factor of two in the year 2040. Projections into the twenty-first century of the net release of CO_2 from changes in the land-use activities mentioned above are difficult to make. The problem lies in the fact that there are large policy-related uncertainties to projections of changes in land-use patterns and their regional differences. However, among other drivers of change, land use has been identified as the major important driver of changes in biodiversity for the year 2100 (Sala et al., 2000) and is the second most important anthropogenic source of CO_2 to the atmosphere (Houghton, 1995). Thus, we show in **Figure 1** the upper and lower bounds for CO_2 emissions from land use. The former is based on projections from Houghton (1983; 1991b; 1995) and Houghton et al. (1998) and the latter is based on emission estimates of about 1.1 ± 0.5 gigatons, i.e., billion tons per year (Gt/year) in 1995 (T. M. L. Wigley, personal communication, 1999). In a later section, we use the anthropogenic CO_2 emission scenarios of **Figure 1** as external forcings in model calculations.

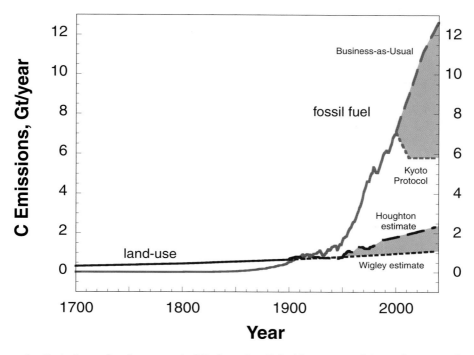

Figure 1 Emissions of anthropogenic CO_2 from fossil-fuel burning and from changes in land-use activities in the past 300 years (1700–1999) and projections to 2040. See text for description and sources of data.

In this paper, we begin analysis of the past and future behavior of the carbon cycle in the year 1700, with the earth-surface system in an assumed quasi-steady state. We analyze the changes in the flows of carbon and its distribution in the major reservoirs of the earth's surface that resulted from the anthropogenic forcings and global temperature rise.

MAJOR FORCINGS AND MODEL ANALYSIS

The analysis of the carbon cycle in the past 300 years is based on an earth system model: the Terrestrial-Ocean-aTmosphere-Ecosystem Model (TOTEM). This is a unique process-based model of the global coupled biogeochemical cycles of carbon, nitrogen, phosphorus, and sulfur (C-N-P-S) that recognizes the dependence of the carbon cycle on the cycles of nitrogen, phosphorus, and sulfur. The essential feature of the model is manifested in the coupling between the C-N-P-S cycles at every biologically mediated transfer process, such as photosynthesis, autorespiration, decay, and burial of organic matter. This provision for the diverse, process-based biogeochemical interactions among the four element cycles in four environmental domains of land, atmosphere, coastal ocean, and open ocean (including the sediments of the latter two domains) distinguishes TOTEM from most other models. The coupled-cycles approach is critical to modeling the responses of biogeochemical systems to global change because an anthropogenic or natural source of one of these elements, such as fossil-fuel burning or humus respiration, is often a source of all three other elements.

The global surface environment, as defined in the model, is comprised of the four major domains and includes thirteen reservoirs **(Figure 2)**: the atmosphere; six terrestrial reservoirs (living biota, humus, inorganic soil, continental soilwater, shallow groundwater, and lakes); three coastal-zone reservoirs (organic matter, water, and sediments); and three open ocean reservoirs (organic matter, surface water, and deep water). The rivers are not defined as a reservoir because the residence time of water in rivers with respect to recharge by atmospheric precipitation is very short, about 20 days (Berner and Berner, 1996). **Figure 2** is the conceptual framework of TOTEM showing the major reservoirs and processes responsible for the transport of C-, N-, P-, and S-bearing components in the system. The initial conditions (reservoir masses, transfer fluxes, and flux constants) of C, N, P, and S in the global reservoirs are given in Appendix A. Model equations and kinetic parameters are listed in Appendix B. Detailed calculations involving derivation of these equations and estimates are provided in Ver (1998) and Ver et al. (1999a). Other details of the model and its applications have been reported elsewhere (Mackenzie et al., 1998a, 1998b, 2000; Mackenzie et al., 1993; Ver et al., 1994, 1999b).

The mathematical structure of the model takes into account the time-dependent external forcings, and combinations of linear and nonlinear transport and reaction kinetics, as in the following generalized mass-balance equation:

$$\frac{dM_i}{dt} = \sum_j F_{ji}(t) - \sum_j F_{ij}(t)$$

where Mi is the mass of an element (C, N, P, or S) in the ith reservoir, t is time, and F_{ji} and F_{ij} are input and output fluxes, respectively, between reservoir i and an adjacent reservoir j that may vary with time. Typically, flux F is a function of ki, a rate parameter for the physical, biological, or chemical removal fluxes from the reservoir, and of physical variables such as temperature, reflecting the complexity of the different processes. In general, rate parameter ki may be either constant or depend on environmental variables, external forcings, or the state of the reservoirs. The coupling between the C-N-P-S cycles is based on the Redfield ratios in each of the biologically mediated transfer processes.

An important major difference between TOTEM and other terrestrial carbon models and global climate models (GCMs) is the treatment of the observed data for atmospheric CO_2 concentration. Whereas this dataset is used as a prescribed input or forcing function in most models of global environmental change (for example, Bruno and Joos, 1997; Cao and Woodward, 1998; Sarmiento et al., 1998), the time course of change of atmospheric CO_2 concentration is not treated as an input function in TOTEM. Rather, the 300-year dataset for the atmospheric C reservoir, as well as that for the other elemental reservoirs in the modeled earth system, are outputs of the model and are a result of iterative model calculations. Numerical results for atmospheric CO_2 concentrations from TOTEM are then compared with the observed dataset.

At the starting time of the model analysis in the year 1700, in the preindustrial earth's environment, the carbon cycle and N, P, and S cycles coupled to it are assumed to be in a quasi-steady state. This assumption is reasonable because the carbon cycle during the preindustrial period, generally defined as the time prior to the year 1800, was in a quasi-steady state under a small but sustained global human disturbance of the environment. For example, the estimated rates of release of CO_2 and CH_4 to the atmosphere associated with the preindustrial human activities of wood-fuel combustion (mostly for food preparation, heating, construction, and biomass burning for metal and clay pottery production) and forest clearing for agricultural land use were almost constant over a period of 500 years prior to 1700. In the year 1200, the land-use derived CO_2 emissions were

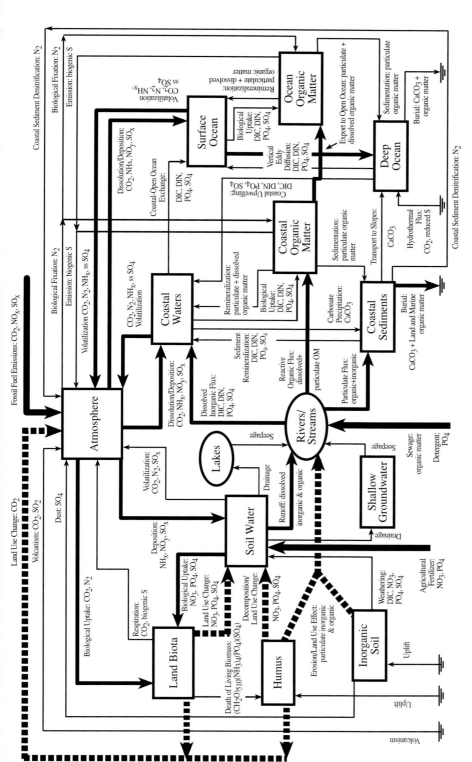

Figure 2 Conceptual diagram of TOTEM, a model of the coupled global biogeochemical cycles of C-N-P-S in the system of land, atmosphere, coastal ocean, and open ocean. Thick arrows are the forcing fluxes of the elements from fossil-fuel combustion, application of agricultural fertilizers, and municipal sewage and wastewater disposal. Thick dashed arrows denote fluxes of the elements from the terrestrial organic and inorganic reservoirs. The medium-sized arrows denote natural fluxes that have been significantly modified by human forcings, whereas the thin arrows represent fluxes that have been less affected by the perturbations. A quasi-steady state is assumed at the initial conditions prior to 1700 (see Appendix A).

estimated to be about 0.4 Gt C/year, and they increased by only 0.06 %/year to the year 1700 (Kammen and Marino, 1993). Fossil-fuel emissions were negligible during this period (Keeling, 1973), the earth system was nearing a period of recovery from the lower global temperatures of the Little Ice Age, and atmospheric CO_2 levels were relatively low (Barnola et al., 1995; Etheridge et al., 1996). Indeed, measurements from ice cores show that the concentration of atmospheric CO_2 over a period of about 1000 years prior to 1800, and including the preindustrial era, varied by no more than 10 ppmv, and imply that the C cycle was approximately in a steady state (Raynaud and Barnola, 1985; Siegenthaler et al., 1988).

In our analysis from the year 1700 on, the system is perturbed by the four forcings listed in the previous section that are representative of the major pathways by which humans have perturbed the C, N, P, and S cycles, and by variations in the climatic variable of temperature (Caraco, 1995; Charlson et al., 1992; Galloway et al., 1995; Houghton, 1996a; 1996b; Smil, 1985). The four perturbations are input functions to the global system, as modeled by TOTEM. Fossil-fuel CO_2 is injected into the atmosphere according to its emissions history (Marland et al., 1999) and becomes part of the well-mixed atmospheric C reservoir. However, the fate of anthropogenic N and S emitted to the atmosphere (Dignon, 1992; Dignon and Hameed, 1989; Hameed and Dignon, 1992) is controlled by the very short lifetimes of these gases and their conversion products in the atmosphere. The nitrogen component does not mix with the large N_2 reservoir but is deposited onto the terrestrial and coastal ocean reservoirs by wet and dry deposition and sedimentation of large particles. This flux is superimposed on the natural fluxes of N out of the atmosphere. Anthropogenic sulfur gases emitted to the atmosphere are also rapidly deposited onto terrestrial and oceanic surfaces.

Land-use activities, including the conversion of land for food production (grazing land, agricultural land), for urbanization (building human settlements, roads, and other structures), for energy development and supply (building dams and hydroelectric plants, and mining fossil fuels), and for resource exploitation (mining metals, harvesting forest hardwood) (Mackenzie, 1998), affect earth's processes by changing the mass and residence time of C, N, P, and S in the reservoirs of terrestrial organic matter. In TOTEM this perturbation is modeled via three mechanisms. First, it is modeled as a direct perturbation when CO_2 is emitted to the atmosphere during the enhanced conversion of organic phytomass and soil humus to its inorganic components. Second, it is modeled as a negative feedback mechanism to rising atmospheric CO_2 concentrations when inorganic nutrients N, P, and S are released to the continental reservoirs accompanying the emission of C to the atmosphere in the same elemental ratios as those in phytomass or soils. The increased availability of nutrients in continental soilwater, coupled with rising atmospheric CO_2 and warming temperatures, stimulates terrestrial productivity and storage of organic carbon in the terrestrial biota, thus increasing the drawdown of atmospheric CO_2. The third mechanism through which earth's processes are affected when humans change the makeup of the land is via the increased transfer of organic and inorganic material from land to the coastal zone owing to soil erosion, mineral dissolution, and surface water runoff.

One of the consequences of changing land-use activities is the release of anthropogenic CO_2 to the atmosphere that is accompanied by the remobilization of inorganic N and P into continental water reservoirs. The resulting fertilization effect on the terrestrial phytomass is manifested by the enhanced photosynthetic uptake, in agreement with the conclusions derived from measured atmospheric CO_2 and O_2/N_2, forest inventories, and global carbon budgets (Houghton et al., 1998; Joos and Bruno, 1998; C. D. Keeling et al., 1996; R. F. Keeling et al., 1996). Our analysis of the land-use perturbation provides a mechanistic solution that explains the concurrent release of CO_2 to the atmosphere and accumulation of C in terrestrial ecosystems (Houghton et al., 1998), and a biochemical basis for the internal fertilization of the terrestrial phytomass. Interestingly, although the effect of land-use activities on nutrient remineralization has been mentioned in the literature (Melillo et al., 1993; Melillo et al., 1996), it has yet to be incorporated in many terrestrial-ecological models.

PARTITIONING OF CARBON

The rise in atmospheric CO_2 during the past 300 years is one of the results of model analysis using TOTEM, and it compares very well with observational data and with results from other models addressing the partitioning of anthropogenic CO_2 **(Figure 3)**. The model-computed values of atmospheric CO_2 concentration over the 300-year time course compare very closely with the combined atmospheric CO_2 data obtained from ice cores (Barnola et al., 1995; Etheridge et al., 1996; Friedli et al., 1986; Neftel et al., 1985) and from Mauna Loa (Keeling and Whorf, 1999).

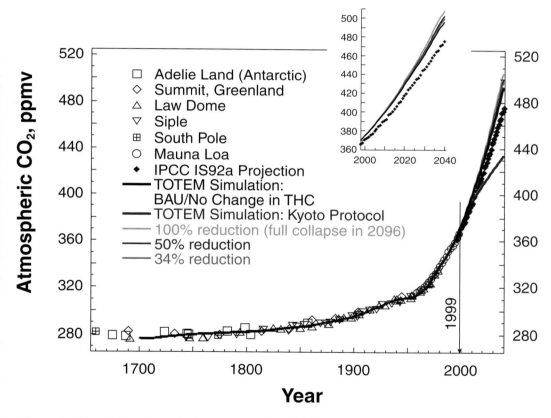

Figure 3 Rise of the atmospheric concentration of CO_2, computed for the period since 1700 to 1999, and projected for an additional 40 years to 2040. Note the agreement of the computed curve with the reported observational data for 1700–1999. Shown are the TOTEM-calculated atmospheric CO_2 rise (solid line), the IPCC IS92a projection (shaded diamonds, Houghton et al., 1996), the projected rise in atmospheric CO_2 for three scenarios of reduction in the oceanic thermohaline circulation by the year 2095 by 34%, 50%, and 100% (red, green, and yellow lines, respectively), and the projected rise under effect of the Kyoto Protocol emissions reduction scenario (blue line). The inset shows in detail the effect of changing oceanic thermohaline circulation on atmospheric CO_2 concentrations.

Figure 4 shows the partitioning of carbon from the two major inputs (fossil-fuel and land-use emissions) among the atmosphere, ocean, land, and coastal oceanic sediments. Our calculated results for the partitioning of anthropogenic CO_2 over the time course from 1850 to present compare very well with results from GCM models (Bruno and Joos, 1997; Sarmiento et al., 1992) and from other models of the global carbon cycle (for example, Hudson et al., 1994). **Table 1** shows calculated values for the carbon sources, sinks, and enhanced fluxes for the years 1700 (preindustrial time), 1860 (early industrial), 1950 (industrial), and 1999 (the present; see also **Figure 4**). The increase in atmospheric CO_2 since the 1940s and 1950s has been faster than its uptake by other environmental reservoirs, indicating that the feedbacks of CO_2 removal from the atmosphere have not been sufficiently strong on this time scale to overcome the rising emissions.

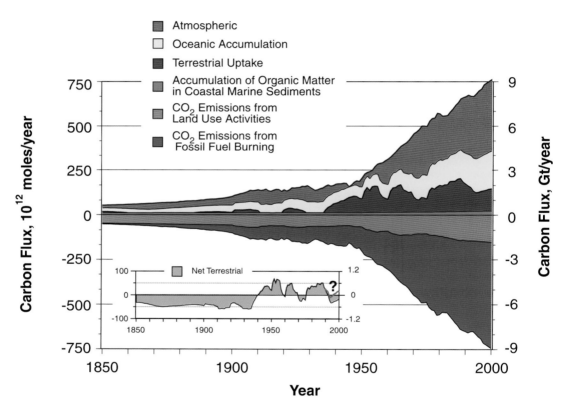

Figure 4 The model-calculated partitioning of the carbon perturbation fluxes (from fossil-fuel burning and land use) to the atmosphere for the period since 1850 to present, in units of 10^{12} moles C/year and Gt C/year. The anthropogenic carbon sources are plotted on the (–) side and the resulting accumulations are plotted on the (+) side. Note that the values for CO_2 emissions from land-use activities (orange field) are plotted separately from those for the terrestrial CO_2 uptake flux (blue field), to highlight their opposing effects on the terrestrial organic reservoirs. The inset shows the net accumulation (or loss) by the terrestrial organic reservoirs and the sensitivity of this process to the assumed magnitude of the land-use flux of anthropogenic CO_2. (See also **Table 1**.) The filled yellow-green area represents results for the standard run using the higher-bound estimates of land-use emissions by Houghton (1983, 1991a, 1991b, 1996b; Houghton et al., 1998); the red line shows TOTEM results using the lower-bound flux estimates by Wigley (personal communication, 1999).

Table 1 Historical summary of carbon sources, sinks, and enhanced fluxes owing to human perturbations and temperature change, as calculated from TOTEM. In column 2 are the initial values prior to the start of the perturbations. In columns 3–5 are the computed magnitudes of change from initial conditions and the percent change relative to initial conditions enclosed in parentheses. See also **Figure 4** for the historical and future trends of carbon sources and sinks.

Carbon fluxes in background (pre-1700) and perturbed (1700–1999) states	Steady State pre-1700	Incremental change since 1700 (Gt C/year)		
		Early Industrial 1860	Industrial 1950	Recent 1999
Major sources of anthropogenic C:				
(1) Fossil-fuel burning [1]	0	0.09 (18%)	1.63 (175%)	7.0 (283%)
(2) Land-use CO_2 emissions [2]	0	0.58 (52%)	0.79 (156%)	1.8 (462%)
(3) Sewage discharges to natural waters	0	0.02	0.05	0.11
Major sinks of anthropogenic C:				
(4) Atmospheric accumulation [3]	0	0.30	0.59	4.79
(5) Oceanic accumulation [4]	0	0.27	0.82	2.49
(6) Organic C in coastal sediments [5]	0	0.03	0.05	0.16
(7) Net terrestrial gain (+) or loss (−) [6]	0	−0.48	0.24	−0.22
Fluxes perturbed by human activities:				
(8) Terrestrial biotic CO_2 uptake [7]	0.60	0.11 (18%)	1.05 (175%)	1.7 (283%)
(9) Oceanic CO_2 uptake (+) or loss (−)	−0.5	0.26 (52%)	0.78 (156%)	2.31 (462%)
(10) Riverine dissolved and particulate organic C	0.41	0.01 (2%)	0.02 (5%)	0.12 (29%)
(11) Riverine dissolved and particulate inorganic C	0.56	0.02 (4%)	0.03 (5%)	0.17 (30%)
(12) Carbonate accumulation	0.25	0.01 (4%)	0.01 (4%)	0.06 (24%)
Long-term, geologic fluxes:				
(13) Organic C burial in sediments	0.13			
(14) Carbonate C burial in sediments	0.4			
(15) Weathering	0.13			
(16) Volcanism and dust	0.10			
(17) Hydrothermal venting	0.06			
(18) Uplift of old organic C	0.37			

[1] See **Figure 4**, CO_2 Emissions from Fossil Fuel Burning, for temporal trends.
[2] See **Figure 4**, CO_2 Emissions from Land Use Activities, for temporal trends.
[3] For values in columns 3–5, Atmospheric accumulation = rows (1) + (2) − (8) − (9); see **Figure 4**, Atmospheric Accumulation for temporal trends.
[4] For values in columns 3–5, Oceanic accumulation = rows (3) + (9) + (10) + (11) − (6) − (12); see **Figure 4**, Oceanic Accumulation for temporal trends.
[5] See Figure 4, Accumulation of Organic Matter in Coastal Marine Sediments, for temporal trends.
[6] For values in columns 3–5, Net terrestrial = rows (8) − (2) − (10); see **Figure 4** inset, Net Terrestrial for temporal trends.
[7] For values in columns 3–5, Terrestrial biotic CO_2 uptake = (Photosynthetic Uptake) − (Plant Respiration) − (Soil Respiration); note that in **Figure 4**, the term *Terrestrial Uptake* includes the loss of organic C to the coastal margin, i.e., Terrestrial Uptake = rows (8) − (10).

There is very close agreement between the TOTEM-calculated fluxes for 1985 **(Figure 4)** and the values reported in the literature as average rates for the period 1980 to 1989. For example, the calculated rate of anthropogenic CO_2 accumulation in the atmosphere (3.1 Gt C/year) is within the range of estimates of direct measurements of atmospheric CO_2 accumulation (average of 3.3 ± 0.2 Gt C/year, Houghton et al., 1996; Sarmiento et al., 1992). In addition, the calculated rate of oceanic accumulation for the same year (1.8 Gt C/year) is consistent with that predicted from ocean-at-mosphere models (1.9 Gt C/year for 1980–1989, Siegenthaler and Sarmiento, 1993) and it is within the range of estimates of the oceanic sink as inferred from changes in oceanic and atmospheric d13C (average of 2.1 ± 0.9 Gt C/year over the period 1970 to 1990, Quay et al., 1992).

R.F. Keeling et al. (1996) estimated the enhanced uptake of CO_2 for the period 1991 to 1994 by the Northern Hemisphere forests (2.0 ± 0.9 Gt C/year) and by the global oceans (1.7 ± 0.9 Gt C/year) from changes in atmospheric CO_2 concentration and atmospheric O_2/N_2 ratio. More recent estimates of Battle et al. (2000) for the decade of the 1990s (1991–1997) give the CO_2 storage on land as 1.4 ± 0.8 Gt C/year and 2.0 ± 0.6 Gt C/year in the oceans. The average rates calculated in the TOTEM simulation for this period fall within the range of these observed values: global terrestrial biotic uptake of 1.6 Gt C/year and global oceanic uptake of 2.1 Gt C/year. Battle et al. (2000), among others, have recently shown that the decade of the 1980s was approximately neutral with respect to CO_2 storage on land. This result is in accord with our model analysis. Estimates of CO_2 storage on land in the coterminous United States for the 1980s and early 1990s vary by a factor of more than three, within a range of approximately 0.1 to 0.35 Gt C/year (Schimel et al., 2000). Although the data for the continental United States cannot be easily extrapolated to the global land reservoir, the differences between the individual estimates demonstrate the present state of uncertainties. The differences in the magnitudes of net CO_2 uptake by land in individual years or decades, as reported by different investigators and our model analysis **(Figure 4,** inset), are attributable to the modeling methodology and uncertainties in the different data bases, as discussed in more detail below.

The issue of the magnitude of the land-use perturbation **(Figure 1)** and its effects on atmospheric CO_2 remains in question. Calculations by Houghton (1983; 1991a; 1991b; 1996b; Houghton et al., 1998) of effects of land-use activities during the period 1850 to 1990 were reported as net CO_2 emissions, including the drawdown or release of CO_2 from such processes as conversion of forest to agricultural land, abandonment of agricultural land, harvest, and regrowth. Houghton's estimate for the early 1990s is 1.4 to 1.7 Gt C/year. More recent estimates by the IPCC are considerably lower (1.1 ± 0.5 Gt C/year in 1995, T. M. L. Wigley, personal communication, 1999), implying that tropical ecosystems were not a strong net source or sink for CO_2 during this period (R. F. Keeling et al., 1996). The net response of the terrestrial organic reservoirs **(Figure 4,** inset) is very sensitive to the assumed magnitude of the land-use perturbation. For example, in the year 2000, the terrestrial organic reservoir is a net sink for anthropogenic CO_2 at the rate of 0.15 Gt C/year when land-use perturbations are within the range estimated by the IPCC. With a land-use flux in the range of Houghton's estimates, the reservoir would be a net source of anthropogenic CO_2 at the rate of 0.2 Gt C/year. Because the magnitude of the fossil-fuel perturbation was overwhelmingly greater than that from land-use activities in recent decades, the atmospheric CO_2 concentration does not show a significant response to changes in the land-use emissions parameter.

For the ocean realm, we confirmed the hypothesis that enhanced organic and carbonate carbon accumulation in coastal marine sediments has been a small sink for anthropogenic CO_2 during recent historical time. Prior to extensive human activities on land, the global coastal margin was likely in a near-steady state of net heterotrophy (i.e., organic matter consumption exceeding production), with an associated net evasion of CO_2 to the atmosphere of about 0.08 Gt C/year (Smith and Hollibaugh, 1993). The accumulation of calcium carbonate in shallow-water, coastal-margin environments was also a source of CO_2 to the atmosphere at that time (Wollast and Mackenzie, 1989). Thus, the preindustrial net flux of CO_2 between the coastal oceans and the atmosphere owing to organic and inorganic processes was a net evasion to the atmosphere of about 0.2 Gt C/year.

It is important to note, however, that various investigators (e.g., Gattuso et al., 1998b; Kempe, 1995; Rabouille et al., submitted; Smith and Hollibaugh, 1993; Ver et al., 1999b; Wollast, 1998) have disagreed as to what extent the coastal zone as a whole is heterotrophic or autotrophic (when organic matter production exceeds consumption), and on what time scale. The problem of the organic carbon balance in the coastal zone and open ocean is certainly not fully resolved at present (see Kempe, 1995, for review). In accepting the estimate of Smith and Hollibaugh (1993), we are cognizant of the fact that this estimate may be subject to further refinements and that the ultimate outcome will significantly influence our estimates of carbon fluxes for the initial steady-state condition of the coastal zone prior to 1700. Regardless of the assumed initial magnitude of heterotrophy in the global coastal zone in the year 1700, the trends over the past 300 years are robust. Sensitivity analysis using TOTEM with different initial values of the trophic status shows that the calculated overall trend toward increasing heterotrophy, particularly since 1940, has been maintained (Ver et al., 1999b). One value of a model of the nature of TOTEM is that it points to processes and fluxes that are not well known or constrained on a global scale.

Our modeling results show that the trophic state of the coastal zone remained almost constant from the initial period of simulation to about the year 1900 (Ver et al., 1999b). However, increasing human perturbations on land and annual temperature variations from 1900 to the present have induced a state of increasing heterotrophy in the coastal margin. Thus, over the period of 300 years of human perturbation on land, the coastal ocean has maintained its state of heterotrophy and its preindustrial role as a net source of CO_2 to the atmosphere. The rate of organic matter imported from land increasingly exceeded the rate of in situ production. The fate of the imported terrestrial organic matter was remineralization to the dissolved inorganic carbon pool (DIC), burial in the coastal sediments, and to a lesser extent, export to the continental slope. Although the riverine input of inorganic nutrients also increased significantly during the same period of time, the incremental new production supported by this input was not sufficient to reverse coastal zone metabolism from net heterotrophy to net autotrophy. Increased heterotrophy and carbonate precipitation in the coastal zone constitute enhanced sources of DIC. Because CO_2 is a component of DIC, its increase in coastal ocean waters tended to oppose the pressure across the air-sea interface from rising atmospheric CO_2 concentration, thereby modulating the response of the coastal ocean to the atmospheric perturbation and modifying the magnitude of the invasion flux of anthropogenic CO_2. The net effect of sustained heterotrophy and increased carbonate precipitation is a reduction in the sink strength of the coastal ocean for anthropogenic CO_2. This conclusion, derived from model analysis, implies that were it not for the accumulation of anthropogenic CO_2 in the atmosphere, the global coastal oceans would be sources of CO_2 to the atmosphere, reflecting the imbalance favoring gross respiration over gross photosynthesis and calcium carbonate accumulation.

The major changes in the carbon fluxes within the land-atmosphere-ocean system, as computed from TOTEM for the conditions of the four forcings and temperature change, are shown in **Table 1**. In the 1700s and early 1800s, the terrestrial reservoirs were a net source of anthropogenic CO_2 to the atmosphere from the clearing of grasslands and extensive deforestation of the Northern Hemisphere (Houghton, 1983). Most of this C accumulated in the atmosphere and oceans. The transfer of perturbation effects to the coastal zone via the rivers was minimal. The period of the Industrial Revolution through early post-World War II (1860–1950) was the time of a growing economy based on fossil fuels (Marland et al., 1999). The period after World War II (1950–2000) is characterized by the extensive use of fossil fuels, synthetic agricultural fertilizers, and phosphate-based detergents (Esser and Kohlmaier, 1991; Smil, 1991). Although the use of phosphate-based detergents decreased in many developed countries in the later part of the twentieth century, such detergents are still being used in developing countries. The accelerating rates of anthropogenic perturbations on land have increased riverine transport of particulate and dissolved organic carbon to the global coastal zone. In the year 1999, the rate of organic C accumulation in coastal marine sediments nearly doubled relative to the geologic long-term rate prior to the year 1700, from 0.13 to 0.29 Gt C/year.

The cumulative changes in the major transfer fluxes and reservoirs are summarized in **Table 2**. Over the past 300 years, the largest perturbation of the earth's surface carbon system was the cumulative emission of about 480 Gt C to the atmosphere: 280 Gt C, or 60%, owing to fossil-fuel burning and cement production, and 200 Gt C, or 40%, from land-use activities. Minor perturbations during this same period included the enhanced cumulative transport to the coastal zone of about 6 Gt organic C and 9 Gt inorganic C owing to land-use activities and the discharge of municipal sewage and wastewater into the coastal zone, adding to this environment a cumulative amount of about 9 Gt C. The cumulative enhanced fluxes of C resulting from these perturbations included the uptake of about 135 Gt C by the global oceans and the fertilization of the terrestrial biotic reservoir of about 140 Gt C, owing to rising atmospheric CO_2, remobilized N and P nutrients from land-use and application of agricultural fertilizers, and warming temperatures. However, the enhanced biotic uptake was greatly exceeded by the loss of mass from land-use activities (200 Gt C emitted to the atmosphere and 6 Gt C transported to the coastal margin), resulting in a net loss of about 66 Gt C from the terrestrial organic C reservoirs. The atmosphere accumulated about 205 Gt of anthropogenic C, the oceans about 145 Gt C, and the coastal organic and inorganic sediments about 14 Gt C.

PROJECTIONS FOR THE FUTURE

The thermohaline circulation of the ocean is a major mechanism of transfer of CO_2 from the surface to the deep ocean (also known as "the oceanic conveyor belt"). Thus changes in the circulation of the ocean that might be caused by global warming will also affect the ocean's role in the sequestration of anthropogenic CO_2. A measure of the circulation is the volume of flow in the meridional direction in the North Atlantic Ocean of the surface waters that flow from the south to the north and sink because of water cooling in the high northern latitudes. The present-day thermohaline circulation is equivalent to about 18 Sv (1 Sverdrup, or 1 Sv, = 10^6 m^3/sec). The intensity of the thermohaline circulation, affected by the global temperature distribution, was estimated by Manabe and Stouffer (1994) for the cases of a doubled (2×) and quadrupled (4×) atmospheric CO_2 concentration. In their 4×-model scenario, by the end of 200 years, the thermohaline circulation in the North Atlantic Ocean would decrease in intensity by about 72%, equivalent to 5 Sv of flow. The ultimate cause of these changes, as projected by Manabe and Stouffer (1994), is the warming of the surface ocean brought about by rising CO_2 concentrations and an enhanced greenhouse effect.

We estimate the effects of such a reduction in global thermohaline circulation on the flows of CO_2 between the atmosphere and coastal and open-ocean surface waters. Consistent with the projections of Manabe and Stouffer (1994), we investigate three scenarios of reduction in thermohaline circulation beginning in 1996: a full collapse (100%), 50%, and 34% reduction by the year 2096 **(Figure 3)**. These scenarios are applied in the model via the effect on the overall mixing time of the deep ocean. When the thermohaline circulation is reduced by 100%, 50%, and 34%, the mixing time is increased by 25%, 10%, and 6%, respectively.

Model results for the period 1996 to 2040 indicate that the greater the reduction in the rate of the thermohaline circulation, the greater the rate of increase of atmospheric CO_2. The reason for the greater concentration of CO_2 in the atmosphere is a decrease in the strength of the high-latitude

Table 2 Cumulative transfers of carbon in the atmosphere, land, coastal ocean, and open ocean in the 300 years (1700–1999) of anthropogenic forcings. Gain or storage and input fluxes are positive numbers; losses and removal are negative. See also **Figures 2** and **4**.

RESERVOIR OR FLUX	MASS (GT C)
Emissions from fossil fuels	280
Emissions from land-use activities	200
TOTAL EMISSIONS	**480**
Atmosphere accumulation	205
CO_2 uptake by land biomass	140
CO_2 uptake by the oceans	135
TOTAL ACCUMULATION AND UPTAKE	**480**
Land	
(1) CO_2 uptake by land biomass	140
(2) CO_2 from land-use activities	−200
(3) Riverine transport of organic C	−6
(4) Riverine transport of inorganic C	−9
(5) Sewage discharge of organic C	−9
TOTAL LAND (ORGANIC C): (1) + (2) + (3)	**−66**
Coastal Ocean	
Riverine input	15
Input by upwelling	1
Sewage discharge	9
CO_2 uptake from atmosphere	10
Storage in sediments	−14
Export to open ocean	−17
TOTAL COASTAL OCEAN (ORGANIC + INORGANIC)	**+4**
Open Ocean	
Import from coastal zone	17
CO_2 uptake from atmosphere	125
Upwelling to coastal waters	−1
Deep-ocean sediments	0.01
TOTAL OCEAN (ORGANIC + INORGANIC)	**+141**

oceanic sink for CO_2, brought about by a decrease in the production and sinking rate of deep water as a consequence of the weaker meridional circulation. The change over these three decades is not substantial, but it represents about 12 ppmv difference by the year 2040 between the cases of an unchanged thermohaline circulation and its complete cessation on a century timescale **(Figure 3)**. In addition, if such a change in thermohaline circulation were to occur and persist, the global coastal ocean would become a stronger sink of anthropogenic CO_2 because of the rising CO_2 concentration in the atmosphere and a decreased supply of waters, rich in inorganic carbon and nutrients, to the coastal zone by upwelling from depth (Mackenzie et al., 2000).

Primary productivity in global coastal margin waters is also affected under scenarios of changing thermohaline circulation. The reduced input of nutrients by upwelling from intermediate oceanic waters not only underscores the importance of this source flux for recycled nutrients but also indicates the stronger role of phosphorus as a nutrient that controls biomass and organic matter accumulation in the coastal zone system. Other biogeochemical processes in the global coastal zone are also affected under scenarios of changing thermohaline circulation; these results are discussed in greater detail in Mackenzie et al. (2000).

We also analyzed the effects of the procedures for decreasing greenhouse gas emissions established by the Kyoto Protocol to the UNFCCC, at the meeting held in Kyoto, Japan, in 1997 (see **Figure 1**, CO_2 emissions scenario). We assume global compliance to the protocol, with future fossil-fuel emissions decreasing linearly to 5.2% below 1990 levels by the year 2012 and maintained at this value until the end of the simulation. If this protocol were to be implemented effective the year 2001, results of our TOTEM experiment show that by the year 2010, the global mean atmospheric CO_2 concentration will have risen by about 20 ppmv, despite a 15% reduction in fossil-fuel emissions relative to that of the year 2000. In comparison, the BAU scenario of the IPCC projects an increase in atmospheric CO_2 by about 25 ppmv, with a 20% increase in emissions relative to those of the year 2000. The sluggish response by the atmospheric CO_2 reservoir to the emissions-reduction protocol may be attributed to the projected continued rise in land-use CO_2 emissions and the 10-year mean residence time of CO_2 in the atmosphere. The effect of the reduction on the atmospheric CO_2 concentration becomes progressively more evident towards the mid-twenty-first century. By the year 2040, fossil-fuel emissions under the Kyoto Protocol scenario would be 55% lower than that of the BAU scenario **(Figure 1)**, and the global mean atmospheric CO_2 concentration is projected to be about 12% (60 ppmv) lower than that of the BAU scenario **(Figure 3)**. It may be anticipated that this reduction in the rate of accumulation of atmospheric CO_2 would be maintained unless other feedback mechanisms act nonlinearly and either increase or decrease the drawdown of CO_2: for example, such processes as changes in the composition of the terrestrial biomes, changes in the rates of terrestrial photosynthesis and respiration, or reduction in the intensity of the oceanic thermohaline circulation. Therefore it appears that the Kyoto Protocol, setting emission limits for greenhouse gases, may be one approach to reducing the rate of increase in atmospheric CO_2, but only if the physical and biogeochemical mechanisms of redistribution of CO_2 among the atmosphere, land, and ocean do not significantly change from the present.

We investigated the sensitivity of the system, particularly that of the terrestrial realm, to the assumed rate of CO_2 emissions from changes in land-use practices. **Figure 5** shows the balance between anthropogenic CO_2 sources and sinks projected to 2040 under two BAU scenarios of the upper and lower bounds of land-use CO_2 emissions shown in **Figure 1**. The inset of **Figure 5** shows the net response (carbon storage or release) of the combined terrestrial reservoirs of phytomass and soil organic C for the same period. Note in particular the results for the period from the late 1990s through the first decade of the twenty-first century. Under the high estimate of land-use CO_2 emissions, the balance between the enhanced terrestrial uptake of CO_2 through fertilization and CO_2 emissions from land-use activities is negative. This implies net release of CO_2 to the atmosphere from the terrestrial organic reservoirs. However, under the lower bounds of land-use CO_2 emissions, the terrestrial reservoirs store organic carbon for much of the 1990s and on into the twenty-first century. This result dramatically demonstrates the role of future land-use activities in the net storage or release of carbon on land.

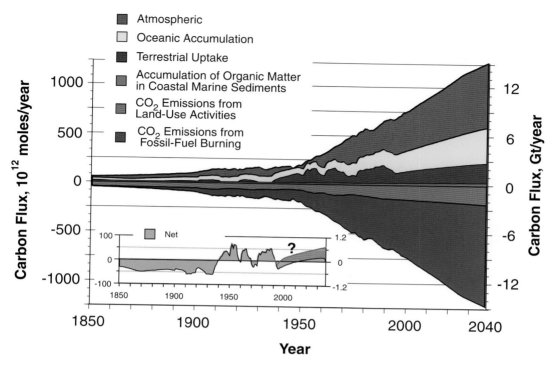

Figure 5 The model-calculated partitioning of the carbon perturbation fluxes (from fossil-fuel burning and land use) to the atmosphere projected to 2040, in units of 10^{12} moles C/year and Gt C/year. The figure composition and legends are as in **Figure 4**. In the inset, the filled yellow-green area represents results for the standard run using the higher-bound estimates of land-use emissions by Houghton (1983, 1991a, 1991b, 1996b) and by Houghton et al., 1998. The red area shows TOTEM results using the lower-bound flux estimates by Wigley (personal communication, 1999).

It is important to keep in mind that although most analyses of the carbon cycle suggest that the land biosphere sequestered carbon in recent decades, this storage might not continue into the twenty-first century. Forests of North America and Europe constituted only about 17% of total forest area in 1995; those of the former USSR, Africa, Latin America, and the Caribbean accounted for about 66% of forest area. It is the Northern Hemispheric forests of parts of North America and Europe that currently have been identified as regions of carbon sequestration, probably because of the combined processes of fertilization and regrowth (see, for example, C.D. Keeling et al., 1996; R.F. Keeling et al., 1996). Major land use change has already occurred in these ecosystems. However, deforestation and other land-use activities are important drivers of change in the developing world and the former USSR. Future policies concerning the management of the forested and other terrestrial ecosystems of these regions will play a strong role in the maintenance of terrestrial carbon sequestration into the future. In addition, continuous rise in global temperature could enhance terrestrial respiration over photosynthesis, leading to a weakening of net ecosystem production on land (Woodwell, 1995).

Another consequence of the rise in atmospheric CO_2 is an increase in dissolved CO_2 concentration in surface waters of the oceanic coastal zone and open ocean. This increase lowers the

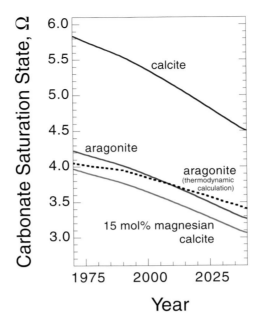

Figure 6 TOTEM-calculated changes in saturation state with respect to carbonate minerals of surface waters of the coastal ocean projected from 1996 to 2040. The saturation index (Ω) for calcite or aragonite ($CaCO_3$) is the ratio of the ion-activity product $IAP = a_{Ca^{2+}} \times a_{CO_3^{2-}}$ in coastal waters to the equilibrium constant K at the in situ temperature (where a is the activity of the aqueous species; for 15 mol % magnesian calcite, the IAP also includes the activity of Mg^{2+}). For comparison, results from thermodynamic calculation for aragonite in surface ocean water by Kleypas et al. (1999) are plotted. Calculations are for the temperature of 25°C.

supersaturation state of surface waters with respect to the common carbonate minerals that form biogenically or inorganically in shallow oceanic environments: calcite, magnesian calcite (e.g., a calcite with 15 mol% Mg), and aragonite. **Figure 6** gives our results that show continuous lowering of the supersaturation state of coastal ocean water with respect to the three carbonate minerals to the year 2040. It should be noted that the primary cause of the lowering of supersaturation is the increased absorption of atmospheric CO_2 by coastal zone waters.

Experimental and observational evidence on the rates of calcification of marine coralline algae and corals and carbonate precipitation suggests that the degree of supersaturation of ocean water is directly related to the rate of mineral deposition (Gattuso et al., 1998a; Gattuso et al., 1998b; Kleypas et al., 1999; Mackenzie and Agegian, 1989; Mackenzie et al., 1983; Morse and Mackenzie, 1990; Smith and Pesret, 1974) . Extending this evidence to the projected conditions of the future, it is conceivable that the rising atmospheric CO_2 might result in slower rates of $CaCO_3$ formation and storage in the shallow waters and the coastal zone of carbonate-producing environments. At the same time, the projected continuation of the increased transport of organic carbon and nutrients N and P from land could result in a greater accumulation of organic matter in coastal sediments, changing the balance between the storage rates of mineral and organic carbon. Perhaps significantly, the phenomenon of a shift from predominantly carbonate to organic matter deposition in industrial time has been documented in lakes found in industrialized countries. Furthermore, the availability of more organic matter for remineralization in sediments might affect the early diagenesis and formation of $CaCO_3$ cements, through release of greater amounts of dissolved CO_2 to sediment pore waters. Although the magnitude and time scale of such changes are not certain, the model analysis indicates at least the direction of future change and the role of the human factor in mineral and organic carbon sedimentation processes.

ACKNOWLEDGMENTS

This research was supported by National Science Foundation Grants EAR93-16133 and ATM00-80878 and by NOAA Office of Global Programs Grant NA37RJ0199. We wish to thank two anonymous reviewers for their constructive comments on an early draft of this paper and Jane S. Tribble and Stephen V. Smith (University of Hawaii) for their critical comments on various parts of this paper. We also wish to express our appreciation to Lee Gerhard and Bill Harrison for their editorial efforts. School of Ocean and Earth Science and Technology Contribution No. 5216. Correspondence and requests for materials should be addressed to FTM: fredm@goldschmidt.soest.hawaii.edu.

APPENDIX A

Table A1   Global reservoir masses of total C, N, P, and S, in units of moles, at the initial assumed steady state prior to the year 1700. Reservoir symbols are used in subsequent tables.

Symbol	Reservoir	Carbon	Nitrogen	Phosphorus	Sulfur
1	Terrestrial phytomass	4.98×10^{16}	3.91×10^{14}	9.76×10^{13}	7.81×10^{13}
2	Reactive humus	2.05×10^{16}	9.76×10^{14}	1.47×10^{14}	1.71×10^{14}
3	Inorganic soil	5.98×10^{16}	1.23×10^{14}	1.15×10^{15}	8.63×10^{19}
4	Coastal water	6.0×10^{15}	6.0×10^{13}	4.50×10^{11}	8.46×10^{16}
5	Coastal organic matter	3.67×10^{14}	3.86×10^{14}	3.5×10^{12}	5.9×10^{12}
6	Coastal sediments	2.07×10^{17}	2.28×10^{15}	9.2×10^{15}	1.21×10^{16}
7	Ocean surface water	6.56×10^{16}	3.5×10^{13}	5.0×10^{13}	9.27×10^{17}
8	Ocean organic matter	3.82×10^{15}	2.13×10^{14}	3.61×10^{13}	6.13×10^{13}
9	Deep ocean water	2.90×10^{18}	5.11×10^{16}	4.53×10^{15}	3.80×10^{19}
	Dissolved inorganic	2.84×10^{18}	4.70×10^{16}	4.0×10^{15}	3.80×10^{19}
	Total organic	6.0×10^{16}	4.1×10^{15}	5.3×10^{14}	9.08×10^{14}
10	Atmosphere	4.90×10^{16}	2.8×10^{20}		9.5×10^{10}
Gw	Shallow groundwater	3.45×10^{15}	2.96×10^{12}	2.7×10^{11}	1.55×10^{18}
La	Lakes	1.04×10^{14}	1.0×10^{11}	8.0×10^{9}	1.25×10^{14}
R	Rivers	3.53×10^{12}	1.1×10^{11}	3.0×10^{10}	4.4×10^{11}
Sw	Soil water	2.48×10^{14}	5.1×10^{12}	1.6×10^{11}	1.34×10^{12}

Table A2 Global transfer fluxes of total C, N, P, and S at the initial assumed steady state prior to the year 1700. Fluxes are in units of 10^{12} moles/year, represented by $XFij$, where X = the element symbol C, N, P, or S, and the subscripts ij = the source reservoir i and sink reservoir j (see **Table A1**). Note that significant figures are retained to preserve material balance at steady state.

Carbon		Nitrogen		Phosphorus		Sulfur	
Fluxes out of the Land Biota (Reservoir 1)							
CF_{12}	5250	NF_{12}	50.18	PF_{12}	10.294	SF_{12}	7.766
CF_{110}	5250					SF_{110}	0.469
Fluxes out of the Humus (Reservoir 2)							
CF_{2R}	16	NF_{2R}	1.50	PF_{2R}	0.258	SF_{2R}	0.133
CF_{2Sw}	5239	NF_{2Sw}	48.68	PF_{2Sw}	10.036	SF_{2Sw}	7.633
Fluxes out of the Inorganic Soil (Reservoir 3)							
						SF_{310}	0.031
CF_{3R}	15	NF_{3R}	0.04	PF_{3R}	0.387	SF_{3R}	0.872
CF_{3Sw}	11	NF_{3Sw}	0.26	PF_{3Sw}	0.290	SF_{3Sw}	0.221
Fluxes out of the Coastal Waters (Reservoir 4)							
CF_{45}	600	NF_{45}	90.71	PF_{45}	5.660	SF_{45}	9.623
CF_{46}	14						
CF_{47}	477	NF_{47}	9.00	PF_{47}	0.375	SF_{47}	10.929
CF_{410}	1055	NF_{410}	1.19			SF_{410}	0.435
Fluxes out of Coastal Organic Matter (Reservoir 5)							
CF_{54}	576	NF_{54}	86.94	PF_{54}	5.430	SF_{54}	9.238
CF_{56}	32	NF_{56}	4.96	PF_{56}	0.208	SF_{56}	0.157
CF_{58}	18	NF_{58}	2.64	PF_{58}	0.170	SF_{58}	0.289
						SF_{510}	0.156
Fluxes out of Coastal Sediments (Reservoir 6)							
$CF_{64\ org}$	31						
$CF_{64\ inorg}$	4	NF_{64}	1.96	PF_{64}	0.136	SF_{64}	0.113
CF_{69}	4						
		NF_{610}	3.57				
$CF_{6out\ inorg}$	21						
$CF_{6out\ org}$	9	NF_{6out}	0.22	PF_{6out}	0.588	SF_{6out}	0.982
Fluxes out of Ocean Surface Waters (Reservoir 7)							
CF_{78}	3600	NF_{78}	543	PF_{78}	33.96	SF_{78}	57.736
CF_{79}	1995	NF_{79}	1.25	PF_{79}	1.590		
CF_{710}	6987	NF_{710}	2.06			SF_{710}	3.939

continued on next page

Carbon		Nitrogen		Phosphorus		Sulfur	
Fluxes out of Ocean Organic Matter (Reservoir 8)							
CF_{87}	3118	NF_{87}	471	PF_{87}	29.42	SF_{87}	50.006
CF_{89}	500	NF_{89}	75.27	PF_{89}	4.71	SF_{89}	8.019
Fluxes out of Deep Ocean Waters (Reservoir 9)							
CF_{94}	465	NF_{94}	10.76	PF_{94}	0.456	SF_{94}	7.814
CF_{97}	2025	NF_{97}	65.68	PF_{97}	5.755		
$CF_{9out\ inorg}$	12						
$CF_{9out\ org}$	2	NF_{9out}	0.08	PF_{9out}	0.089	SF_{9out}	0.218
Fluxes out of Atmosphere (Reservoir 10)							
CF_{101}	10500	NF_{101}	9				
CF_{104}	1038	NF_{104}	0.92			SF_{104}	0.281
		NF_{105}	2.37				
CF_{107}	6962	NF_{107}	0.63			SF_{107}	0.74
		NF_{108}	0.63				
		NF_{10Sw}	1.69			SF_{10Sw}	4.197
Flux out of Shallow Groundwater (Reservoir Gw)							
CF_{GwR}	1.9	NF_{GwR}	0.01	PF_{GwR}	0.0002	SF_{GwR}	0.288
Flux out of Lakes (Reservoir La)							
CF_{LaR}	8.7	NF_{LaR}	0.01	PF_{LaR}	0.001	SF_{LaR}	1.094
Fluxes out of Rivers (Reservoir R)							
CF_{R4}	32	NF_{R4}	0.32	PF_{R4}	0.013	SF_{R4}	3.541
$CF_{R5\ diss}$	18	$NF_{R5\ diss}$	0.71	$PF_{R5\ diss}$	0.019	$SF_{R5\ diss}$	0.150
$CF_{R5\ part}$	8	$NF_{R5\ part}$	0.75	$PF_{R5\ part}$	0.129	$SF_{R5\ part}$	0.067
$CF_{R6\ inorg}$	15	$NF_{R6\ inorg}$	0.04	$PF_{R6\ inorg}$	0.387	$SF_{R6\ inorg}$	0.872
$CF_{R6\ org}$	8	$NF_{R6\ org}$	0.75	$PF_{R6\ org}$	0.129	$SF_{R6\ org}$	0.067
Fluxes out of Continental Soil Water (Reservoir Sw)							
		NF_{Sw1}	41.18	PF_{Sw1}	10.294	SF_{Sw1}	8.235
CF_{Sw10}	5200	NF_{Sw10}	8.42			SF_{Sw10}	0.125
CF_{SwGw}	1.9	NF_{SwGw}	0.01	PF_{SwGw}	0.0002	SF_{SwGw}	0.288
CF_{SwLa}	8.7	NF_{SwLa}	0.01	PF_{SwLa}	0.001	SF_{SwLa}	1.094
CF_{SwR}	39.4	NF_{SwR}	1.01	PF_{SwR}	0.031	SF_{SwR}	2.309
External Input Fluxes							
CF_{out2}	5						
CF_{out3}	26	NF_{out3}	0.30	PF_{out3}	0.677	SF_{out3}	1.124
CF_{out10}	8					SF_{out10}	0.063
CF_{hydro}	5					SF_{hydro}	0.013

Table A3 Flux constants adopted in TOTEM at the initial condition of an assumed steady state prior to 1700. Flux constants are in units of 1/year and are represented by $kXij$ (where X = C, N, P, or S) and the subscripts ij denote the source reservoir i and sink reservoir j (see **Table A1**).

Carbon	Nitrogen	Phosphorus	Sulfur
$kC_{12} = 5250/(4.98 \times 10^4)$	$kN_{12} = 50.18/391$	$kP_{12} = 10.294/97.65$	$kS_{12} = 7.766/78.1$
$kC_{110} = 5250/(4.98 \times 10^4)$			$kS_{110} = 0.469/78.1$
$kC_{2Sw} = 5239/(2.05 \times 10^4)$	$kN_{2Sw} = 48.68/976$	$kP_{2Sw} = 10.036/147$	$kS_{2Sw} = 7.633/170.5$
			$kS_{310} = 0.031/(8.63 \times 10^7)$
$kC_{3Sw} = 11/(5.98 \times 10^4)$	$kN_{3Sw} = 0.26/123$	$kP_{3Sw} = 0.29/(1.15 \times 10^3)$	$kS_{3Sw} = 0.221/(8.63 \times 10^7)$
$kC_{46} = 14/(6.0 \times 10^3)$			
$kC_{47} = 477/(6.0 \times 10^3)$	$kN_{47} = 9.00/60$	$kP_{47} = 0.375/0.45$	$kS_{47} = 10.929/(8.46 \times 10^4)$
	$kN_{410} = 1.19/60$		$kS_{410} = 0.435/(8.46 \times 10^4)$
$kC_{54} = 576/367$	$kN_{54} = 86.94/386$	$kP_{54} = 5.43/3.5$	$kS_{54} = 9.238/5.9$
$kC_{56} = 32/367$	$kN_{56} = 4.96/386$	$kP_{56} = 0.2080/3.5$	$kS_{56} = 0.1565/5.9$
$kC_{58} = 18/367$	$kN_{58} = 2.64/386$	$kP_{58} = 0.17/3.5$	$kS_{58} = 0.289/5.9$
			$kS_{510} = 0.156/5.9$
$kC_{64\,org} = 31/(2.07 \times 10^5)$	$kN_{64} = 1.96/(2.28 \times 10^3)$	$kP_{64} = 0.136/(9.2 \times 10^3)$	$kS_{64} = 0.113/(1.21 \times 10^4)$
$kC_{64\,inorg} = 4/(2.07 \times 10^5)$			
$kC_{69} = 4/(2.07 \times 10^5)$			
$kC_{6out\,inorg} = 21/(2.07 \times 10^5)$	$kN_{610} = 3.57/(2.28 \times 10^3)$		
	$kN_{79} = 1.25/35$	$kP_{79} = 1.59/50$	$kS_{710} = 3.939/(9.27 \times 10^5)$
	$kN_{710} = 2.06/35$		
$kC_{87} = 3118/(3.82 \times 10^3)$	$kN_{87} = 471/213$	$kP_{87} = 29.42/36.1$	$kS_{87} = 50.006/61.3$
$kC_{89} = 500/(3.82 \times 10^3)$	$kN_{89} = 75.27/213$	$kP_{89} = 4.71/36.1$	$kS_{89} = 8.019/61.3$
$kC_{94} = 465/(2.90 \times 10^6)$	$kN_{94} = 10.76/(5.11 \times 10^4)$	$kP_{94} = 0.456/(4.53 \times 10^3)$	$kS_{94} = 7.814/(3.8 \times 10^7)$
$kC_{9out\,inorg} = 12/(2.90 \times 10^6)$	$kN_{97} = 65.68/(5.11 \times 10^4)$	$kP_{97} = 5.755/(4.53 \times 10^3)$	

Table A3 *(continued)*

Carbon	Nitrogen	Phosphorus	Sulfur
$kC_{101} = 10500/(4.98 \times 10^4)$			
	$kN_{104} = 0.92/(2.80 \times 10^8)$		$kS_{104} = 0.281/5.218$
	$kN_{107} = 0.63/(2.80 \times 10^8)$		$kS_{107} = 0.74/5.218$
	$kN_{10Sw} = 1.69/(2.80 \times 10^8)$		$kS_{10Sw} = 4.197/5.218$
$kC_{GwR} = 1.9/(3.45 \times 10^3)$	$kN_{GwR} = 0.01/2.96$	$kP_{GwR} = 0.0002/0.27$	$kS_{GwR} = 0.288/(1.55 \times 10^6)$
$kC_{LaR} = 8.7/104$	$kN_{LaR} = 0.01/0.10$	$kP_{LaR} = 0.001/0.008$	$kS_{LaR} = 1.094/125$
$kC_{R4} = 32/3.53$	$kN_{R4} = 0.32/0.11$	$kP_{R4} = 0.013/0.03$	$kS_{R4} = 3.541/0.44$
$kC_{R5\ diss} = 18/3.53$	$kN_{R5\ diss} = 0.71/0.11$	$kP_{R5\ diss} = 0.019/0.03$	$kS_{R5\ diss} = 0.15/0.44$
$kC_{R5\ part} = 8/3.53$	$kN_{R5\ part} = 0.75/0.11$	$kP_{R5\ part} = 0.129/0.03$	$kS_{R5\ part} = 0.0665/0.44$
$kC_{R6\ org} = 8/3.53$	$kN_{R6\ org} = 0.75/0.11$	$kP_{R6\ org} = 0.129/0.03$	$kS_{R6\ org} = 0.0665/0.44$
$kC_{R6\ inorg} = 15/3.53$	$kN_{R6\ inorg} = 0.04/0.11$	$kP_{R6\ inorg} = 0.387/0.03$	$kS_{R6\ inorg} = 0.872/0.44$
$kC_{Sw10} = 5200/248$	$kN_{Sw10} = 8.42/5.1$		$kS_{Sw10} = 0.125/1.34$
$kC_{SwGw} = 1.9/248$	$kN_{SwGw} = 0.01/5.1$	$kP_{SwGw} = 0.0002/0.16$	$kS_{SwGw} = 0.288/1.34$
$kC_{SwLa} = 8.7/248$	$kN_{SwLa} = 0.01/5.1$	$kP_{SwLa} = 0.001/0.16$	$kS_{SwLa} = 1.094/1.34$
$kC_{SwR} = 39.4/248$	$kN_{SwR} = 1.01/5.1$	$kP_{SwR} = 0.0308/0.16$	$kS_{SwR} = 2.309/1.34$

APPENDIX B

Numerical or letter subscripts denote each of the 13 reservoirs and the rivers and streams **(Table A1)**. The term C_1, for example, is the mass of carbon in terrestrial phytomass. In the flux term XF_{ij}, $X =$ C (carbon), N (nitrogen), P (phosphorus), or S (sulfur); the subscripts i = originating reservoir and j = receiving reservoir. Net exchange fluxes of carbon between the atmosphere and surface waters and between the surface and deep ocean are indicated as CF_{iji} (e.g., CF_{4104}, CF_{7107}, CF_{797}). Other subscripts used are: atmex = exchange between the atmosphere and surface waters; Detergent = detergent consumption; diss = dissolved; Fert = inorganic fertilizers; Fert cons = fertilizer consumption; FF = fossil-fuel and cement manufacturing; hydro = hydrothermal; inorg = inorganic; LU = land use; org = organic; out = outside of defined model boundaries (i.e., external); part = particulate; photo = photosynthesis; react = reactive; resp = respiration; Sewage = sewage discharge; unr = unreactive. Redfield ratios are defined in section *IIIE* and denoted in the equations as, for example, $(C{:}P)_{1\,photo}$. Reservoir masses are in units of 10^{12} moles of the element; fluxes are in units of 10^{12} moles of element/year; rate constants are in units of 1/year.

I. EQUATIONS OF THE MODEL TOTEM

IA. Carbon Equations

IA1. Mass Balance Equations

$$C_1(t) = C_1(t - dt) + (CF_{101} - CF_{12} - CF_{110} - CF_{1LU}) \times dt$$
$$C_2(t) = C_2(t - dt) + (CF_{12} + CF_{out2} - CF_{2Sw} - CF_{2LU} - CF_{2R}) \times dt$$
$$C_3(t) = C_3(t - dt) + (CF_{out3} - CF_{3Sw} - CF_{3R}) \times dt$$
$$C_4(t) = C_4(t - dt) + (CF_{54} + CF_{64\,inorg} + CF_{64\,org} + CF_{94} + CF_{R4} + CF_{4104} - CF_{45} - CF_{46} - CF_{47}) \times dt$$
$$C_5(t) = C_5(t - dt) + (CF_{45} + CF_{R5\,diss} + CF_{R5\,part} + CF_{Sewage} - CF_{54} - CF_{56} - CF_{58}) \times dt$$
$$C_6(t) = C_6(t - dt) + (CF_{46} + CF_{56} + CF_{R6\,org} + CF_{R6\,inorg} - CF_{64\,inorg} - CF_{64\,org} - CF_{6out\,org} - CF_{6out\,inorg} - CF_{69}) \times dt$$
$$C_7(t) = C_7(t - dt) + (CF_{47} + CF_{7107} + CF_{87} - CF_{78} - CF_{797}) \times dt$$
$$C_8(t) = C_8(t - dt) + (CF_{58} + CF_{78} - CF_{87} - CF_{89}) \times dt$$
$$C_9(t) = C_9(t - dt) + (CF_{69} + CF_{797} + CF_{89} + CF_{hydro} - CF_{94} - CF_{9out\,org} - CF_{9out\,inorg}) \times dt$$
$$C_{10}(t) = C_{10}(t - dt) + (CF_{110} + CF_{out10} + CF_{Sw10} + CF_{1LU} + CF_{2LU} + C_{FF} - CF_{101} - CF_{4104} - CF_{7107}) \times dt$$
$$C_{Gw}(t) = C_{Gw}(t - dt) + (CF_{SwGw} - CF_{GwR}) \times dt$$
$$C_{La}(t) = C_{La}(t - dt) + (CF_{SwLa} - CF_{LaR}) \times dt$$
$$C_R(t) = C_R(t - dt) + (CF_{SwR} + CF_{GwR} + CF_{2R} + CF_{3R} + CF_{LaR} - CF_{R4} - CF_{R5\,diss} - CF_{R5\,part} - CF_{R6\,org} - CF_{R6\,inorg}) \times dt$$
$$C_{Sw}(t) = C_{Sw}(t - dt) + (CF_{2Sw} + CF_{3Sw} - CF_{Sw10} - CF_{SwR} - CF_{SwGw} - CF_{SwLa}) \times dt$$

IA2. Flux Equations

$$CF_{12} = kC_{12} \times C_1(t)$$
$$CF_{2Sw} = kC_{2Sw} \times C_2(t) \times f_T$$
$$CF_{3Sw} = kC_{3Sw} \times C_3(t)$$

$$CF_{110} = kC_{110} \times C_1(t) \times f_T$$
$$CF_{2LU} = (1 - f_{LU}) \times C_{LU}$$
$$CF_{3R} = CF_{3R\,t=0} \times (1 + f_{LUR})$$

$$CF_{1LU} = f_{LU} \times C_{LU}$$
$$CF_{2R} = CF_{2R\,t=0} \times (1 + f_{LUR})$$

$$CF_{45} = CF_{45\,t=0} \times f_{N4} \times f_{P4} \qquad CF_{46} = kC_{46} \times \mathbf{C}_4(t) \qquad CF_{47} = kC_{47} \times \mathbf{C}_4(t)$$

$$CF_{54} = kC_{54} \times \mathbf{C}_5(t) \qquad CF_{56} = kC_{56} \times \mathbf{C}_5(t) \qquad CF_{58} = kC_{58} \times \mathbf{C}_5(t)$$

$$CF_{64\,org} = kC_{64\,org} \times \mathbf{C}_6(t) \qquad CF_{64\,inorg} = kC_{64\,inorg} \times \mathbf{C}_6(t) \qquad CF_{69} = kC_{69} \times \mathbf{C}_6(t)$$

$$CF_{6out\,org} = 9 \qquad CF_{6out\,inorg} = kC_{6out\,inorg} \times \mathbf{C}_6(t)$$

$$CF_{78} = CF_{78\,t=0} \times f_{N7} \times f_{P7}$$

$$CF_{797} = -CF_{797\,t=0} + [(\mathbf{C}_7(t) - \mathbf{C}_{7\,t=0})/(h_7 \times \tau_9)] - [(\mathbf{C}_9(t) - \mathbf{C}_{9\,t=0})/(h_9 \times t_9)]$$

$$CF_{87} = kC_{87} \times \mathbf{C}_8(t) \qquad CF_{89} = kC_{89} \times \mathbf{C}_8(t)$$

$$CF_{94} = kC_{94} \times \mathbf{C}_9(t) \qquad CF_{9out\,inorg} = kC_{9out\,inorg} \times \mathbf{C}_9(t) \qquad CF_{9out\,org} = 2$$

$$CF_{101} = kC_{101} \times \mathbf{C}_1(t) \times K_{photo} = GPP$$

$$CF_{4104} = (d\mathbf{C}_4/dt)_{atmex} - \{CF_{R4} + CF_{54} + CF_{64\,org} + CF_{64\,inorg} + CF_{94} - CF_{45} - CF_{46} - CF_{47}\}$$

$$CF_{7107} = (d\mathbf{C}_7/dt)_{atmex} - \{CF_{47} + CF_{87} - CF_{78} - CF_{797}\}$$

$$CF_{GwR} = kC_{GwR} \times \mathbf{C}_{Gw}(t)$$

$$CF_{LaR} = kC_{LaR} \times \mathbf{C}_{La}(t)$$

$$CF_{R4} = kC_{R4} \times \mathbf{C}_R(t) \qquad CF_{R5\,diss} = kC_{R5\,diss} \times \mathbf{C}_R(t) \qquad CF_{R5\,part} = kC_{R5\,part} \times \mathbf{C}_R(t)$$

$$CF_{R6\,org} = kC_{R6\,org} \times \mathbf{C}_R(t) \qquad CF_{R6\,inorg} = kC_{R6\,inorg} \times \mathbf{C}_R(t)$$

$$CF_{Sw10} = kC_{Sw10} \times \mathbf{C}_{Sw}(t) \times f_T \qquad CF_{SwR} = kC_{SwR} \times \mathbf{C}_{Sw}(t) \qquad CF_{SwGw} = kC_{SwGw} \times \mathbf{C}_{Sw}(t)$$

$$CF_{SwLa} = kC_{SwLa} \times \mathbf{C}_{Sw}(t)$$

$$CF_{out2} = 5 \qquad CF_{out3} = 26 \qquad CF_{hydro} = 5$$

$$CF_{out10} = 8$$

IB. Nitrogen Equations

IB1. Mass Balance Equations

$$\mathbf{N}_1(t) = \mathbf{N}_1(t - dt) + (NF_{Sw1} + NF_{101} - NF_{12} - NF_{1LU}) \times dt$$

$$\mathbf{N}_2(t) = \mathbf{N}_2(t - dt) + (NF_{12} - NF_{2Sw} - NF_{2LU} - NF_{2R}) \times dt$$

$$\mathbf{N}_3(t) = \mathbf{N}_3(t - dt) + (NF_{out3} - NF_{3Sw} - NF_{3R}) \times dt$$

$$\mathbf{N}_4(t) = \mathbf{N}_4(t - dt) + (NF_{54} + NF_{64} + NF_{94} + NF_{104} + N_{FF4} + NF_{R4} - NF_{45} - NF_{47} - NF_{410}) \times dt$$

$$\mathbf{N}_5(t) = \mathbf{N}_5(t - dt) + (NF_{45} + NF_{105} + NF_{R5\,diss} + NF_{R5\,part} + NF_{Sewage} - NF_{54} - NF_{56} - NF_{58}) \times dt$$

$$\mathbf{N}_6(t) = \mathbf{N}_6(t - dt) + (NF_{56} + NF_{R6\,org} + NF_{R6\,inorg} - NF_{64} - NF_{6out} - NF_{610}) \times dt$$

$$\mathbf{N}_7(t) = \mathbf{N}_7(t - dt) + (NF_{47} + NF_{87} + NF_{97} + NF_{107} - NF_{78} - NF_{79} - NF_{710}) \times dt$$

$$\mathbf{N}_8(t) = \mathbf{N}_8(t - dt) + (NF_{58} + NF_{78} + NF_{108} - NF_{87} - NF_{89}) \times dt$$

$$\mathbf{N}_9(t) = \mathbf{N}_9(t - dt) + (NF_{79} + NF_{89} - NF_{94} - NF_{97} - NF_{9out}) \times dt$$

$$\mathbf{N}_{10}(t) = \mathbf{N}_{10}(t - dt) + (NF_{410} + NF_{610} + NF_{710} + NF_{Sw10} - NF_{101} - NF_{104} - NF_{105} - NF_{107} - NF_{108} - NF_{10Sw}) \times dt$$

$$\mathbf{N}_{Gw}(t) = \mathbf{N}_{Gw}(t - dt) + (NF_{SwGw} - NF_{GwR}) \times dt$$

$$\mathbf{N}_{La}(t) = \mathbf{N}_{La}(t - dt) + (NF_{SwLa} - NF_{LaR}) \times dt$$

$$\mathbf{N}_R(t) = \mathbf{N}_R(t - dt) + (NF_{LaR} + NF_{GwR} + NF_{FertR} + NF_{SwR} + NF_{2R} + NF_{3R} - NF_{R4} - NF_{R5\,diss} - NF_{R5\,part} - NF_{R6\,org} - NF_{R6\,inorg}) \times dt$$

$$\mathbf{N}_{Sw}(t) = \mathbf{N}_{Sw}(t - dt) + (NF_{2Sw} + NF_{3Sw} + NF_{10Sw} + NF_{1LU} + NF_{2LU} + N_{FFSw} + NF_{Fert\,leach} - NF_{Sw1} - NF_{Sw10} - NF_{SwGw} - NF_{SwLa} - NF_{SwR}) \times dt$$

IB2. Flux Equations

$NF_{12} = kN_{12} \times \mathbf{N}_2(t)$

$NF_{2Sw} = kN_{2Sw} \times \mathbf{N}_2(t) \times f_T$

$NF_{3Sw} = kN_{3Sw} \times \mathbf{N}_3(t)$

$NF_{45} = CF_{45}/(\mathbf{C:N})_{5\ photo}$

$NF_{54} = kN_{54} \times \mathbf{N}_5(t)$

$NF_{64} = kN_{64} \times \mathbf{N}_6(t)$

$NF_{78} = CF_{78}/(\mathbf{C:N})_{8\ photo}$

$NF_{87} = kN_{87} \times \mathbf{N}_8(t)$

$NF_{94} = kN_{94} \times \mathbf{N}_9(t)$

$NF_{101} = 9.0$

$NF_{107} = kN_{107} \times \mathbf{N}_{10}(t)$

$NF_{GwR} = kN_{GwR}(t) \times \mathbf{N}_{Gw}(t)$

$NF_{LaR} = kN_{LaR}(t) \times \mathbf{N}_{La}(t)$

$NF_{R4} = kN_{R4} \times \mathbf{N}_R(t)$

$NF_{R6\ org} = kN_{R6\ org} \times \mathbf{N}_R(t)$

$NF_{Sw1} = CF_{101}/(\mathbf{C:N})_{1\ photo}$

$NF_{SwLa} = kN_{SwLa} \times \mathbf{N}_{Sw}(t)$

$NF_{out3} = 0.30$

$NF_{1LU} = CF_{1LU}/(\mathbf{C:N})_{1\ resp}$

$NF_{2LU} = CF_{2LU}/(\mathbf{C:N})_{2\ resp}$

$NF_{3R} = NF_{3R\ t=0} \times (1 + f_{LUR})$

$NF_{47} = kN_{47} \times \mathbf{N}_4(t)$

$NF_{56} = kN_{56} \times \mathbf{N}_5(t)$

$NF_{610} = kN_{610} \times \mathbf{N}_6(t)$

$NF_{79} = kN_{79} \times \mathbf{N}_7(t)$

$NF_{89} = kN_{89} \times \mathbf{N}_8(t)$

$NF_{97} = kN_{97} \times \mathbf{N}_9(t)$

$NF_{104} = kN_{104} \times \mathbf{N}_{10}(t)$

$NF_{108} = 0.63$

$NF_{R5\ diss} = kN_{R5\ diss} \times \mathbf{N}_R(t)$

$NF_{R6\ inorg} = kN_{R6\ inorg} \times \mathbf{N}_R(t)$

$NF_{Sw10} = kN_{Sw10} \times \mathbf{N}_{Sw}(t) \times f_T$

$NF_{SwR} = kN_{SwR} \times \mathbf{N}_{Sw}(t)$

$NF_{2R} = NF_{2R\ t=0} \times (1 + f_{LUR})$

$NF_{410} = kN_{410} \times \mathbf{N}_4(t)$

$NF_{58} = kN_{58} \times \mathbf{N}_5(t)$

$NF_{6out} = 0.22$

$NF_{710} = kN_{710} \times \mathbf{N}_7(t)$

$NF_{9out} = 0.08$

$NF_{105} = 2.37$

$NF_{10Sw} = kN_{10Sw} \times \mathbf{N}_{10}(t)$

$NF_{R5\ part} = kN_{R5\ part} \times \mathbf{N}_R(t)$

$NF_{SwGw} = kN_{SwGw} \times \mathbf{N}_{Sw}(t)$

IC. Phosphorus Equations

IC1. Mass Balance Equations

$\mathbf{P}_1(t) = \mathbf{P}_1(t - dt) + (PF_{Sw1} - PF_{12} - PF_{1LU} - PF_{1LU\ unr}) \times dt$

$\mathbf{P}_2(t) = \mathbf{P}_2(t - dt) + (PF_{12} - PF_{2Sw} - PF_{2LU} - PF_{2LU\ unr} - PF_{2R}) \times dt$

$\mathbf{P}_3(t) = \mathbf{P}_3(t - dt) + (PF_{out3} + PF_{1LU\ unr} + PF_{2LU\ unr} + PF_{Fert\ unr} - PF_{3Sw} - PF_{3R}) \times dt$

$\mathbf{P}_4(t) = \mathbf{P}_4(t - dt) + (PF_{54} + PF_{64} + PF_{94} + PF_{R4} + PF_{Detergent} - PF_{45} - PF_{47}) \times dt$

$\mathbf{P}_5(t) = \mathbf{P}_5(t - dt) + (PF_{45} + PF_{R5\ diss} + PF_{R5\ part} + PF_{Sewage} - PF_{54} - PF_{56} - PF_{58}) \times dt$

$\mathbf{P}_6(t) = \mathbf{P}_6(t - dt) + (PF_{56} + PF_{R6\ org} + PF_{R6\ inorg} - PF_{64} - PF_{6out}) \times dt$

$\mathbf{P}_7(t) = \mathbf{P}_7(t - dt) + (PF_{47} + PF_{87} + PF_{97} - PF_{78} - PF_{79}) \times dt$

$\mathbf{P}_8(t) = \mathbf{P}_8(t - dt) + (PF_{58} + PF_{78} - PF_{87} - PF_{89}) \times dt$

$\mathbf{P}_9(t) = \mathbf{P}_9(t - dt) + (PF_{79} + PF_{89} - PF_{94} - PF_{97} - PF_{9out}) \times dt$

$\mathbf{P}_{Gw}(t) = \mathbf{P}_{Gw}(t - dt) + (PF_{SwGw} - PF_{GwR}) \times dt$

$\mathbf{P}_{La}(t) = \mathbf{P}_{La}(t - dt) + (PF_{SwLa} - PF_{LaR}) \times dt$

$\mathbf{P}_R(t) = \mathbf{P}_R(t - dt) + (PF_{SwR} + PF_{GwR} + PF_{LaR} + PF_{FertR} + PF_{2R} + PF_{3R} - PF_{R4} - PF_{R5\ diss} - PF_{R5\ part} - PF_{R6\ org} - PF_{R6\ inorg}) \times dt$

$\mathbf{P}_{Sw}(t) = \mathbf{P}_{Sw}(t - dt) + (PF_{2Sw} + PF_{3Sw} + PF_{1LU} + PF_{2LU} + PF_{Fert\ leach} - PF_{Sw1} - PF_{SwGw} - PF_{SwLa} - PF_{SwR}) \times dt$

IC2. Flux Equations

$$PF_{12} = kP_{12} \times \mathbf{P}_1(t) \qquad\qquad PF_{1LU} = CF_{1LU} \times f\,P_{LU\ react}/(\mathbf{C:P})_{1\ resp}$$

$$PF_{1LU\ unr} = CF_{1LU} \times (1 - f\,P_{LU\ react})/(\mathbf{C:P})_{1\ resp}$$

$$PF_{2Sw} = kP_{2Sw} \times \mathbf{P}_2(t) \times f_T \qquad PF_{2LU} = CF_{2LU} \times f\,P_{LU\ react}/(\mathbf{C:P})_{2\ resp}$$

$$PF_{2R} = PF_{2R\ t=0} \times (1 + f_{LUR}) \qquad PF_{2LU\ unr} = CF_{2LU} \times (1 - f\,P_{LU\ react})/(\mathbf{C:P})_{2\ resp}$$

$$PF_{3Sw} = kP_{3Sw} \times \mathbf{P}_3(t) \times f_T \qquad PF_{3R} = PF_{3R\ t=0} \times (1 + f_{LUR}) \times f_T$$

$$PF_{45} = CF_{45}/(\mathbf{C:P})_{5\ photo} \qquad PF_{47} = kP_{47} \times \mathbf{P}_4(t)$$

$$PF_{54} = kP_{54} \times \mathbf{P}_5(t) \qquad\qquad PF_{56} = kF_{56} \times \mathbf{P}_5(t) \qquad PF_{58} = kP_{58} \times \mathbf{P}_5(t)$$

$$PF_{64} = kP_{64} \times \mathbf{P}_6(t) \qquad\qquad PF_{6out} = 0.59$$

$$PF_{78} = CF_{78}/(\mathbf{C:P})_{8\ photo} \qquad PF_{79} = kP_{79} \times \mathbf{P}_7(t)$$

$$PF_{87} = kP_{87} \times \mathbf{P}_8(t) \qquad\qquad PF_{89} = kP_{89} \times \mathbf{P}_8(t)$$

$$PF_{94} = kP_{94} \times \mathbf{P}_9(t) \qquad\qquad PF_{97} = kP_{97} \times \mathbf{P}_9(t) \qquad PF_{9out} = 0.089$$

$$PF_{GwR} = kP_{GwR} \times \mathbf{P}_{Gw}(t)$$

$$PF_{LaR} = kP_{LaR} \times \mathbf{P}_{La}(t)$$

$$PF_{R4} = kP_{R4} \times \mathbf{P}_R(t) \qquad PF_{R5\ diss} = kP_{R5\ diss} \times \mathbf{P}_R(t) \qquad PF_{R5\ part} = kP_{R5\ part} \times \mathbf{P}_R(t)$$

$$PF_{R6\ org} = kP_{R6\ org} \times \mathbf{P}_R(t) \qquad PF_{R6\ inorg} = kP_{R6\ inorg} \times \mathbf{P}_R(t)$$

$$PF_{Sw1} = CF_{101}/(\mathbf{C:P})_{1\ photo} \qquad PF_{SwGw} = kP_{SwGw} \times \mathbf{P}_{Sw}(t) \qquad PF_{SwLa} = kP_{SwLa} \times \mathbf{P}_{Sw}(t)$$

$$PF_{SwR} = kP_{SwR} \times \mathbf{P}_{Sw}(t)$$

$$PF_{out3} = 0.677$$

ID. Sulfur Equations

ID1. Mass Balance Equations

$$\mathbf{S}_1(t) = \mathbf{S}_1(t - dt) + (SF_{Sw1} - SF_{12} - SF_{110} - SF_{1LU}) \times dt$$

$$\mathbf{S}_2(t) = \mathbf{S}_2(t - dt) + (SF_{12} - S_{2LU} - SF_{2R} - SF_{2Sw}) \times dt$$

$$\mathbf{S}_3(t) = \mathbf{S}_3(t - dt) + (SF_{out3} - SF_{310} - SF_{3R} - SF_{3Sw}) \times dt$$

$$\mathbf{S}_4(t) = \mathbf{S}_4(t - dt) + (SF_{54} + SF_{64} + SF_{94} + SF_{104} + SF_{R4} + S_{FF4} - SF_{45} - SF_{47} - SF_{410}) \times dt$$

$$\mathbf{S}_5(t) = \mathbf{S}_5(t - dt) + (SF_{45} + SF_{R5\ diss} + SF_{R5\ part} - SF_{54} - SF_{56} - SF_{58} - SF_{510}) \times dt$$

$$\mathbf{S}_6(t) = \mathbf{S}_6(t - dt) + (SF_{56} + SF_{R6\ inorg} + SF_{R6\ org} - SF_{64} - SF_{6out}) \times dt$$

$$\mathbf{S}_7(t) = \mathbf{S}_7(t - dt) + (SF_{47} + SF_{87} + SF_{107} - SF_{78} - SF_{710}) \times dt$$

$$\mathbf{S}_8(t) = \mathbf{S}_8(t - dt) + (SF_{58} + SF_{78} - SF_{87} - SF_{89}) \times dt$$

$$\mathbf{S}_9(t) = \mathbf{S}_9(t - dt) + (SF_{89} + SF_{hydro} - SF_{94} - SF_{9out}) \times dt$$

$$\mathbf{S}_{10}(t) = \mathbf{S}_{10}(t - dt) + (SF_{110} + SF_{310} + SF_{410} + SF_{510} + SF_{710} + SF_{Sw10} + SF_{out10} - SF_{104} - SF_{107} - SF_{10Sw}) \times dt$$

$$\mathbf{S}_{Gw}(t) = \mathbf{S}_{Gw}(t - dt) + (SF_{SwGw} - SF_{GwR}) \times dt$$

$$\mathbf{S}_{La}(t) = \mathbf{S}_{La}(t - dt) + (SF_{SwLa} - SF_{LaR}) \times dt$$

$$\mathbf{S}_R(t) = \mathbf{S}_R(t - dt) + (SF_{2R} + SF_{3R} + SF_{GwR} + SF_{LaR} + SF_{SwR} - SF_{R4} - SF_{R5\ diss} - SF_{R5\ part} - SF_{R6\ inorg} - SF_{R6\ org}) \times dt$$

$$\mathbf{S}_{Sw}(t) = \mathbf{S}_{Sw}(t - dt) + (SF_{1LU} + SF_{2LU} + SF_{2Sw} + SF_{3Sw} + SF_{10Sw} + S_{FFSw} - SF_{Sw1} - SF_{Sw10} - SF_{SwGw} - SF_{SwLa} - SF_{SwR}) \times dt$$

ID2. Flux Equations

$SF_{12} = kS_{12} \times \mathbf{S}_1(t)$

$SF_{110} = kS_{110} \times \mathbf{S}_1(t)$

$SF_{1LU} = CF_{1LU}/(\mathbf{C{:}S})_{1\,resp}$

$SF_{2LU} = CF_{2LU}/(\mathbf{C{:}S})_{2\,resp}$

$SF_{2R} = SF_{2R\,t=0} \times (1 + f_{LUR})$

$SF_{2Sw} = kS_{2Sw} \times \mathbf{S}_2(t) \times f_T$

$SF_{310} = kS_{310} \times \mathbf{S}_3(t)$

$SF_{3R} = SF_{3R\,t=0} \times (1 + f_{LUR})$

$SF_{3Sw} = kS_{3Sw} \times \mathbf{S}_3(t)$

$SF_{45} = CF_{45}/(\mathbf{C{:}S})_{5\,photo}$

$SF_{47} = kS_{47} \times \mathbf{S}_4(t)$

$SF_{410} = kS_{410} \times \mathbf{S}_4(t)$

$SF_{54} = kS_{54} \times \mathbf{S}_5(t)$

$SF_{56} = kS_{56} \times \mathbf{S}_5(t)$

$SF_{58} = kS_{58} \times \mathbf{S}_5(t)$

$SF_{510} = kS_{510} \times \mathbf{S}_5(t)$

$SF_{64} = kS_{64} \times \mathbf{S}_6(t)$

$SF_{6out} = 0.982$

$SF_{78} = CF_{78}/(\mathbf{C{:}S})_{8\,photo}$

$SF_{710} = kS_{710} \times \mathbf{S}_7(t)$

$SF_{87} = kS_{87} \times \mathbf{S}_8(t)$

$SF_{89} = kS_{89} \times \mathbf{S}_8(t)$

$SF_{94} = kS_{94} \times \mathbf{S}_9(t)$

$SF_{9out} = 0.218$

$SF_{104} = kS_{104} \times \mathbf{S}_{10\,Input}$

$SF_{107} = kS_{107} \times \mathbf{S}_{10\,Input}$

$SF_{10Sw} = kS_{10Sw} \times \mathbf{S}_{10\,Input}$

$\mathbf{S}_{10\,Input} = SF_{110} + SF_{310} + SF_{410} + SF_{510} + SF_{710} + SF_{Sw10} + SF_{out10}$

$SF_{GwR} = kS_{GwR} \times \mathbf{S}_{Gw}(t)$

$SF_{LaR} = kS_{LaR} \times \mathbf{S}_{La}(t)$

$SF_{R4} = kS_{R4} \times \mathbf{S}_R(t)$

$SF_{R5\,diss} = kS_{R5\,diss} \times \mathbf{S}_R(t)$

$SF_{R5\,part} = kS_{R5\,part} \times \mathbf{S}_R(t)$

$SF_{R6\,inorg} = kS_{R6\,inorg} \times \mathbf{S}_R(t)$

$SF_{R6\,org} = kS_{R6\,org} \times \mathbf{S}_R(t)$

$SF_{Sw1} = CF_{101}/(\mathbf{C{:}S})_{1\,photo}$

$SF_{Sw10} = kS_{Sw10} \times \mathbf{S}_{Sw}(t) \times f_T$

$SF_{SwGw} = kS_{SwGw} \times \mathbf{S}_{Sw}(t)$

$SF_{SwLa} = kS_{SwLa} \times \mathbf{S}_{Sw}(t)$

$SF_{SwR} = kS_{SwR} \times \mathbf{S}_{Sw}(t)$

$SF_{out3} = 1.124$

$SF_{hydro} = 0.013$

$SF_{out10} = 0.063$

II. ANTHROPOGENIC FLUXES

IIA. Fossil-Fuel Burning

C_{FF} = total CO_2 emissions from fossil-fuel burning and cement manufacturing (Marland et al., 1999)

$N_{FF4} = N_{FF} \times fN_{FF4}$

$N_{FFSw} = N_{FF} \times (1 - fN_{FF4})$

$N_{FF} = N_{FF4} + N_{FFSw}$; total anthropogenic NO_x emissions (Brown et al., 1997; Dignon, 1992; Dignon and Hameed, 1989; Hameed and Dignon, 1992)

$fN_{FF4} = 0.379$; fraction of NO_x emissions that is deposited back onto the coastal water reservoir

$S_{FF4} = S_{FF} \times fS_{FF4}$

$S_{FFSw} = S_{FF} \times (1 - fS_{FF4})$

$S_{FF} = S_{FF4} + S_{FFSw}$; total anthropogenic SO_x emissions (Brown et al., 1997; Dignon, 1992; Dignon and Hameed, 1989; Hameed and Dignon, 1992)

$fS_{FF4} = 0.5$; fraction of SO_x emissions that is deposited back onto the coastal water reservoir

IIB. Changes in Land-Use Activities

$CF_{1LU} = f_{LU} \times C_{LU}$; emissions of CO_2 from land-use disturbance of terrestrial biota

$CF_{2LU} = (1 - f_{LU}) \times C_{LU}$; emissions of CO_2 from land-use disturbance of humus

$$NF_{1LU} = CF_{1LU}/(\textbf{C:N})_{1\,resp}$$
$$NF_{2LU} = CF_{2LU}/(\textbf{C:N})_{2\,resp}$$
$$PF_{1LU} = CF_{1LU} \times fP_{LU\,react}/(\textbf{C:P})_{1\,resp}$$
$$PF_{1LU\,unr} = CF_{1LU} \times (1 - fP_{LU\,react})/(\textbf{C:P})_{1\,resp}$$
$$PF_{2LU} = CF_{2LU} \times fP_{LU\,react}/(\textbf{C:P})_{2\,resp}$$
$$PF_{2LU\,unr} = CF_{2LU} \times (1 - fP_{LU\,react})/(\textbf{C:P})_{2\,resp}$$
$$SF_{1LU} = CF_{1LU}/(\textbf{C:S})_{1\,resp}$$
$$SF_{2LU} = CF_{2LU}/(\textbf{C:S})_{1\,resp}$$

$C_{LU} = C_{1LU} + C_{2LU}$; total CO_2 emissions from land-use activities (Houghton, 1983, 1991a, 1991b; Kammen and Marino, 1993)

$f_{LU} = 0.5 - 0.25$; fraction of land-use CO_2 emissions from terrestrial biota

$fP_{LU\,react} = 0.6$; $1700 < t < 1935$
$= 1.0$; $t > 1935$; fraction of remobilized P that remains in the reactive inorganic form

$$CF_{2R} = CF_{2R\,t=0} \times (1 + f_{LUR})$$
$$CF_{3R} = CF_{3R\,t=0} \times (1 + f_{LUR})$$
$$NF_{2R} = NF_{2R\,t=0} \times (1 + f_{LUR})$$
$$NF_{3R} = NF_{3R\,t=0} \times (1 + f_{LUR})$$
$$PF_{2R} = PF_{2R\,t=0} \times (1 + f_{LUR})$$
$$PF_{3R} = PF_{3R\,t=0} \times (1 + f_{LUR}) \times f_T$$
$$SF_{2R} = SF_{2R\,t=0} \times (1 + f_{LUR})$$
$$SF_{3R} = SF_{3R\,t=0} \times (1 + f_{LUR})$$

$f_{LUR} = 0.008 \times (C_{LU} - C_{LU\,t=0})$; effect of land-use activities on the transport of C, N, P, and S to coastal margins via runoff.

IIC. Agricultural Fertilizer Application

$NF_{Fert\,crops} = NF_{Fert\,cons} \times fN_{Fert\,crops}$; applied N fertilizer that is assimilated into crops

$NF_{Fert\,leach} = NF_{Fert\,cons} \times fNP_{Fert\,leach}$; applied N fertilizer that leaches into soilwater reservoir

$NF_{FertR} = NF_{Fert\,cons} \times fN_{FertR}$; applied N fertilizer that is lost in runoff

$NF_{FertVol} = NF_{Fert\,cons} - NF_{Fert\,crops} - NF_{Fert\,leach} - NF_{FertR}$; applied N fertilizer that is volatilized

$PF_{Fert\,crops} = NF_{Fert\,crops}/(\textbf{N:P})_{Crops}$; applied P fertilizer that is assimilated into crops

$PF_{Fert\,leach} = PF_{Fert\,cons} \times fNP_{Fert\,leach}$; applied P fertilizer that leaches into soilwater reservoir

$PF_{FertR} = PF_{Fert\,cons} \times fP_{FertR}$; applied P fertilizer that is lost in runoff

$PF_{Fert\,unr} = PF_{Fert\,cons} - PF_{Fert\,crops} - PF_{Fert\,leach} - PF_{FertR}$; applied P fertilizer that is transformed into the biochemically unreactive inorganic form

$NF_{Fert\,cons} = NF_{Fert\,crops} + NF_{Fert\,leach} + NF_{FertR} + NF_{FertVol}$; global consumption of inorganic N fertilizer (United Nations FAO, various years)

$PF_{Fert\,cons} = PF_{Fert\,crops} + PF_{Fert\,leach} + PF_{FertR} + PF_{Fert\,unr}$; global consumption of inorganic P fertilizer (United Nations FAO, various years)

$fN_{FertR} = 0.25$; fraction of applied N fertilizer that is lost in runoff (Smil, 1991)

$fN_{Fert\,crops} = 0.45$; fraction of applied N fertilizer that is assimilated into crops (Smil, 1991)

$f\text{NP}_{\text{Fert leach}}$ = 0.3; fraction of applied N and P fertilizer that leaches into soilwater reservoir (Smil, 1991)

$f\text{P}_{\text{FertR}}$ = 0.1; fraction of applied P fertilizer that is lost in runoff (Smil, 1991)

IID. Sewage Disposal

$\text{CF}_{\text{Sewage}} = f\text{C}_{\text{Sewage}} \times \text{Pop}_{\text{global}}(t)$; organic C from sewage discharged into coastal zone

$\text{NF}_{\text{Sewage}} = f\text{N}_{\text{Sewage}} \times \text{Pop}_{\text{global}}(t)$; total N from sewage discharged into coastal zone

$\text{PF}_{\text{Sewage}} = f\text{P}_{\text{Sewage}} \times \text{Pop}_{\text{global}}(t)$; total P from sewage discharged into coastal zone

$\text{PF}_{\text{Detergent}} = f\text{P}_{\text{Detergent}} \times \text{Pop}_{\text{urban}}(t)$; total inorganic P from detergent consumption

$\text{Pop}_{\text{global}}(t)$ = global population (United Nations Population Division, 1995)

$\text{Pop}_{\text{urban}}(t)$ = urban industrialized population (United Nations Population Division, 1995)

$f\text{C}_{\text{Sewage}}$ = 1521 moles C/person/yr (Billen, 1993)

$f\text{N}_{\text{Sewage}}$ = 261 moles N/person/yr (Billen, 1993)

$f\text{P}_{\text{Sewage}}$ = 16.5 moles P/person/yr (Billen, 1993; Caraco, 1995)

$f\text{P}_{\text{Detergent}}$ = 30.6 moles P/urban industrial person/yr (Billen, 1993; Caraco, 1995)

III. CONSTANTS AND PARAMETERS

IIIA. Scaling Factors in Terrestrial Photosynthetic Flux Equation

$K_{\text{photo}} = f_{C10} \times f_{Nsw} \times f_{Psw} \times f_T$

$f_{C10} = (R_{\text{max},C}/\text{CF}_{101\ t=0}) \times [\mathbf{C}_{10}(t)/(k_C + \mathbf{C}_{10}(t))]$; atmospheric CO_2 response function, dimensionless

$R_{\text{max},C} = 1.445 \times 10^5$ mol/yr

$k_C = 6.25 \times 10^5$ moles

$f_{Nsw} = (R_{\text{max},N}/\text{NF}_{Sw1\ t=0}) \times [\mathbf{N}_{Sw}(t)/(k_N + \mathbf{N}_{Sw}(t))]$; soil-water N response function, dimensionless

$R_{\text{max},N} = 566.9$ mol/yr

$k_N = 65.11$ moles

$f_{Psw} = (R_{\text{max},P}/\text{PF}_{Sw1\ t=0}) \times [\mathbf{P}_{Sw}(t)/(k_P + \mathbf{P}_{Sw}(t))]$; soil-water P response function, dimensionless

$R_{\text{max},P} = 141.7$ mol/yr

$k_P = 2.04$ moles

$f_T = Q_{10}^{(\Delta T/10)}$, temperature response function, dimensionless

$Q_{10} = 2$ (Ver, 1998)

$\Delta T = T(t) - T_{1700}$ global mean temperature change since the year 1700, in °C (Houghton et al., 1996; Nicholls et al., 1996; UCAR/OIES, 1991)

IIIB. Atmosphere-Ocean Exchange Flux

$(d\mathbf{C}_4/dt)_{\text{atmex}} = [\mathbf{C}_{4\ t=0}/\{(R_0 \times \mathbf{C}_{10\ t=0})+[d \times (\mathbf{C}_{10}(t) - \mathbf{C}_{10\ t=0})]\}] \times [1 - \{d \times (\mathbf{C}_{10}(t) - \mathbf{C}_{10\ t=0})/\{(R_0 \times \mathbf{C}_{10\ t=0})+[d \times (\mathbf{C}_{10}(t) - \mathbf{C}_{10\ t=0})]\}\}] \times d\mathbf{C}_{10}/dt$

$(d\mathbf{C}_7/dt)_{\text{atmex}} = [\mathbf{C}_{7\ t=0}/\{(R_0 \times \mathbf{C}_{10\ t=0})+[d \times (\mathbf{C}_{10}(t) - \mathbf{C}_{10\ t=0})]\}] \times [1 - \{d \times (\mathbf{C}_{10}(t) - \mathbf{C}_{10\ t=0})/\{(R_0 \times \mathbf{C}_{10\ t=0})+[d \times (\mathbf{C}_{10}(t) - \mathbf{C}_{10\ t=0})]\}\}] \times d\mathbf{C}_{10}/dt$

$R_0 = 9$; Revelle factor, dimensionless (Revelle and Munk, 1977)

$d = 4$; Revelle constant, dimensionless (Revelle and Munk, 1977)

IIIC. Surface Ocean–Deep Ocean Waters Exchange

$$CF_{797} = -CF_{797\,t=0} + [(\mathbf{C}_7(t) - \mathbf{C}_{7\,t=0})/(h_7 \times (\tau_9))] - [(\mathbf{C}_9(t) - \mathbf{C}_{9\,t=0})/(h_9 \times (\tau_9))]$$

h_9 = 3900 meters; mean deep ocean water depth

h_7 = 100 meters; mean surface ocean mixed layer depth

τ_9 = 500 yrs; time constant to disperse an ocean surface perturbation into the bulk oceans (Revelle and Munk, 1977)

IIID. Scaling Factors in Marine Photosynthetic Flux Equation

$$CF_{45} = CF_{45\,t=0} \times f_{N4} \times f_{P4}$$
$$CF_{78} = CF_{78\,t=0} \times f_{N7} \times f_{P7}$$

$$f_{N4} = \mathbf{N}_4(t)/\mathbf{N}_{4\,t=0}$$
$$f_{P4} = \mathbf{P}_4(t)/\mathbf{P}_{4\,t=0}$$
$$f_{N7} = \mathbf{N}_7(t)/\mathbf{N}_{7\,t=0}$$
$$f_{P7} = \mathbf{P}_7(t)/\mathbf{P}_{7\,t=0}$$

IIIE. Redfield Ratios

$(\mathbf{C}{:}\mathbf{N})_{1\,photo}$ = 10500/41.18	$(\mathbf{C}{:}\mathbf{P})_{1\,photo}$ = 10500/10.294	$(\mathbf{C}{:}\mathbf{S})_{1\,photo}$ = 10500/8.235
$(\mathbf{C}{:}\mathbf{N})_{1\,resp}$ = 5250/41.18	$(\mathbf{C}{:}\mathbf{P})_{1\,resp}$ = 5250/10.294	$(\mathbf{C}{:}\mathbf{S})_{1\,resp}$ = 5250/8.235
$(\mathbf{C}{:}\mathbf{N})_{2\,resp}$ = 2.05×10^4/976	$(\mathbf{C}{:}\mathbf{P})_{2\,resp}$ = 2.05×10^4/147	$(\mathbf{C}{:}\mathbf{S})_{2\,resp}$ = 2.05×10^4/170.5
$(\mathbf{C}{:}\mathbf{N})_{5\,photo}$ = 600/90.71	$(\mathbf{C}{:}\mathbf{P})_{5\,photo}$ = 600/5.66	$(\mathbf{C}{:}\mathbf{S})_{5\,photo}$ = 600/9.623
$(\mathbf{C}{:}\mathbf{N})_{8\,photo}$ = 3600/543	$(\mathbf{C}{:}\mathbf{P})_{8\,photo}$ = 3600/33.96	$(\mathbf{C}{:}\mathbf{S})_{8\,photo}$ = 3600/57.736
$(\mathbf{N}{:}\mathbf{P})_{Crops}$ = 20		
$(\mathbf{C}{:}\mathbf{N}{:}\mathbf{P})_{Sewage}$ = 32/5.5/1		

REFERENCES CITED

Barnola, J. M., M. Anklin, J. Procheron, D. Reynaud, J. Schwander, and B. Stauffer, 1995, CO$_2$ evolution during the last millennium as recorded by Antarctic and Greenland ice: Tellus, v. 47B, p. 264–272.

Battle, M., M. L. Bender, P. P. Tans, J. W. C. White, J. T. Ellis, T. Conway, and R. J. Francey, 2000, Global carbon sinks and their variability inferred from atmospheric O$_2$ and $\delta^{13}C$: Science, v. 287, p. 2467–2470.

Berner, E. A., and R. A. Berner, 1996, Global Environment: Water, Air and Geochemical Cycles: Upper Saddle River, N.J., Prentice Hall, 376 p.

Billen, G., 1993, The PHISON river system: A conceptual model of C, N and P transformations in the aquatic continuum from land to sea, in R. Wollast, F. T. Mackenzie, and L. Chou, eds., Interactions of C, N, P and S Biogeochemical Cycles and Global Change, Berlin, Springer-Verlag, p. 141–161.

Brown, L. R., M. Renner, and C. Flavin, 1997, The Environmental Trends that are Shaping Our Future: Vital Signs 1997: New York, W.W. Norton & Company, Inc., 165 p.

Bruno, M., and F. Joos, 1997, Terrestrial carbon storage during the past 200 years: A Monte Carlo analysis of CO$_2$ data from ice core and atmospheric measurements: Global Biogeochemical Cycles, v. 11, p. 111–124.

Cao, M., and F. I. Woodward, 1998, Dynamic responses of terrestrial ecosystem carbon cycling to global climate change: Nature, v. 393, p. 249–252.

Caraco, N. F., 1995, Influence of human populations on P transfers to aquatic systems: A regional scale study using large rivers, in H. Tiessen, ed., Phosphorus in the Global Environment: Transfers, Cycles and Management. SCOPE 54, Chichester, England, John Wiley & Sons, Ltd., p. 235–244.

Charlson, R. J., T. L. Anderson, and R. E. McDuff, 1992, The sulfur cycle, in S. S. Butcher, R. J. Charlson, G. H. Orians, and G. V. Wolfe, eds., Global Biogeochemical Cycles, London, Academic Press, p. 285–300.

Dignon, J., 1992, NO_x and SO_x emissions from fossil fuels: A global distribution: Atmospheric Environment, v. 26, p. 1157–1163.

Dignon, J., and S. Hameed, 1989, Global emissions of nitrogen and sulfur oxides from 1860 to 1980: Journal Air Pollution Control Association, v. 39, p. 180–186.

Esser, G., and G. H. Kohlmaier, 1991, Modeling terrestrial sources of nitrogen, phosphorus, sulphur and organic carbon to rivers, in E. T. Degens, S. Kempe, and J. E. Richey, eds., Biogeochemistry of Major World Rivers. SCOPE 42, Chichester, John Wiley & Sons, p. 297–322.

Etheridge, D. M., L. P. Steele, R. L. Langenfelds, R. J. Francey, J.-M. Barnola, and V. I. Morgan, 1996, Natural and anthropogenic changes in atmospheric CO_2 over the last 1000 years from air in Antarctic ice and firn: Journal of Geophysical Research, v. 101, p. 4115–4128.

Friedli, H., H. Lotscher, H. Oeschger, U. Siegenthaler, and B. Stauffer, 1986, Ice core record of the $^{13}C/^{12}C$ ratio of atmospheric carbon dioxide in the past two centuries: Nature, v. 324, p. 237–238.

Galloway, J. N., W. H. Schlesinger, H. Levy II, A. Michaels, and J. L. Schnoor, 1995, Nitrogen fixation: Anthropogenic enhancement-environmental response: Global Biogeochemical Cycles, v. 9, p. 235–252.

Gattuso, J. P., M. Frankignoulle, and R. W. Buddemeier, 1998a, Effect of calcium carbonate saturation of seawater on coral calcification: Global and Planetary Change, v. 18, p. 37–46.

Gattuso, J. P., M. Frankignoulle, and R. Wollast, 1998b, Carbon and carbonate metabolism in coastal aquatic ecosystems: Annual Review of Ecology and Systematics, v. 29, p. 405–434.

Hameed, S., and J. Dignon, 1992, Global emissions of nitrogen and sulfur oxides in fossil fuel combustion 1970–1986: Journal of the Air and Waste Management Association, v. 42, p. 159–163.

Houghton, J., L. G. M. Filho, B. A. Callander, N. Harris, A. Kattenberg, and K. Maskell, 1996, Climate Change 1995: The Science of Climate Change, Cambridge, Cambridge University Press, p. 572.

Houghton, R. A., 1983, Changes in the carbon content of terrestrial biota and soils between 1860–1980: A net release of CO_2 to the atmosphere: Ecological Monographs, v. 53, p. 235–262.

Houghton, R. A., 1991a, Release of carbon to the atmosphere from degradation of forests in tropical Asia: Canadian Journal of Forest Research, v. 21, p. 132–142.

Houghton, R. A., 1991b, Tropical deforestation and atmospheric carbon dioxide: Climatic Change, v. 19, p. 99–118.

Houghton, R. A., 1995, Effects of land-use change, surface temperature and CO_2 concentration on terrestrial stores of carbon, in G. M. Woodwell and F. T. Mackenzie, eds., Biotic Feedbacks in the Global Climatic System: Will the Warming Feed the Warming?, Oxford University Press, p. 333–350.

Houghton, R. A., 1996a, Land-use change and terrestrial carbon: The temporal record, in M. J. Apps and D. T. Proce, eds., Forest Ecosystems, Forest Management and the Global Carbon Cycle, Berlin, Springer-Verlag, p. 117–134.

Houghton, R. A., 1996b, Terrestrial sources and sinks of carbon inferred from terrestrial data: Tellus, v. 48B, p. 420–432.

Houghton, R. A., E. A. Davidson, and G. M. Woodwell, 1998, Missing sinks, feedbacks, and understanding the role of terrestrial ecosystems in the global carbon balance: Global Biogeochemical Cycles, v. 12, p. 25–34.

Hudson, R. J. M., S. A. Gherini, and R. A. Goldstein, 1994, Modeling the global carbon cycle: Nitrogen fertilization of the terrestrial biosphere and the "missing" CO_2 sink: Global Biogeochemical Cycles, v. 8, p. 307–333.

Joos, F., and M. Bruno, 1998, Long-term variability of the terrestrial and oceanic carbon sinks and the budgets of the carbon isotopes ^{13}C and ^{14}C: Global Biogeochemical Cycles, v. 12, p. 277–295.

Kammen, D., and B. Marino, 1993, On the origin and magnitude of pre-industrial anthropogenic CO_2 and CH_4 emissions: Chemosphere, v. 26, p. 69–86.

Keeling, C. D., 1973, Industrial production of carbon dioxide from fossil fuels and limestone: Tellus, v. 25, p. 174–198.

Keeling, C. D., J. F. S. Chin, and T. P. Whorf, 1996, Increased activity of northern vegetation inferred from atmospheric CO_2 measurements: Nature, v. 382, p. 146–149.

Keeling, C. D., and T. P. Whorf, 1999, Atmospheric CO_2 records from sites in the SIO air sampling network, Mauna Loa Observatory, Hawaii, 1958–1998, Trends: A compendium of data on global change, http://cdiac.esd.ornl.gov/ftp/maunaloa-co2/maunaloa.co2, Oak Ridge, Tennessee, Carbon Dioxide Information Analysis Center, Oak Ridge National Laboratory.

Keeling, R. F., S. C. Piper, and M. Heimann, 1996, Global and hemispheric CO_2 sinks deduced from changes in atmospheric O_2 concentration: Nature, v. 381, p. 218–221.

Kempe, S., 1995, Coastal seas: A net source or sink of atmospheric carbon dioxide?, Land-Ocean Interactions in the Coastal Zone Core Project of the IGBP, p. 27.

Kleypas, J. A., R. W. Buddemeier, D. Archer, J.-P. Gattuso, C. Langdon, and B. N. Opdyke, 1999, Geochemical consequences of increased atmospheric carbon dioxide on coral reefs: Science, v. 284, p. 118–120.

Mackenzie, F. T., 1998, Our Changing Planet: An Introduction to Earth System Science and Global Environmental Change: Upper Saddle River, New Jersey, Prentice-Hall, Inc., 486 p.

Mackenzie, F. T., and C. Agegian, 1989, Biomineralization and tentative links to plate tectonics, in R. E. Crick, ed., Origin, Evolution and Modern Aspects of Biomineralization in Plants and Animals, New York, Plenum Press, p. 11–28.

Mackenzie, F. T., W. B. Bishoff, F. C. Bishop, M. Loijens, J. Schoonmaker, and R. Wollast, 1983, Magnesian calcites: Low temperature occurrence, solubility, and solid-solution behavior, in R. J. Reeder, ed., Carbonates: Mineralogy and Chemistry, Chelsea, Michigan, Mineralogical Society of America, p. 91–144.

Mackenzie, F. T., A. Lerman, and L. M. Ver, 1998a, Role of the continental margin in the global carbon balance during the past three centuries: Geology, v. 26, p. 423–426.

Mackenzie, F. T., L. M. Ver, and A. Lerman, 1998b, Coupled biogeochemical cycles of carbon, nitrogen, phosphorus, and sulfur in the land-ocean-atmosphere system, in J. N. Galloway and J. M. Melillo, eds., Asian Change in the Context of Global Change, Cambridge, Cambridge University Press, p. 42–100.

Mackenzie, F. T., L. M. Ver, and A. Lerman, 2000, Coastal-zone biogeochemical dynamics under global warming: International Geology Review, v. 42, p. 193–206.

Mackenzie, F. T., L. M. Ver, C. Sabine, M. Lane, and A. Lerman, 1993, C, N, P, S global biogeochemical cycles and modeling of global change, in R. Wollast, F. T. Mackenzie, and L. Chou, eds., Interactions of C, N, P and S Biogeochemical Cycles and Global Change, Berlin, Springer-Verlag, p. 1–62.

Manabe, S., and R. J. Stouffer, 1994, Multiple-century response of a coupled ocean-atmosphere model to increase of atmospheric carbon dioxide: Journal of Climate, v. 7, p. 5–23.

Marland, G., T. A. Boden, R. J. Andres, A. L. Brenkert, and C. A. Johnston, 1999, Global, regional, and national CO_2 emissions, NDP-030/R8, Trends: A compendium of data on global change, http://cdiac.esd.ornl.gov/trends/emis/tre_glob.htm, Oak Ridge, Tennessee, Carbon Dioxide Information Analysis Center, Oak Ridge National Laboratory.

Melillo, J. M., A. D. McGuire, D. W. Kicklighter, B. Moore III, C. J. Vorosmarty, and A. L. Schloss, 1993, Global climate change and terrestrial net primary production: Nature, v. 363, p. 234–240.

Melillo, J. M., I. C. Prentice, G. D. Farquhar, E. D. Schulze, and O. E. Sala, 1996, Terrestrial biotic responses to environmental change and feedbacks to climate, in J. Houghton, L. G. M. Filho, B. A. Callander, N. Harris, A. Kattenberg, and K. Maskell, eds., Climate Change 1995: The Science of Climate Change, Cambridge, Cambridge University Press, p. 445–481.

Morse, J. W., and F. T. Mackenzie, 1990, Geochemistry of Sedimentary Carbonates: Amsterdam, The Netherlands, Elsevier Science Publishing Co., 679 p.

Neftel, A., E. Moor, H. Oeschger, K. K. Turekian, and R. E. Dodge, 1985, Evidence from polar ice cores for the increase in atmospheric CO_2 in the past two centuries: Nature, v. 315, p. 45–47.

Nicholls, N., G. V. Gruza, J. Jouzel, T. R. Karl, L. A. Ogallo, and D. E. Parker, 1996, Observed climate variability and change, in J. Houghton, L. G. M. Filho, B. A. Callander, N. Harris, A. Kattenberg, and K. Maskell, eds., Climate Change 1995: The Science of Climate Change, Cambridge University Press, p. 133–192.

Quay, P. D., B. Tilbrook, and C. S. Wong, 1992, Oceanic uptake of fossil fuel CO_2: Carbon-13 evidence: Science, v. 256, p. 74–79.

Rabouille, C., Mackenzie, F. T., and Ver, L. M. (submitted) Influence of the human perturbation on carbon, nitrogen, and oxygen biogeochemical cycles in the global coastal ocean: Geochimica et Cosmochimica Acta.

Raynaud, D., and J. M. Barnola, 1985, An Antarctic ice core reveals atmospheric CO_2 variations over the past few centuries: Nature, v. 315, p. 309–311.

Revelle, R., and W. Munk, 1977, The carbon dioxide cycle and the biosphere, in N. G. S. Committee, ed., Energy and Climate, Washington, D.C., National Academy Press, p. 140–158.

Sala, O. E., F. S. Chapin III, J. J. Armesto, E. Berlow, J. Bloomfield, R. Dirzo, E. Huber-Sanwald, L. F. Huenneke, R. B. Jackson, A. Kinzig, R. Leemans, D. M. Lodge, H. A. Mooney, M. Oesterheld, N. L. Poff, M. T. Sykes, B. H. Walker, M. Walker, and D. H. Wall, 2000, Global biodiversity scenarios for the year 2100: Science, v. 287, p. 1770–1774.

Sarmiento, J. L., T. M. C. Hughes, R. J. Stouffer, and S. Manabe, 1998, Simulated response of the ocean carbon cycle to anthropogenic climate warming: Nature, v. 393, p. 245–249.

Sarmiento, J. L., J. C. Orr, and U. Siegenthaler, 1992, A perturbation simulation of CO_2 uptake in an ocean general circulation model: Journal of Geophysical Research, v. 97, p. 3621–3645.

Schimel, D., J. Melillo, H. Tian, A. D. McGuire, D. Kicklighter, T. Kittel, N. Rosenbloom, S. Running, P. Thornton, D. Ojima, W. Parton, R. Kelly, M. Sykes, R. Neilson, and B. Rizzo, 2000, Contribution of increasing CO_2 and climate to carbon storage by ecosystems in the United States: Science, v. 287, p. 2004–2006.

Siegenthaler, U., H. Friedli, H. Loetscher, E. Moor, A. Neftel, H. Oeschger, and B. Stauffer, 1988, Stable-isotope ratios and concentration of CO_2 in air from polar ice cores: Annals of Glaciology, v. 10, p. 1–6.

Siegenthaler, U., and J. L. Sarmiento, 1993, Atmospheric carbon dioxide and the ocean: Nature, v. 365, p. 119–125.

Smil, V., 1985, Carbon, Nitrogen and Sulfur: Human Interference in Grand Biospheric Cycles: New York, Plenum Press, 459 p.

Smil, V., 1991, Population growth and nitrogen: An exploration of a critical existential link: Population and Development Review, v. 17, p. 569–601.

Smith, S. V., and J. T. Hollibaugh, 1993, Coastal metabolism and the oceanic organic carbon balance: Reviews of Geophysics, v. 31, p. 75–89.

Smith, S. V., and F. Pesret, 1974, Processes of carbon dioxide flux in the Fanning Island lagoon: Pacific Science, v. 28, p. 225–245.

UCAR/OIES, 1991, Changes in Time in the Temperature of the Earth: Boulder, Colorado, UCAR/OIES.

United Nations FAO, various years, Annual Fertilizer Review: Rome, Italy: Food and Agriculture Organization of the United Nations.

United Nations Population Division, 1995, World Population Prospects: The 1994 Revision: New York, United Nations, 884 p.

Ver, L. M., 1998, Global Kinetic Models of the Coupled C, N, P, and S Biogeochemical Cycles: Implications for Global Environmental Change: Ph.D. dissertation, University of Hawaii, Honolulu, 681 p.

Ver, L. M., F. T. Mackenzie, and A. Lerman, 1994, Modeling preindustrial C-N-P-S biogeochemical cycling in the land-coastal margin system: Chemosphere, v. 29, p. 855–887.

Ver, L. M., F. T. Mackenzie, and A. Lerman, 1999a, Biogeochemical responses of the carbon cycle to natural and human perturbations: Past, present, and future: American Journal of Science, v. 299, p. 762–801.

Ver, L. M., F. T. Mackenzie, and A. Lerman, 1999b, Carbon cycle in the coastal zone: Effects of global perturbations and change in the past three centuries: Chemical Geology, v. 159, p. 283–304.

Wollast R., 1998, Evaluation and comparison of the global carbon cycle in the coastal zone and in the open ocean, in The sea: The global coastal ocean, Vol. 10 (ed. K. H. Brink and A. R. Robinson), John Wiley & Sons, p. 213–252.

Wollast, R., and F. T. Mackenzie, 1989, Global biogeochemical cycles and climate, in A. Berger, S. Schneider, and J. C. Duplessy, eds., Climate and Geo-Sciences, Kluwer Academic Publishers, p. 453–473.

Woodwell, G. M., 1995, Biotic feedbacks from the warming of the Earth, in G. M. Woodwell and F. T. Mackenzie, eds., Biotic Feedbacks in the Global Climatic System: Will the Warming Feed the Warming?, New York, Oxford University Press, p. 3–21.

Broecker, W. S., Are we headed for a thermohaline catastrophe?, 2001, *in* L. C. Gerhard, W. E. Harrison, and
B. M. Hanson, eds., Geological perspectives of
global climate change, p. 83–95.

4 | Are We Headed for a Thermohaline Catastrophe?

Wallace S. Broecker
Lamont-Doherty Earth Observatory
of Columbia University
Palisades, New York, U.SA.

ABSTRACT

If we continue along the business-as-usual fossil-fuel-use track, we run the risk, late in the twenty-first century, of triggering an abrupt reorganization of the ocean's thermohaline circulation. This conclusion is based on evidence stored in Greenland ice, continental-margin sediments, and mountain moraines that tells us that the large and abrupt global climate changes during the last period of glaciation were associated with sudden reorganizations of the ocean's thermohaline circulation and on simulations carried out in joint atmosphere-ocean models that suggest that raising the greenhouse capacity of the atmosphere to the carbon-dioxide equivalent of 750 ppm would cripple the ocean's conveyor circulation. However, the recent discovery by Gerard Bond that the 1500-year cycle which paced these glacial disruptions continued in a muted form during times of interglaciation casts a new light on this situation. It leads me to suspect that the large and rapid atmospheric changes of glacial time were driven by a sea-ice amplifier. If so, then, because little sea ice will remain at the time of a greenhouse-induced thermohaline reorganization, perhaps the threat will be far smaller than I had previously envisioned.

INTRODUCTION

A debate rages regarding the significance of the changes in climate that will occur during the next century as the result of the ongoing buildup of greenhouse gases in the earth's atmosphere. A large majority of scientists involved in atmosphere and ocean research support the conclusions of the Intergovernmental Panel on Climate Change (IPCC) report, which state that if unabated, this buildup will result in a significant warming of the planet with consequent changes in rainfall, storminess, and soil moisture. These changes, which will intensify over the next century, may adversely impact the production of food and will certainly pose an additional threat to the already stressed wildlife on our planet.

One prominent atmospheric scientist, Richard Lindzen of the Massachusetts Institute of Technology, strongly opposes this view. He correctly points out that a significant warming will occur only if the primary forcing (by CO_2, CH_4, N_2O, and CFCs) is amplified by an increase in the atmosphere's water-vapor content. In the absence of this water-vapor feedback, a tripling of the atmosphere's CO_2 content would lead to only a 1.8°C global warming. The general circulation models (GCMs) employed by atmospheric scientists have in common that the water-vapor content of the atmosphere rises in proportion to the vapor pressure of water (i.e., about 7% per °C) and thereby, through its large infrared absorption capacity, generates an amplification of primary warming by a factor of about 2.5, thus vaulting the warming for a CO_2 tripling to about 4.5°C. Lindzen et al. (1982) are convinced that existing GCMs do not distribute water vapor correctly and, in particular, that the water-vapor content of the air descending over the desert regions of the world will decrease rather than increase. Because these regions constitute the atmosphere's major radiator, this decrease would tend to null the primary CO_2 warming.

My view lies at the opposite pole from that of Lindzen. I believe that the results obtained using GCMs are the best guide we have to the future. Further, in addition to the gradual warming predicted by these GCMs, I fear that if the earth's response to the buildup of greenhouse gases is as large as the GCMs predict and if the buildup were to triple the preindustrial CO_2 content (i.e. 3×280 or 840 ppm [parts per million]), there is a distinct possibility that warming and increased precipitation in the polar regions will lead to a disruption of the ocean's large-scale thermohaline circulation. By analogy to events recorded in ice cores, in mountain moraines, and in rapidly accumulating marine sediments, such a disruption could bring about a large and abrupt change in the climate of our planet.

EXPECTED MAGNITUDE OF THE GREENHOUSE BUILDUP

Before discussing the evidence in support of a possible greenhouse-triggered reorganization, a few words regarding the expected magnitude of the anthropogenic buildup of the atmosphere's greenhouse capacity are in order. Although it must be kept in mind that currently the impact of excess methane, nitrous oxide, and CFCs roughly matches that of CO_2, it is CO_2 emissions that pose the major future hazard because they will prove to be the most difficult to rein in. We now emit 6.8 gigatons of carbon (GtC) as CO_2 per annum. To this must be added about 1 GtC per year resulting from deforestation. With an expected increase in population to 9 billion or 10 billion and with the expected increase in the standard of living of people in developing countries, if fossil fuels were to remain our primary source of energy, then it is likely that the CO_2 emissions will rise well above 10 GtC per annum before 2050 A.D. As shown in **Figure 1**, if emissions were to average 10 GtC over the entire course of the twenty-first century, then in the absence of a significant enhancement of storage in terrestrial biomass (i.e., trees and soil humus), the CO_2 content would reach well into the danger zone. By this I mean that when the greenhouse contributions of CH_4, N_2O, and CFCs are taken into account, the greenhouse capacity of the atmosphere will surpass the level at which models suggest that thermohaline circulation would be seriously impacted (see Manabe and Stouffer, 1993, and Stocker and Schmittner, 1997). Even if by the end of the twenty-first century enhanced storage of carbon in the terrestrial biosphere reaches the generous magnitude of 200 GtC, the greenhouse content of the atmosphere will still reach into the danger zone. In

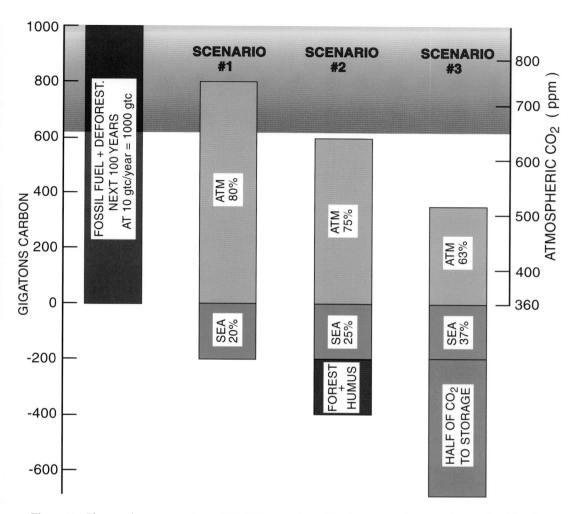

Figure 1 If, over the next century, 1000 GtC are released to the atmosphere as the result of fossil-fuel burning and deforestation, then in the absence of significant greening of the terrestrial biosphere, roughly 80% will remain in the atmosphere, raising its CO_2 content to about 750 ppm. If we are lucky and greening driven by fixed nitrogen and carbon dioxide increases storage in the terrestrial biosphere by as much as 200 GtC, CO_2 would increase to about 640 ppm. The only way by which the CO_2 content could be held below 500 ppm would be to capture and store more than half of the CO_2 generated or to substitute nonfossil-fuel energy systems for at least half of the conventional sources. When the contribution of CH_4, N_2O, and the CFCs is taken into account, a CO_2 buildup to more than 600 ppm puts us in the danger zone with regard to a shutdown of thermohaline circulation.

order to avoid entry into this zone, we would either have to turn to nonfossil fuel sources to supply roughly half of the energy needed in the twenty-first century or we would have to sequester roughly half of the CO_2 generated by fossil fuel burning over the course of the twenty-first century.

THE CLIMATIC RECORD

The basis for the claim that the earth's climate system is capable of jumping from one mode of operation to another comes from records stored in polar ice (Dansgaard et al., 1993), in rapidly deposited marine sediments (Behl and Kennett, 1996; Schulz et al., 1998), and in the moraines formed by mountain glaciers (Denton and Hendy, 1994). Taken together, these records provide convincing evidence that these switches were abrupt, strong, and global. By "abrupt," I mean that they were completed in two to four decades. Furthermore, during the transition interval, climate flickered (Taylor et al., 1993) much as fluorescent lights do when they are turned on. By "strong," I mean that they eclipsed, by far, any climate change experienced during historic time. By "global," I mean that their impacts were felt everywhere on the planet.

Because I have published several papers summarizing this evidence (see Broecker, 1997b), I will present here only a brief recapitulation. During the course of the last glacial period, the earth experienced 20 or so millennial-duration oscillations in climate between a very cold state and an intermediately cold state. The last of these major chills has been dubbed the Younger Dryas (Y.D.). Its abrupt ending came 11,500 years ago. Since then, except for one century of cooling centered at 8200 years ago (Alley et al., 1997), the earth's climate system appears to have remained locked in its interglacial mode. The birth of agriculture occurred early in the present interglacial period(and likely in response to the reduction in middle eastern rainfall brought about by the demise of glacial climates). Hence, civilization as we know it developed during a period of unusual climate quiescence. No mode change has marred its progress.

Due to a combination of results of annual layer counting in Greenland's ice cores, radiocarbon dating of moraines and sediments, and thorium-uranium dating of corals and speliothems, we can speak with confidence concerning the chronology of the events of the last 130,000 years. These results provide a firm absolute chronology covering the entire duration of the last major glacial-interglacial cycle. In particular, the radiocarbon method allows us to correlate events during the last 40,000 years across the entire planet.

Much of the information comes from two 3-kilometer-long ice cores located close to the geographic center of the Greenland ice cap. Drilled at sites separated by 30 kilometers, these cores provide records agreeing to the finest detail back to 110,000 years ago (i.e., back to roughly the middle of the last interglacial period). The oxygen isotope record in the ice itself provides the pattern of the temperature changes (**Figure 2**). The structure of the temperature profile measured in the borehole itself provides a means of calibrating the isotope record. It shows that the mean air temperature over Greenland's ice plateau was on the average about 15°C colder during glacial time than now (Cuffey et al., 1994). The dust content in the ice varied in concert with the isotopes (Mayewski et al., 1994). It ranged up to 50 times higher than today's concentration. Because isotope fingerprinting (via radiogenic daughter nuclides of the elements lead, strontium, and neodymium) demonstrates that this dust originated in Asia's Gobi Desert (Biscaye et al., 1997), this concordance requires that the storminess over Asia underwent jumps in frequency and intensity in exact concert with Greenland's air temperature changes. Furthermore, the input of dust clearly flickered during the transitions (Taylor et al., 1993). Finally, the methane content of air trapped in bubbles in the ice shows sharp changes (Chappellaz et al., 1993; Brook et al., 1996) which have been shown to be synchronous with the dust and temperature changes (Severinghaus et al., 1998). Because the major source of methane during glacial time was likely to have been tropical swamps, these water bodies must have become warmer and wetter at the times of the abrupt cold to warm transitions in Greenland.

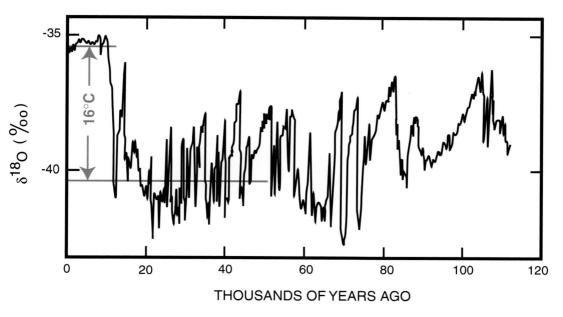

Figure 2 The pattern of temperature change over the last 110,000 years as recorded by the [18]O to [16]O ratio in Greenland ice (Dansgaard et al., 1993). The absolute temperature range has been independently determined from the temperature profile measured in the borehole (Cuffey et al., 1994).

Further evidence for the widespread occurrence of Greenland's millennial-duration events comes from marine sediments from areas of high sedimentation rate (i.e., >50 cm/1000 yrs.). Three such sites, one in the Santa Barbara Basin off California (Behl and Kennett, 1996), one in the Arabian Sea off India (Schulz et al., 1998), and one in the Cariaco Trench off Venezuela (K. Hughen, personal communication, 1999), are currently bathed in oxygen-poor intermediate-depth water (see **Figure 3**). During the present interglacial period, the sediments at these sites are annually layered for the most part, demonstrating that oxygen is absent in the sediment pore waters. This absence prevents worms from stirring the sediment and thereby erasing the layering. At all three locations, the O_2 content of the sediment must have been considerably higher during the Younger Dryas and at the times of millennial-duration cold extremes. This alternation in thermocline O_2 content makes it clear that some combination of the rate of ventilation by O_2-rich surface waters and the rate of rain of organic matter from the overlying water at these cold times must have maintained higher pore water O_2 contents. Although it is not possible to say exactly how the operation of the upper oceans changed, the fact that the same temporal pattern is seen in the Pacific, Indian, and Atlantic Oceans suggests that the changes were global in scale.

Detailed measurements of alkenone ratios (an excellent paleothermometer) on rapidly accumulating sediment from a site near Bermuda reveal that during the time interval 30,000 to 60,000 years ago, surface water temperatures in the Sargasso Sea underwent 4° to 5°C changes in concert with the temperature swings observed in the Greenland ice core (Sachs and Lehman, 1999).

Only for the period between the onset of the Bolling-Allerod warm period, which brought to an end the last glacial period and the subsequent Younger Dryas cold lapse (**Figure 4**), do we have

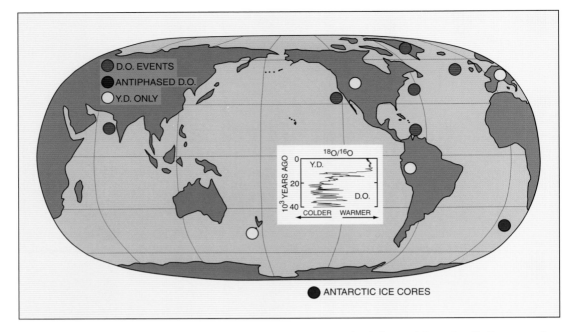

Figure 3 The red dots depict localities on the globe where the full set of Greenland's Dansgaard Oeschger events has been identified (Behl and Kennett, 1996; Schulz et al., 1998). The yellow dots depict localities where radiometric ages confirm a Younger Dryas age for mountain moraines (Gosse et al., 1995; Denton and Hendy, 1994; Ivy-Ochs et al., 1996) and for ^{18}O cold anomalies in Andean ice cores (Thompson et al., 1998). The blue dots depict southern sites where these events are antiphased with respect to those for the rest of the globe (Blunier et al., 1997, 1998).

extensive coverage for the continents. The pollen records contained in bog and lake sediments clearly demonstrate that a profound global vegetation change occurred at the time of the onset of the Bolling-Allerod warm. The vastness of this climate change is also recorded by closed basin lakes in desert regions. Prior to the onset of the Bolling-Allerod, Lake Victoria, which straddles the equator in East Africa, was bone-dry (Johnson et al., 1996). The lake reappeared early in the Bolling-Allerod. The nearby Red Sea had become so saline during late glacial time that planktonic foraminifera could no longer survive (Hemleben et al., 1996). Then suddenly at the onset of the Bolling-Allerod warm, planktonic foraminifera reappeared. In contrast, the situation in North America's Great Basin was exactly the opposite. During late glacial time, its closed basin lakes, Bonneville and Lahontan, were as much as 10 times larger in area than those of the present-day remnant lakes (Benson, 1981, 1993). The shutdown of the water supply that maintained these large lakes came at the onset of the Bolling-Allerod. Hence, not only was this event marked by a pronounced global warming but also by a major redistribution of rainfall (see Broecker, 1998).

ABRUPT-CHANGE TRIGGERS

Only one part of the climate system, the Atlantic Ocean's conveyor circulation, has been demonstrated to have clearly defined alternate modes of operation **(Figure 5)**. Model studies demonstrate

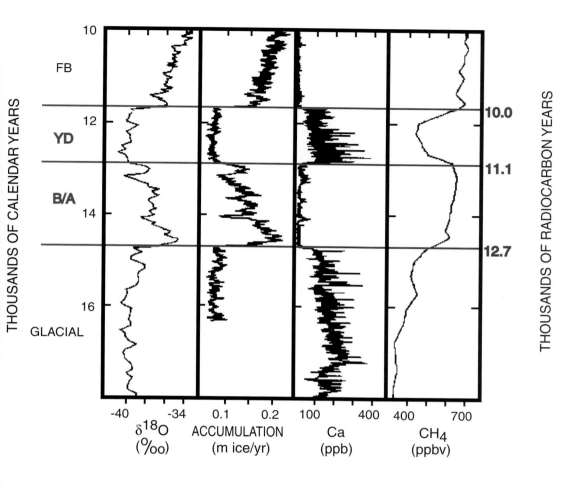

Figure 4 Oxygen isotope (Grootes et al., 1993), snow accumulation rate (Alley et al., 1993), calcium component of dust (Mayewski et al., 1994), and methane content of trapped air (Chappellaz et al., 1993) of Summit Greenland ice cores for the deglaciation period including the Bolling-Allerod (B/A) warm and the Younger Dryas (YD) cold.

that the ocean's large-scale thermohaline (*thermo* = cold, *haline* = salty) circulation can lock into more than one pattern (Marotske and Willebrand, 1991; Stocker et al., 1992; Rahmstorf, 1994, 1995). The existence of these alternate patterns is possible because the deep sea can be ventilated from both high northern and southern latitudes. Which particular source locale dominates depends on the distribution of salt in surface ocean waters (the density of seawater rises with increasing salt content and with decreasing temperature). Today's ocean surface waters of the northern Pacific are so low in salt content that even when at their freezing point (i.e., −1.8°C) they do not descend more than a few hundred meters. By contrast, waters in the northern Atlantic are particularly high in salt content. When cooled to only +2°C, they become dense enough to sink to the bottom and flood southward into the circumpolar raceway surrounding the Antarctic continent. Although the surface waters in the Southern Ocean are, like those in the northern Pacific, too fresh to permit them to

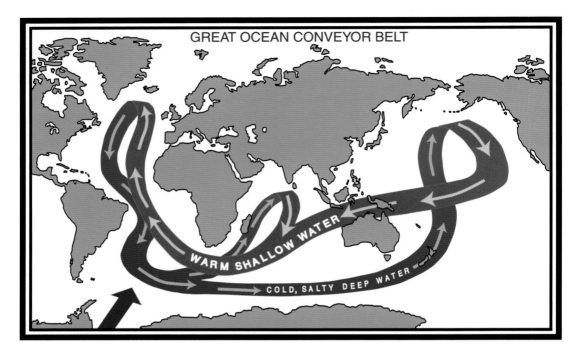

Figure 5 A conveyorlike circulation in the Atlantic Ocean carries an enormous amount of heat to the vicinity of Iceland. Here, winter cooling densifies this already salty water, allowing it to sink to the bottom. This newly formed deep water flows to the south, passing around the southern tip of Africa, where it is joined by deep waters formed in the Weddell Sea. These waters mix to form a circumpolar deep current, which swirls around the Antarctic continent. This water eventually peels off and floods the deep Indian and Pacific Oceans (Broecker, 1991).

sink to the abyss, brines released beneath the sea ice pack covering the narrow continental margin of Antarctica during the winter expansion of the ice densify the subice waters to the point that they spill off the shelves and sink to the abyss. Currently, these waters join those emanating from the Atlantic to form a mix which flows northward into the deep Pacific and Indian Oceans.

A variety of evidence from marine sediments tells us that the currently important role of deep waters formed in the northern Atlantic was greatly lessened during glacial time. Further, the fact that reorganizations of the ocean's thermohaline circulation were intimate parts of the abrupt change process is demonstrated by the record of carbon-14 to carbon (^{14}C/C) ratios in the upper ocean and atmosphere (Hughen et al., 1998). These ratios have been reconstructed by making radiocarbon measurements on carbon from materials of known calendar age. Such calendar ages can be obtained by counting annual layers (rings produced by trees and annual layers deposited in standing water). Radiocarbon measurements on planktonic foraminifera from varved Cariaco Trench sediments revealed that during the first 200 years of the Y.D., the ^{14}C/C ratio in the surface waters of the Caribbean Sea rose by an astounding 5% (Hughen et al., 1998). Then, during the remaining 1000 years of Y.D. time, the ^{14}C/C ratio gradually declined to the pre-Y.D. value. Because the onset of the ^{14}C rise coincides exactly with the onset of the Y.D. (marked by a distinct and sharp color change in Cariaco sediment), the only reasonable explanation for its occurrence is that the formation of new deep water in the northern Atlantic came to a halt. Because today the conveyorlike circulation in the Atlantic supplies most of the deep sea's radiocarbon and because deep sea waters house about two-thirds of the radiocarbon present on our planet, this cutoff would lead to a backlogging in the atmosphere and upper ocean of the new radiocarbon atoms produced by the bombardment of our planet by cosmic rays. Assuming that these upper reservoirs contain about one-third of the ocean-atmosphere carbon, their $^{14}C/C$ ratio would rise by 5% in 200 years. Correspondingly, the $^{14}C/C$ ratio in the deep sea would fall by 2.5% in 200 years. So the radiocarbon record from the Cariaco Trench provides a smoking gun with regard to the intimate involvement of thermohaline circulation. The most reasonable cause for the shutdown of the Atlantic's conveyor circulation at the onset of the Y.D. is a sudden influx of fresh water into the region where deep waters are formed. Indeed, we now know that such a flood did take place. Drainage from a large lake impounded in front of the re-treating Canadian ice sheet suddenly shifted from the Mississippi River to the Saint Lawrence (Teller and Thorleifson, 1983). This shift was heralded by a drop in the lake's outlet, permitting a large portion of the stored water to flood directly into the region where deep waters are formed.

In addition, there is evidence for a second meltwater flood close to the onset of a brief Y.D.-like cold event at 8200 years ago (Barber et al., 1999). This flood occurred when meltwater tunneled its way through the retreating Canadian ice sheet into Hudson Bay, once again flooding the area where deep water forms. As was the case for the Y.D., we surmise that this brief cold episode came to an end when the Atlantic's conveyor popped back into action.

While these two cold events which punctuated the present interglacial period were very likely triggered by floods of fresh water, it is difficult to believe that each of the 20 or so glacial flip-flops recorded in the glacial portion of the Greenland ice record was triggered by a meltwater flood. Rather, I envision that they were driven by an oscillation in the salt content of the Atlantic Ocean (Broecker et al., 1990). During the periods of intense cold, precipitation over Canada and Scandinavia became locked in the growing ice sheets allowing the Atlantic to become more salty. Eventually, the density of surface waters became high enough to permit the conveyor to snap back into action. The heat transported by the upper limb of the conveyor then changed the situation. Net growth of the ice gave way to net melting. The consequent input of fresh water drove down the salinity of Atlantic waters until the conveyor once again shut down. This cycle repeated itself over and over on a millennial timescale.

THE ELUSIVE GLOBAL CONNECTION

Although reorganizations of the ocean's thermohaline circulation appear to have served as the trigger for abrupt change, the link to global climate remains obscure. Studies conducted in joint ocean-atmosphere models suggest that the climate changes associated with a shutdown or rejuvenation of the Atlantic's conveyor should be confined in latitude to a belt north of Gibraltar and in longitude by the western margin of the Atlantic and roughly the Eurasian boundary. Winter temperatures in this region would be 6°–10°C warmer when the conveyor is operating than when it is shut down. However, as we have seen, the impacts of the abrupt changes were global. The lowering of mountain snowlines during the Y.D. was about the same at 40°S as at 40°N (**Figure 3**). In order to produce such large and hemispherically symmetrical changes, it is necessary to impact the great equatorial convective systems which supply moisture to much of the world's atmosphere. But alas, to date, no one has been able to articulate even a first-order conceptual model designed to link the tropics to the Atlantic's conveyor. So difficult does this task seem that some dynamists are tempted to turn the connection around and somehow trigger a change in the dynamics of the tropical atmosphere that would then propagate to the thermohaline circulation.

PROGNOSIS FOR THE FUTURE

Clearly, in the absence of an understanding of the mechanism which allows the earth's climate to undergo radical changes, there is no way to predict whether we put the earth at risk by adding greenhouse gases to the atmosphere. However, if as I suspect, these mode switches are triggered by reorganizations of the ocean's thermohaline circulation, then one avenue we can follow is to assess at what point such a buildup might impact the production of new deep water. Clearly, as the planet warms, the density of surface waters in the polar regions will decline. In addition to the thermal effect, there will be a dilution of the salt content of these waters, for on a warmer planet, there will be more precipitation. In areas poleward of 40° where precipitation exceeds evaporation, this excess will result in increased wetting. Indeed, joint ocean-atmosphere models (Manabe and Stauffer, 1993; Stocker and Schmittner, 1997) suggest that if the greenhouse capacity of our atmosphere (i.e., the joint impact of excess CO_2, CH_4 N_2O, and CFCs) were to reach the CO_2 equivalent of 750 ppm, then a substantial weakening of the Atlantic's conveyor circulation would be brought about. By analogy to the events of glacial time, such a weakening might lead to a jump to one of the earth's climate system's alternate modes. Because these jumps were heralded by a several-decades-long series of flickers, such an event would perhaps stress our ability to feed the 9 billion or so people likely to be present on the planet a century from now.

A discovery made by my colleague, Gerard Bond, has in recent months led me to downgrade this rather dire prediction. His studies of the ice-rafted debris in sediment cores from the northern Atlantic (Bond et al., 1997) reveals that the same 1500-year cycles in the ratio of iron-stained to clean quartz grains he observed in glacial-age sediments continue during Holocene time and also during the last interglacial. As the 1500-year cycles of glacial time appear to have been a response to the seesawing of deepwater formation, it seems likely that the interglacial cycles are as well. Indeed, Bond has shown that the last of these cycles appears to correspond to the Medieval Warm–Little Ice Age climate oscillation which played an important role in both the establishment and demise of the Vikings' Greenland colony. If so, then the question immediately comes to mind as to why, during times of glaciation, these seesaws in thermohaline circulation led to very large climate swings, but during interglacial time, to only modest ones. In a paper recently submitted to *Nature* magazine, I propose a possible explanation that involves the extent of planetary ice cover.

During glacial time, when extent of ice cover was very large, the amount of dust and sea salt transported through the atmosphere was an order of magnitude larger than today's. I link the two. The greater the ice extent, the stronger the latitudinal thermal gradient and hence the higher the frequency of intense storms. It is these intense storms which loft the dust and sea salt. As dust grains and sea salt aerosols act as solar reflectors and act to increase the number of nuclei available for raindrop formation, they tend to cool the planet.

With this background in mind, it is easy to see why the seesawing of deepwater formation between the northern Atlantic and the Southern Ocean might lead to larger climate changes during glacials than during interglacials. The key is that during glacials, the extent of sea ice in the northern Atlantic would have been on the average much larger than now. Because the time required to form or to melt away sea ice is at most a decade, were the strength of conveyor circulation in the Atlantic to undergo sudden changes, then so also would the extent of sea-ice cover, and hence storminess. By contrast, during interglacials, the magnitude of the sea-ice changes associated with these seesaws would have been much smaller and, were the ongoing greenhouse warming to lead to a disruption of thermohaline circulation, they would be even smaller yet. It is for this reason that I must back off a bit regarding my previous warning. Perhaps the climate beast is far less angry during times when our planet is relatively ice free.

Figure 6   The earth's climate system has proved itself to be an angry beast. We are now poking this beast with greenhouse gases. Unfortunately, we lack the wherewithal to predict how the beast will respond.

CONCLUSIONS

The record kept in Greenland ice and in rapidly accumulating marine sediments speaks clearly to me that earth's climate system has at times been an angry beast. Through the addition of greenhouse gases to the atmosphere, we are poking this beast **(Figure 6)**. But because we have yet to fully understand the rules governing the beast's behavior, we cannot predict its response to our poke. If, as suggested above, climate is far more responsive during times of large ice cover, then perhaps the disruption resulting from a greenhouse-induced thermohaline reorganization would be more akin to a Little Ice Age than to a Younger Dryas. Nevertheless, prudence dictates that we take firm steps to cut back the release of CO_2 into the atmosphere. These steps will involve a combination of energy conservation, the implementation of nonfossil-fuel-based energy sources, and the capture and sequestration of CO_2 from stationary power plants.

REFERENCES CITED

Alley, R. B., Mayewski, P. A., Sowers, T., Stuiver, M., Taylor, K. C., and Clark, P. U., 1997, Holocene climatic instability: A prominent, widespread event 8200 years ago: Geology, v. 25, p. 483–486.

Alley, R. B., Meese, D. A., Shuman, C. A., Gow, A. J., Taylor, K. C., Grootes, P. M., White, J. W. C., Ram, M., Waddington, E. D., Mayewski, P. A., and Zielinski, G. A., 1993, Abrupt increase in Greenland snow accumulation at the end of the Younger Dryas event: Nature, v. 362, p. 527–529.

Barber, D. C., Dyke, A., Hillaire-Marcel, C., Jennings, A. E., Andrews, J. T., Kerwin, M. W., Bilodeau, G., Mc-Neely, R., Southon, J., Morehead, M. D., and Gagnon, J.-M., 1999, Forcing of the cold event 8200 years ago by outburst drainage of Laurentide lakes, Nature, v. 400, p. 344–348.

Behl, R. J., and Kennett, J. P., 1996, Brief interstadial events in the Santa Barbara basin, NE Pacific, during the past 60 kyr: Nature, v. 379, p. 243–246.

Benson, L. V., 1981, Paleoclimatic significance of lake-level fluctuations in the Lahontan Basin: Quaternary Research, v. 16, p. 390–403.

Benson, L., 1993, Factors affecting ^{14}C ages of lacustrine carbonates: Timing and duration of the last Highstand Lake in the Lahontan Basin: Quaternary Research, v. 39, p. 163–174.

Biscaye, P. E., Grousset, F. E., Revel, M., Van der Gaast, S., Zielinski, G. A., Vaars, A., and Kukla, G., 1997, Asian provenance of glacial dust (stage 2) in the Greenland Ice Sheet Project 2 Ice Core, Summit, Greenland: Journal of Geophysical Research, v. 102, p. 26, 765–26, 781.

Blunier, T., Schwander, J., Stauffer, B., Stocker, T., Dallenbach, A., Indermühle, A., Tschumi, J., Chappellaz, J., Raynaud, D., and Barnola, J.-M., 1997, Timing of the Antarctic cold reversal and the atmospheric CO_2 increase with respect to the Younger Dryas event: Geophysical Research Letters, v. 24, p. 2683–2686.

Blunier, T., Chappellaz, J., Schwander, J., Dallenbach, A., Stauffer, B., Stocker, T. F., Raynaud, D., Jouzel, J., Clausen, H. B., Hammer, C. U., and Johnsen, S. J., 1998, Asynchrony of Antarctic and Greenland climate change during the last glacial period: Nature, v. 394, p. 739–743.

Bond, G., Showers, W., Cheseby, M., Lotti, R., Almasi, P., deMenocal, P., Priore, P., Cullen, H., Hajdas, I., and Bonani, G., 1997, A pervasive millennial-scale cycle in North Atlantic Holocene and Glacial Climates: Science, v. 278, p. 1257–1266.

Broecker, W. S., 1991, The great ocean conveyor: Oceanography, v. 4, p. 79–89.

Broecker, W. S., 1997a, Will our ride into the greenhouse future be a smooth one?: GSA Today, v. 5, p. 1–7.

Broecker, W. S., 1997b, Thermohaline circulation, The Achilles heel of our climate system: Will manmade CO_2 upset the current balance?: Science, v. 278, p. 1582–1588.

Broecker, W. S., 1998, Paleocean circulation during the last deglaciation: A bipolar seesaw?: Paleoceanography, v. 13, p. 119–121.

Broecker, W. S., Bond, G., Klas, M., Bonani, G., and Wolfli, W., 1990, A salt oscillator in the glacial North Atlantic? 1. The Concept: Paleoceanography, v. 5, p. 469–477.

Broecker, W. S., Peteet, D., Hajdas, I., Lin, J., and Clark, E., 1998, Antiphasing between rainfall in Africa's Rift Valley and North America's Great Basin: Quaternary Research, v. 50, p. 12–20.

Brook, E. J., Sowers, T., and Orchardo, J., 1996, Rapid variations in atmospheric methane concentration during the past 110,000 years: Science, v. 273, p. 1087–1091.

Chappellaz, J., Blunier, T., Raynaud, D., Branola, J. M., Schwander, J., and Stauffer, B., 1993, Synchronous changes in atmospheric CH_4 and Greenland climate between 40 and 8 kyr BP: Nature, v. 366, p. 443–445.

Cuffey, K. M., Alley, R. B., Grootes, P. M., Bolzan, J. M., and Anandakrishnan, S., 1994, Calibration of the $\delta^{18}O$ isotopic paleothermometer for central Greenland, using borehole temperatures: Journal of Glaciology, v. 40, p. 341–349.

Dansgaard, W., Johnsen, S. J., Clausen, H. B., Dahl-Jensen, D., Gundestrup, N. S., Hammer, C. U., Hvidberg, C. S., Steffensen, J. P., Sveinbjornsdottir, A. E., Jouzel, J., and Bond, G., 1993, Evidence for general instability of past climate from a 250-kyr ice-core record: Nature, v. 364, p. 218–220.

Denton, G. H., and Hendy, C. H., 1994, Younger Dryas age advance of Franz Josef Glacier in the Southern Alps of New Zealand: Science, v. 264, p. 1434–1437.

Gosse, J. C., Klein, J., Evenson, E. B., Lawn, B., and Middleton, R., 1995, Beryllium-10 dating of the duration and retreat of the last Pinedale glacial sequence: Science, v. 268, 1329–1333.

Grootes, P. M., Stuiver, M., White, J. W. C., Johnsen, J. J., and Jouzel, J., 1993, Comparison of oxygen isotope records from the GISP2 and GRIP Greenland ice cores: Nature, v. 366, p. 552–554.

Hemleben, C., Meischner, D., Zahn, R., Almogi-Labin, A., Erlenkeuser, H., and Hiller, B., 1996, Three hundred eighty thousand year long stable isotope and faunal records from the Red Sea: Influence of global sea level change on hydrography: Paleoceanography, v. 11, p. 147–156.

Hughen, K. A., Overpeck, J. T., Lehman, S. J., Kashgarian, M., Southon, J., Peterson, L. C., Alley, R., and Sigman, D. M., 1998, Deglacial changes in ocean circulation form an extended radiocarbon calibration: Nature, v. 391, 65–68.

IPCC, 1995, Climate Change 1994, Radiative Forcing of Climate Change: Intergovernmental Panel on Climate Change, Cambridge Press, 339 pp.

Ivy-Ochs, S., Schluchter, C., Kubik, P. W., Synal, H.-A., Beer, J., and Kerschner, H., 1996, The exposure age of an Egesen moraine at Julier Pass, Switzerland, measured with the cosmogenic radionuclides [10]Be, [26]Al and [36]Cl: Eclogae Geologicae Helvetiae, v. 89, p. 1049–1063.

Johnson, T. C., Scholz, C. A., Talbot, M. R., Kelts, K., Ricketts, R. D., Ngobi, G., Beuning, K., Ssemmanda, I., and McGill, J. W., 1996, Late Pleistocene desiccation of Lake Victoria and rapid evolution of Cichlid fishes: Science, v. 273, p. 1091–1093.

Lindzen, R. S., Hou, A. Y., and Farrel, B. F., 1982, The role of convective model choice in calculating the climate impact of doubling CO_2: Journal of Atmospheric Science, v. 39, p. 1189–1205.

Manabe, S., and Stouffer, R. J., 1993, Century-scale effects of increased atmospheric CO_2 on the ocean-atmosphere system: Nature, v. 364, p. 215–218.

Marotzke, J., and Willebrand, J., 1991, Multiple equilibria of the global thermohaline circulation: Journal of Physical Oceanography, v. 21, p. 1372–1385.

Mayewski, P. A., Meeker, L. O., Whitlow, S., Twickler, M. S., Morrison, M. C., Bloomfield, P., Bond, G. C., Alley, R. B., Gow, A. J., Grootes, P. M., Meese, D. A., Ram, M., Taylor, K. C., and Wumkes, W., 1994, Changes in atmospheric circulation and ocean ice cover over the North Atlantic during the last 41,000 years: Science, v. 263, p. 1747–1751.

Rahmstorf, S., 1994, Rapid climate transitions in a coupled ocean-atmosphere model: Nature, v. 372, p. 82–85.

Rahmstorf, S., 1995, Bifurcations of the Atlantic thermohaline circulation in response to changes in the hydrological cycle: Nature, v. 378, p. 145–149.

Rahmstorf, S., 1996, On the freshwater forcing and transport of the Atlantic thermohaline circulation: Climate Dynamics, v. 12, p. 799–811.

Sachs, J. P., and Lehman, S. J., 1999, Subtropical North Atlantic temperatures 60,000 to 30,000 years ago: Science, v. 286, p. 756–759.

Schulz, H., von Rad, U., and Erlenkeuser, H., 1998, Correlation between Arabian Sea and Greenland climate oscillations of the past 110,000 years: Nature, v. 393, p. 54–57.

Severinghaus, J. P., Sowers, T., Brook, E. J., Alley, R. B., and Bender, M. L., 1998, Timing of abrupt climate change at the end of the Younger Dryas interval from thermally fractionated gases in polar ice: Nature, v. 391, p. 141–146.

Stocker, T. F., Wright, D. G., and Broecker, W. S., 1992, The influence of high-latitude surface forcing on the global thermohaline circulation: Paleoceanography, v. 7, p. 529–541.

Stocker, T. F., and Schmittner, A., 1997, Influence of CO_2 emission rates on the stability of the thermohaline circulation: Nature, v. 388, p. 862–865.

Taylor, K. C., Lamorey, G. W., Doyle, G. A., Alley, R. B., Grootes, P. M., Mayewski, P. A., White, J. W. D., and Barlow, L. K., 1993, The "flickering switch" of late Pleistocene climate change: Nature, v. 361, p. 432–436.

Teller, J. T., and Thorleifson, L. H., 1983, The Lake Agassiz–Lake Superior connection, in Teller, J. T., and Clayton, L., eds., Glacial Lake Agassiz: Geological Association of Canada Special Paper 26, p. 261–290.

Thompson, L. G., Davis, M. E., Mosley-Thompson, E., Sowers, T. A., Henderson, K. A., Zagorodnov, V. S., Lin, P.-N., Mikhalenko, V. N., Campen, R. K., Bolzan, J. F., Cole-Dai, J., and Francou, B., 1998, A 25,000-year tropical climate history from Bolivian ice cores: Science, v. 282, p. 1858–1864.

Part II | Methods of Estimating Ancient Temperature

To understand climate changes of the past, proxies for actual temperature measurements are required. Geologists routinely interpret conditions of the past, including climate. Several techniques for evaluating ancient temperatures have been developed recently. Some of these newer proxies bring to the forefront valuable information on past climate changes.

Thompson, L. G., Stable isotopes and their relationship
to temperature as recorded in low-latitude ice cores,
2001, *in* L. C. Gerhard, W. E. Harrison, and B. M.
Hanson, eds., Geological perspectives of global
climate change, p. 99–119.

5 | Stable Isotopes and their Relationship to Temperature as Recorded in Low-Latitude Ice Cores

Lonnie G. Thompson

Department of Geological Sciences and
Byrd Polar Research Center
The Ohio State University, Columbus, Ohio, U.S.A.

ABSTRACT

The potential of stable isotopic ratios ($^{18}O/^{16}O$ and $^{2}H/^{1}H$) in mid- to low-latitude glaciers as a modern tool for paleoclimate reconstruction is reviewed. To interpret quantitatively the ice-core isotopic records, the response of the isotopic composition of precipitation to long-term fluctuations of key climatic parameters (temperature, precipitation amount, relative humidity) over the given area should be known. Furthermore, it is important to establish the transfer functions that relate the climate-induced changes of the isotopic composition of precipitation to the isotope record preserved in the glacier. This paper will present long-term perspectives of isotopic composition variations in ice cores spanning the last 25,000 years from the mid- to low-latitude glaciers. The $\delta^{18}O$ records from the far western Tibetan Plateau suggest temperatures as warm as today occurred approximately 3000 years ago. However, $\delta^{18}O$ records from the Himalayas and the eastern side of the Tibetan Plateau confirm that the twentieth century is the warmest period in the last 12,000 years. In the South American Andes on Huascarán, $\delta^{18}O$ records suggest temperatures as warm as those of today occurred 5000 years ago.

All the tropical glaciers for which data exist are disappearing. The evidence for recent and rapid warming in the low latitudes is presented and possible reasons for this warming are examined. The isotopic composition of precipitation should be viewed not only as a powerful proxy indicator of climate, but also as an additional parameter for understanding climate-induced changes in the water cycle, on both regional and global scales.

INTRODUCTION

An overview is presented of some of the oxygen isotopic ($\delta^{18}O$) records from low-latitude, high-altitude glaciers of the world. This paper addresses the question of how faithfully these archives record temperature and then examines 25,000 years of documented climatic history. Finally, an overview of the disappearance of these tropical archives in the last 30 years is presented.

The work presented here reflects the combined efforts of a team of scientists and engineers at the Byrd Polar Research Center of the Ohio State University (OSU), and our colleagues in China, Peru, and Bolivia. These records have been recovered over the previous two decades as part of an ongoing program to reconstruct earth's past climate and environment from a global array of ice cores, as shown in **Figure 1**. The latest addition to this array is the cores drilled on Kilimanjaro in Tanzania. Here, six cores were drilled to bedrock in 2000 from the wasting ice fields on the top of the mountain. Drilling the glaciers was accomplished by using two systems, electrical/mechanical and

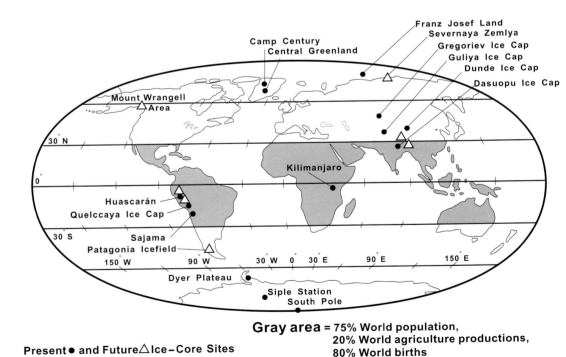

Figure 1 This map shows the ice-core sites where ice cores have already been recovered, as well as planned future sites. The gray area between 30°N and 30°S represents the tropical regions of the world that contain 50% of the surface area, 75% of the population, and 80% of the new births, yet produce only 20% of the agricultural products.

thermal-alcohol. In all our drilling programs, at least two and usually three ice cores are recovered so that the records in the archive can be verified by duplication. The cores are always transported to the Byrd Polar Research Center frozen and are analyzed for stable isotopes, dust, chemistry, and radioactivity produced by atmospheric thermonuclear testing.

A glacier is formed when fresh snow accumulates and survives the ablation season. The snow is buried and becomes firn; eventually, as the density continues to increase, it becomes ice. When we investigate ice-core parameters such as oxygen isotopic ratios ($\delta^{18}O$), we try to reconstruct the atmospheric input signal of temperatures, although there are postdepositional processes that can potentially alter that signal as densification takes place. Stable isotopes can be altered by vapor transfer through the snowpack as temperatures change both diurnally and seasonally. The relative importance of this process can be evaluated by drilling multiple cores at each site.

Whether a glacier is polar or tropical, three key factors determine the length of the record preserved within it. The first consideration is the annual accumulation rate, i.e., the amount of snow that falls each year on the ice cap; the second factor is the thickness of the glacier (most tropical glaciers are less than 300 meters thick); and the third and most important factor is the temperature at the ice-bedrock contact. If a glacier is frozen to bedrock and has remained frozen through its life, then time cannot be lost from the base. Thus, very old archives of climate and environmental variability can be preserved in quite thin ice caps.

The annual climate of the tropics is often dominated by a distinct wet (summer) and dry (winter) season. The glaciers are nourished during the wet season, and during the dry season a prominent dust layer is deposited on the ice surface. It often forms distinct visible layers such as those exposed along the vertical margin of the Quelccaya ice cap, pictured in **Figure 2**, which illustrates the strong seasonal nature of the tropical climatic and environmental record archived in these glaciers. All the records discussed in this paper were recovered from areas characterized by distinct wet and dry seasons. Ice cores provide unique archives of the past, not only because they can provide well-dated histories of climatic and environmental change as recorded in the layers of ice, but also because one can examine possible climatic forcings such as greenhouse gases, changes in solar activity, and volcanic eruptions. The emphasis here is on the stable isotope archive as a proxy for temperature in six ice cores (low-latitude, high-altitude: 400 to 500 millibar atmospheric pressure), three from the Tibetan Plateau and three from the tropical Andes of South America.

THE TIBETAN PLATEAU

One of the most important questions concerning a stable isotopic record is whether it represents a proxy indicator of temperature for the lower troposphere. We have collaborated with our Chinese colleagues at the Lanzhou Institute of Glaciology and Geocryology (LIGG), who have collected precipitation samples and measured temperatures at six meteorological stations across the Tibetan Plateau. At the Delingha station, which is the closest to the Dunde ice cap (150 km to southeast), the correlation coefficient (R^2) is 0.86 (see **Table 1**), showing a very strong relationship between temperature and $\delta^{18}O$ (Yao et al., 1996). These data suggest that for the period of measurement, as well as for longer time periods (Thompson, 2000; Thompson et al., 2000a), the $\delta^{18}O$ faithfully records temperature in this section of the Plateau.

Since 1987, ice core records have been recovered from three sites along the perimeter of the Tibetan Plateau **(Figure 3)**. The Dunde ice cap, drilled in 1987, lies in the north-central part of the plateau and covers about 60 km². The scale of this ice cap is illustrated in **Figure 4**, showing (in center) the Mongolian ponies used to transport the drilling equipment to the drill site. **Figure 5**

Figure 2 A 50-m ice cliff near the margin of the Quelccaya ice cap in the southeastern Peruvian Andes about 5650 m above sea level. The individual layers represent annual increments of accumulation.

illustrates the $\delta^{18}O$ record from Dunde, plotted as 50-year averages, that extends back approximately 12,000 years. The histograms illustrate periods that are warmer and colder than the mean. The projection (shaded) of the most recent 50-year period (1938–1987) back 12,000 years indicates that the most recent 50-year period is the most isotopically enriched 50-year period.

Table 1 Linear regression coefficients between $\delta^{18}O$ and air temperature for both individual precipitation events and monthly averages of precipitation collected at three meteorological stations on the north-central Tibetan Plateau. Delingha is the meteorological station closest to the Dunde ice cap.

Station	Long.	Lat.	Alt. (m)	Averages for Individual Events		Regression			
						Individual Events		Monthly Average	
				$\delta^{18}O$ %	Air Temp. °C	Regression	Coeff.	Regression	Coeff.
Delingha	97°22'E	37°222'N	2981.5	−8.66	7.40	$\delta^{18}O(\%) =$ 0.67T−13.59	$R^2 =$ 0.69	$\delta^{18}O(\%) =$ 0.76T−14.29	$R^2 =$ 0.86
Tuotuohe	97°22'E	37°222'N	2981.5	−8.66	7.40	$\delta^{18}O(\%) =$ 0.36T−11.2	$R^2 =$ 0.13	$\delta^{18}O(\%) =$ 0.48T−11.09	$R^2 =$ 0.45
Xining	97°22'E	37°222'N	2981.5	−8.66	7.40	$\delta^{18}O(\%) =$ 0.29T−8.51	$R^2 =$ 0.14	$\delta^{18}O(\%) =$ 0.49T−10.51	$R^2 =$ 0.60

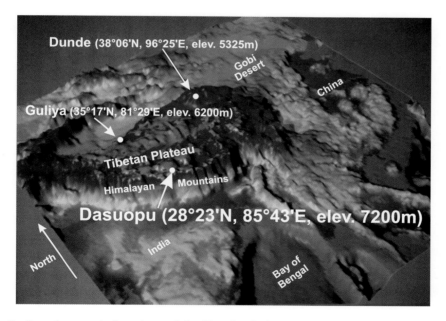

Figure 3 Locations and elevations of the Dunde, Guliya, and Dasuopu ice-core sites on the Tibetan Plateau.

Figure 4 The eastern margin of the Dunde ice cap in the Qilian Mountains on the northeastern margin of the Qinghai-Tibetan Plateau. The ice cap covers 60 km² and attains a summit elevation of 5325 m.

Figure 6 illustrates 1000 years of decadally averaged $\delta^{18}O$ for Dunde, along with similar results from two more recent sites: the Guliya ice cap in the far western Kunlun Mountains and the Dasuopu Glacier at 7200 meters in the Himalayas on the southern margin of the Tibetan Plateau (Thompson et al., 2000a). The overall characteristics of the profiles reveal major differences throughout the last millennium, possibly due to different precipitation sources and postdepositional processes. However, over the last 200 years, all three cores clearly show strong isotopic enrichment, suggesting a recent warming across the entire Tibetan Plateau.

THE ANDES OF SOUTH AMERICA

Figure 7 illustrates the three sites in the tropical Andes from which ice cores have been recovered. For all sites, the moisture source is the tropical Atlantic Ocean by way of the Amazon Basin. This source has remained constant since the Last Glacial Maximum (LGM) (about 20,000 years ago), as revealed by the fact that the mountain snowlines at that time were tilted toward the Amazon Basin as they are at present (Klein et al., 1995).

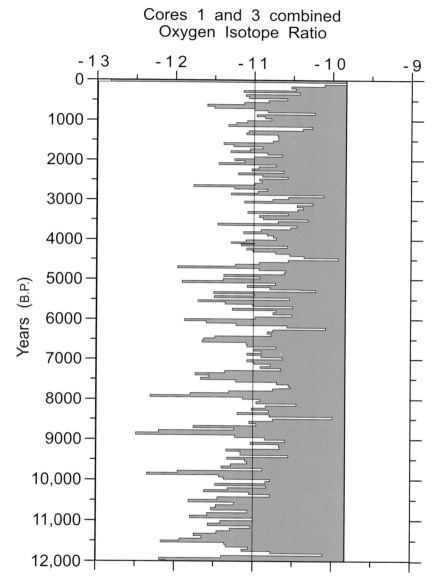

Figure 5 Fifty-year averages of $\delta^{18}O$ from the combination of cores 1 and 3 for the last 12,000 years from the Dunde ice cap, China. The reference line at -11% represents the long-term average of the records, and projections into the gray area indicate 50-year periods with warmer than average $\delta^{18}O$. Note that the most recent 50-year period (1937–1987) is the warmest since the end of the last glacial stage.

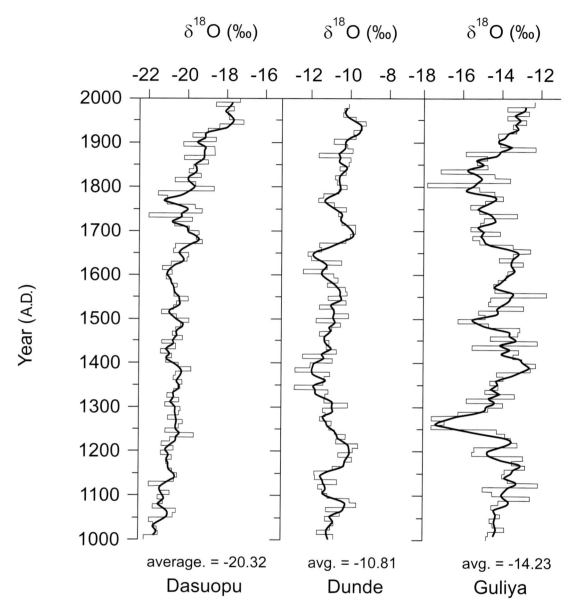

Figure 6 Decadal averages of $\delta^{18}O$ for Dasuopu, Dunde, and Guliya are shown for the last 1000 years. The dark lines represent three-decade running means.

Huascarán (9°S, 77°W, 6048 m asl [meters above sea level]), the highest tropical mountain on earth, is located between the Amazon Basin to the east and the center of the El Niño-Pacific warm pool to the west. The ice cores were drilled with the use of solar power **(Figure 8)** in the col between the north and south peaks at an elevation of 6050 m. Two $\delta^{18}O$ profiles from cores drilled 100 m

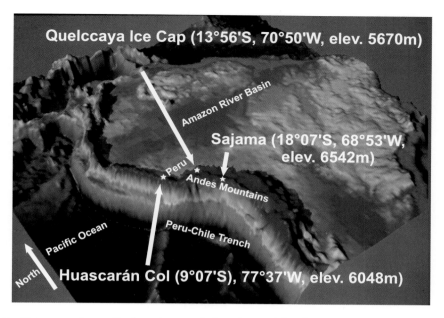

Figure 7 This map shows the locations and elevations of the three tropical ice-core sites in the Andes of South America.

Figure 8 An array of 60 photovoltaic cells was used on the col of Huascarán in 1993 to produce four kilowatts of power which allowed recovery of two ice cores to bedrock (over 166 m in depth). This reflects an environmentally correct way to recover ice cores from pristine tropical environments.

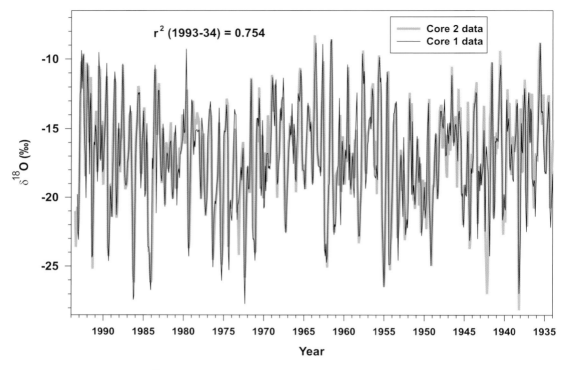

Figure 9 Individual $\delta^{18}O$ samples are plotted for the two cores drilled on the col of Huascarán. The sites are separated by a 100-m horizontal distance. The similarity of these profiles demonstrates that drifting snow and other surface disturbances have little effect on the reproducibility of the record from site to site.

apart are shown in **Figure 9**. Core 1 was brought back from the field as bottled water samples and core 2 was transported to OSU frozen. This diagram illustrates that stable isotope records are spatially reproducible across the glacier. Of concern, however, is the issue of temporal reproducibility, or how much the input signal changes with time as it becomes buried in the ice cap. **Figure 10** illustrates the $\delta^{18}O$ signal in a firn core drilled in 1980 with a mean value of –17.56‰ and compares the same period of time covered in the deep ice cores drilled in 1993 to bedrock. Even after 13 years, the mean isotopic value is –17.33‰, which is nearly identical. This clearly illustrates that at least under present-day climate conditions, these ice cores faithfully archive and preserve the isotopic signature at this tropical site.

Currently, two models have been used to explain the $\delta^{18}O$ composition of ice cores in the tropical Andes. The model developed by Grootes et al. (1989) is a hydrological model of moisture transport **(Figure 11)**. The isotopic values are initially determined by the composition of ocean water, and as the vapor moves across the Amazon Basin, it is recycled in thunderstorms. Every time condensation takes place, the heavier isotopes are preferentially removed. This moisture falls from clouds and is transported out of the Amazon Basin by the river system in the wet season. In the dry season most of the water that falls in the Amazon is reevaporated, and thus very little isotopic fractionation takes place **(Figure 11)**. However, by the time the moisture

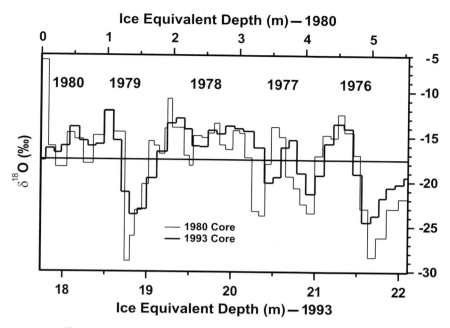

Figure 10 The δ¹⁸O record from a 10-m core drilled in 1980 on the Huascarán col is compared with that representing the same time period but drilled 13 years later (1993) as Huascarán core 2. Comparison of these records demonstrates that very little alteration of the input signal has occurred with time and depth.

reaches the base of the Andes in the wet season, it has a mean isotopic value of –20‰. When the air masses are forced to rise over the Andes (above 5000 m) an additional 10‰ depletion takes place.

The second model brings into consideration another important factor by which interpretation of stable isotopes in the tropics differs from interpretations in the higher latitudes. In the Andes, the snowfall comes from thunderstorms, i.e., convective cells with great vertical extent. Therefore, the mean level of condensation in the tropics is higher than that in the polar regions. More importantly, the location and height of the mean condensation level changes from the wet to the dry season. In the wet season, Huascarán is in the center of the region affected by maximum deep convection, but in the dry season this activity moves to the north. The condensation level during the height of the wet season is roughly 2 km higher where the temperatures are cooler, while in the dry season condensation occurs at a lower, warmer level in the atmosphere. The isotopic composition of precipitation may well reflect these changes (Thompson et al., 2000b).

Figure 12 illustrates a record of δ¹⁸O and the concentrations of dust and nitrate (NO_3^-) from Huascarán over the most recent 100 years of the record. Because of the preservation of the very distinct annual signal in the upper sections of these ice cores, they can be used to provide records of El Niño–Southern Oscillation (ENSO) variations (Henderson et al., 1999).

A profile of centennial averages of δ¹⁸O over the last 20,000 years **(Figure 13)** illustrates that the isotopic depletion of 6.3‰ during the LGM is very similar to that in polar regions (Thompson

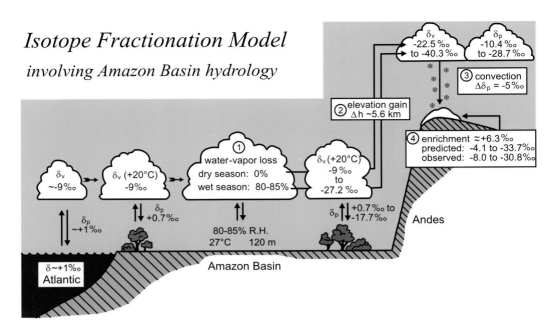

Figure 11 This schematic traces the $\delta^{18}O$ composition of water vapor and precipitation along a transect from the Atlantic Ocean to the top of the Andes (Quelccaya ice cap). This figure was modified from Grootes et al. (1989).

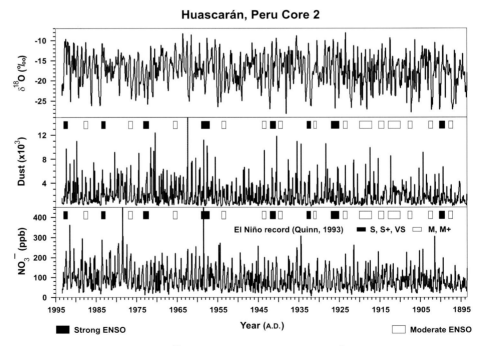

Figure 12 Seasonal variations in $\delta^{18}O$ and concentrations of (NO_3^-) and insoluble dust measured in Huascarán core 2 for the last 100 years. Also shown are the El Niño events identified by Quinn (1993). Dust concentrations are for particles with diameters > 2.0 μm per milliliter sample.

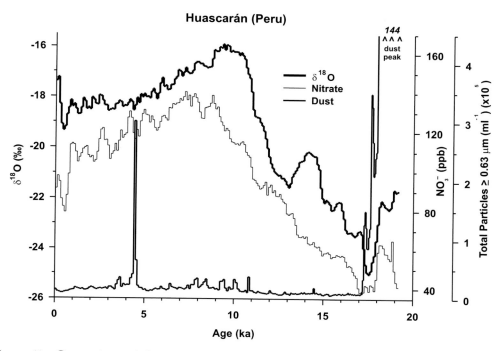

Figure 13 Comparison of the 100-year averages of $\delta^{18}O$ and concentrations NO_3^- and dust particles with diameters > 2.0 μm from Huascarán over the last 20,000 years.

et al., 1995). The enrichment of the last 200 years is unprecedented in the most recent 5000 years of this record. Between 11,000 and 6000 years B.P. (before the present), however, $\delta^{18}O$ values are the highest of the entire climatic history presented by this ice-core record. This observation suggests that this part of the world was even warmer in the early Holocene than it is today.

The Quelccaya ice cap (14°S) was drilled in 1983 and has provided an annually resolved record of climatic variation extending back to A.D. 470 (Thompson et al., 1985; 1986; 1989). The reproducibility of the $\delta^{18}O$ records between two cores over the last 500 years is demonstrated in **Figure 14**. Also demonstrated is how well the $\delta^{18}O$ records reflect the Northern Hemisphere temperature record as reconstructed by Mann et al. (1998). In fact, the Quelccaya ice cap has provided the clearest $\delta^{18}O$ record of the "Little Ice Age" of any ice core recovered to date.

Further to the south is Sajama, Bolivia (18°S). It is a 6550-m asl extinct volcano on which two cores were drilled to bedrock in 1997 (Thompson et al., 1998). The ice cores from this site are very unusual in that they contain both insect and plant remains in sufficient supply to allow accelerated mass spectrometer (AMS) carbon-14 (^{14}C) dating by two separate laboratories. Thus, unusually good absolute time control is possible for the Sajama records. **Figure 15** illustrates the 25,000 calendar years of reconstructed climate history. Not only does the $\delta^{18}O$ show a 5.4‰ decrease in glacial conditions, similar to Huascarán and the polar ice cores, but it also demonstrates that this area which is very dry today was very wet during the LGM. Moreover, the good time control for the Sajama record allows a detailed comparison with the GISP II (Greenland Ice Sheet Project 2) core (Grootes et al., 1993) from central Greenland **(Figure 16)**. The Younger Dryas (in Greenland) or deglaciation cold reversal (in Bolivia) are very clear in both these cores, each showing a depletion

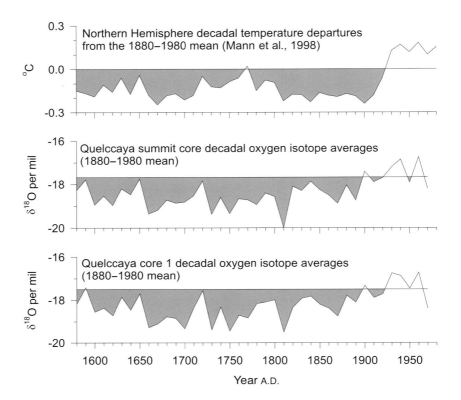

Figure 14 Decadal temperature departures (from the 1881–1980 mean) in the Northern Hemisphere (Mann et al., 1998) from 1580 to 1980, compared with decadally averaged $\delta^{18}O$ for both Quelccaya ice cores. The solid line is the 1881–1980 mean.

of 5.2‰. Additionally, the abruptness of the termination of the event, as far as it can be constrained by the timescales, appears to be the same in the tropics as in the high latitudes. These data indicate that the natural climate system may actually operate in modes and be capable of very rapid abrupt changes at magnitudes which have not been documented by human observations.

These new ice-core records challenge existing paradigms of the earth's climate system because whether we look at ice core records from Antarctica, Greenland, or the low- to mid-latitudes, the Holocene period (last 10,000 years) shows quite different trends. On the other hand, the isotopic shift from the glacial maximum to early Holocene is very similar around the world, suggesting a global cooling during glacial maximum of 5° to 6°C. We believe that the recent noble gas work on groundwater (Stute et al., 1995; Weyhenmeyer et al., 2000), pollen (Colinvaux et al., 1996), tropical corals (Guilderson et al., 1994; Beck et al., 1997), and marine sediment pore fluids (Shrag et al., 1996) all supports the temperature interpretation placed on these tropical ice-core stable isotopic results.

THE MELTING OF THE TROPICAL GLACIER CLIMATE ARCHIVES

There is mounting evidence for a recent, strong warming in the tropics, which is signaled by the rapid retreat and even disappearance of ice caps and glaciers at high elevations. Indeed, we have

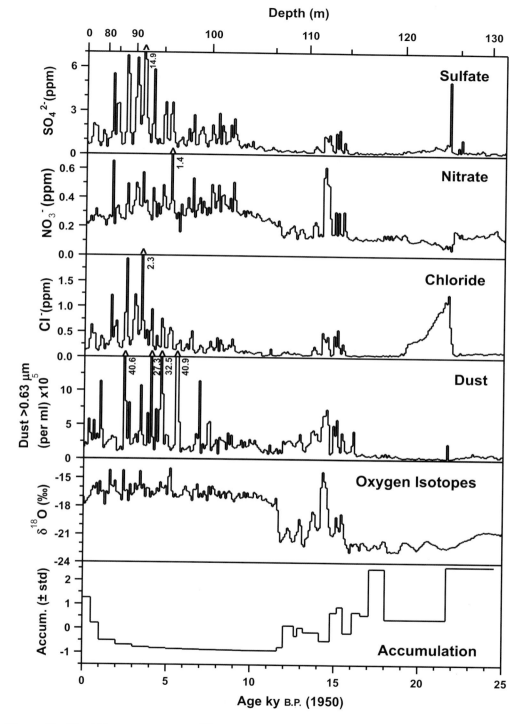

Figure 15 The 100-year averages of $\delta^{18}O$ and concentrations of dust, chloride (Cl^-), nitrate (NO_3^-), and sulfate (SO_4^{2-}) from Sajama core 1 are shown for the past 25,000 years. The accumulation record is the average of cores 1 and 2 (Thompson et al., 1998). ky B.P. = thousand years before the present.

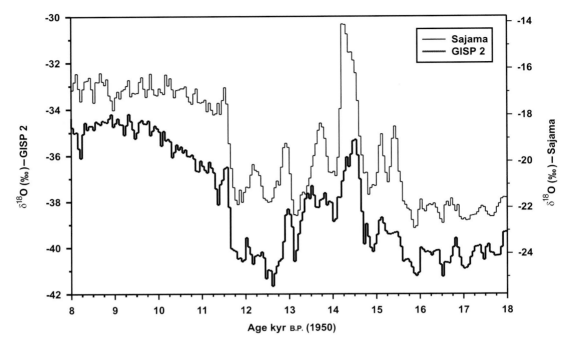

Figure 16 The 50-year averages of $\delta^{18}O$ since the deglaciation as recorded in the tropical Sajama, Bolivia, and polar GISP 2 Greenland ice cores.

extensively documented these changes on Quelccaya, one of the best-studied tropical ice caps. $\delta^{18}O$ profiles from shallow cores drilled on Quelccaya in 1976 from different elevations along the summit of the ice cap are illustrated in **Figure 17**. At the highest elevation site an annual cycle in $\delta^{18}O$ existed and this isotopic signal allowed dating of the most recent 600 years of the ice cores drilled to bedrock in 1983. Cores also recovered in 1976 from lower elevation sites on Quelccaya show a loss of annual $\delta^{18}O$ cyclicity with depth due to melting and movement of water through the firn column. Since 1976, Quelccaya has been visited repeatedly for monitoring. In 1991, another shallow core was drilled at the summit. It revealed that the seasonally resolved paleoclimatic record, formerly preserved as $\delta^{18}O$ variations, was no longer being retained within the accumulating snowfall. The percolation of meltwater throughout the snowpack was vertically homogenizing the $\delta^{18}O$, enriching the mean isotopic value by 2‰ between 1976 and 1991, indicating a strong recent warming.

The general warming of the twentieth century is now well documented. Although not all regions of the earth have warmed, the globally averaged temperature has increased 0.7°C since the end of the nineteenth century (Hansen et al., 1999). Annual global land-temperature anomalies, as measured by meteorological stations, indicate that 1998 was the warmest year on record, and the 1990s have been the warmest decade on record. The best interpretation of proxy records from borehole temperatures, stable isotopes from ice cores, tree-ring data, and other proxy data suggests that the decade of the 1990s was the warmest in the last 1000 years (Mann et al., 1999).

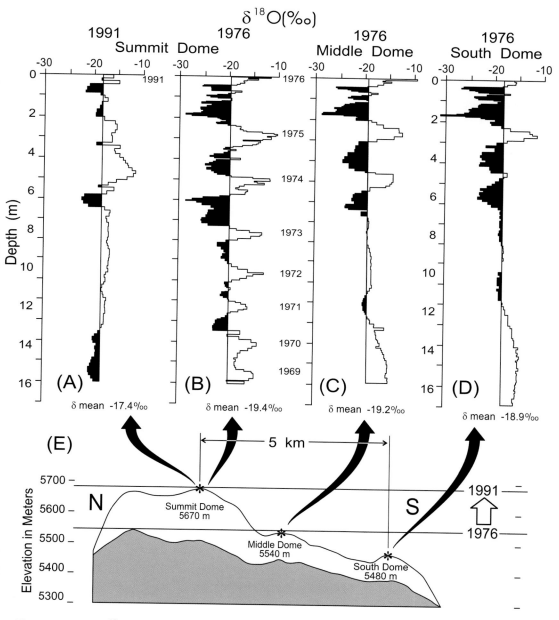

Figure 17   The δ¹⁸O profiles drilled in 1976 along a north-south transect (summit, middle, and south domes) on the Quelccaya ice cap, Peru, are compared with a 1991 record from the summit dome. Also shown is the rise in the 0°C melting line between 1976 and 1991.

Tropical ice masses are particularly sensitive to small changes in ambient temperatures because they already exist very close to the melting point. The retreat of Quelccaya has been monitored since 1978 by terrestrial photogrammetry of a valley glacier called Qori Kalis on the western side of the ice cap (Brecher and Thompson, 1993; Thompson et al., 1993). The movement of the ice front and the volume loss have been determined for six time intervals, and these observations document a drastic retreat of the glacier that has accelerated with time. The latest photographic evidence from 1998 indicates that this retreat continues to accelerate. It was twice as great in the latest three-year period, between 1995 and 1998, as it was in the previous two-year period, between 1993 and 1995 (48.7 vs. 28.7 m a^{-1} (meters per year)). Currently, the retreat rate is an order of magnitude greater than it was from 1963 to 1978, when comparison between the position from the first terrestrial photograph and that from the aerial photographs gave a value of 4.9 m a^{-1}. The volume loss has accelerated at an even greater rate. The graph in the lower right of **Figure 18** illustrates the retreat of the terminus. The sketch map shows the position of the glacier terminus for each of the seven determinations since 1963. Note that the small proglacial lake that first appeared in the 1991 photograph has continued to grow with the retreat of the ice front, and in 1998 it was twice as large as in 1991 **(Figure 18)**. The retreat of Qori Kalis is consistent with the records from two ice cores drilled on the col of Huascarán whose $\delta^{18}O$ records (discussed earlier) indicate that the nineteenth and twentieth centuries were the warmest in the last 5000 years.

Additional glaciological evidence for tropical warming exists. Hastenrath and Kruss (1992) report that the total ice cover on Mount Kenya decreased by 40% between 1963 and 1987 and continues to diminish today. The Speke Glacier in the Ruwenzori Range of Uganda has retreated substantially since it was first observed in 1958 (Kaser and Noggler, 1991). Today the total ice mass on Kilimanjaro is roughly 25% of that which existed in 1912 (Hastenrath and Greischar, 1997). Schubert (1992) noted that in the Sierra Nevada de Merida of Venezuela, at least three glaciers have disappeared since 1972. The shrinking of these ice masses in the high mountains of Africa and South America is consistent with similar observations throughout most of the world.

We believe that warming is being enhanced through the tropical hydrological system at the higher-elevation sites in the tropics. At these latitudes, water evaporates from the oceans and heat energy is released from the latent heat of condensation at the higher elevations. A change in the tropical lapse rate may have occurred in the late 1970s (Gaffen et al., 2000). The authors found, if anything, a cooling since 1979 relative to the surface in the records from both satellite and the radiosonde data. Over the longer record from about 1960 to present, they found evidence of differential temperature changes, with the lower troposphere warming somewhat more. This is in opposition to the glacier records. However, a small temperature rise at the surface evidently is being enhanced through latent heat release at these higher elevations associated with enhanced deep convection in the tropics, to account for not only the retreat but also the acceleration in the rate of retreat of tropical glaciers. Water vapor, pumped into the atmosphere at low latitudes, is the most important greenhouse gas on earth. Thus, the increases in CO_2 leading to surface warming may be enhanced through positive feedbacks and generate an increasing global inventory of water vapor. Two of the factors we must try to understand are how the tropical hydrological system has varied through time and how the water-vapor content in the atmosphere has changed in response. Because the forcing is driven from the tropics, changes in the global water-vapor content can have dramatic impacts on the climates of both the Northern and Southern Hemispheres. Clearly, much more research is needed in this area. We must realize that the tropical glaciers may be the "canary in the cage" for the earth's climate system.

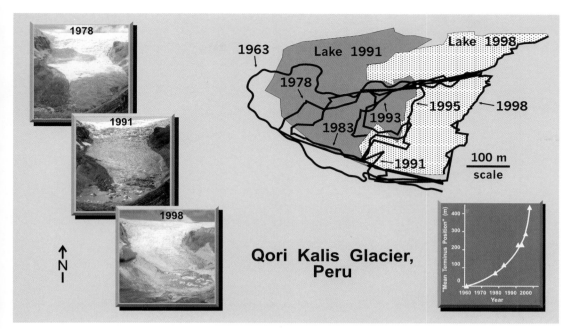

Figure 18 Recent warming in the Peruvian Andes is illustrated by the mapped and photographed changes in the extent of the tropical outlet glacier, Qori Kalis, from 1963 to 1998. The right-hand graph illustrates the exponential increase in the rate of retreat of the Qori Kalis margin since 1963.

ACKNOWLEDGMENTS

I would like to thank the outstanding members of the paleoclimate ice-core group at the Byrd Polar Research Center: Jihong Cole-Dai, Mary Davis, Ross Edwards, Keith Henderson, Ping-Nan Lin, Ellen Mosley-Thompson, and Victor Zagorodnov. Special thanks are extended to Yao Tandong, Lanzhou Institute of Glaciology and Geocryology (LIGG), and Vladimir Mikhalenko, Institute of Geography (IG). We acknowledge the contributions of Bruce Koci, University of Wisconsin, who has been our drilling engineer on all these projects, and Henry Brecher for producing the maps for retreat of the Qori Kalis Glacier. We thank the many scientists, technicians, graduate students and support personnel from the Ohio State University, LIGG, and IG and the mountain guides of the Casa de Guias, Huaraz, Peru and the Sherpas of Nepal. Special thanks to Mary Davis and Ellen Mosley-Thompson for editing this manuscript. This work has been supported by the National Science Foundation and the National Oceanic and Atmospheric Administration. This is contribution 1173 of the Byrd Polar Research Center.

REFERENCES CITED

Beck, J. W., Récy, J., Taylor, F., Edwards, R. L., and Cabioch, G. 1997. Abrupt changes in early Holocene tropical sea surface temperature derived from coral records. Nature, 385, 705–707.

Brecher, H. H., and Thompson, L. G. 1993. Measurement of the retreat of Qori Kalis in the tropical Andes of Peru by terrestrial photogrammetry. Photogrammetric Engineering and Remote Sensing, 59, 1017–1022.

Colinvaux, P. A., DeOliveira, P. E., Moreno, J. E., Miller, M. C., and Bush, M. B. 1996. A long pollen record from lowland Amazonia forest and cooling in glacial times. Science, 274, 85–88.

Gaffen, D. J., Santer, B. D., Boyle, J. S., Christy, J. R., Graham, N. E., and Ross, R. J. 2000. Multidecadal changes in the vertical temperature structure of the tropical troposphere. Science, 287, 1242–1245.

Grootes, P. M., Stuiver, M., Thompson, L. G., and Mosley-Thompson, E. 1989. Oxygen isotope changes in tropical ice, Quelccaya, Peru. Journal of Geophysical Research, 94, 1187–1194.

Grootes, P. M., Stuvier, M., White, J. W. C., Johnsen, S., and Jouzel, J. 1993. Comparison of oxygen isotope records from the GISP2 and GRIP Greenland ice cores. Nature, 366, 552–554.

Guilderson, T. P., Fairbanks, R. G., and Rubenstein, J. L. 1994. Tropical temperature variations since 20,000 years ago: Modulating interhemispheric climate change. Science, 263, 663–665.

Hansen, J., Ruedy, R., Glascoe, J., and Sato, M. 1999. GISS analysis of surface temperature change. Journal of Geophysical Research, 104, 30, 997–31, 022.

Hastenrath, S., and Greischar, L. 1997. Glacier recession on Kilimanjaro, East Africa, 1912–1989. Journal of Glaciology, 43, 455–459.

Hastenrath, S., and Kruss, P. D. 1992. The dramatic retreat of Mount Kenya's glaciers between 1963 and 1987. Annals of Glaciology, 16, 127–133.

Henderson K. A., Thompson, L. G., and Lin, P.-N. 1999. Recording of El Niño in ice core $\delta^{18}O$ records from Nevado Huascarán, Peru. Journal of Geophysical Research, 104, 31, 053–31, 065.

Kaser, G., and Noggler, B. 1991. Observations on Speke Glacier, Ruwenzori Range, Uganda. Journal of Glaciology, 37, 315–318.

Klein, A. G., Isacks, B. L., and Bloom, A. L. 1995. Modern and Last Glacial Maximum snowlines in Peru and Bolivia: Implications for regional climatic change. Bulletin de l'Institut Français d'Études Andines, 24, 607–617.

Mann, M. E., Bradley, R. S., and Hughes, M. K. 1998. Global-scale temperature patterns and climate forcing over the past six centuries. Nature, 392, 779–787.

Mann, M. E., Bradley, R. S., and Hughes, M. K. 1999. Northern Hemisphere temperatures during the past millennium: Inferences, uncertainties, and limitations. Geophysical Research Letters, 26, 759–762.

Quinn, W. H. 1993. The large-scale ENSO event, the El Niño and other important regional patterns. Bulletin de l'Institut Francais d'Éstudes Andines, 22, 13–34.

Schrag, D. P., Hampt, G., and Murray, D. W. 1996. Pore fluid constraints on the temperature and oxygen isotopic composition of the glacial ocean. Science, 272, 1930–1932.

Schubert, C. 1992. The glaciers of the Sierra Nevada de Mérida (Venezuela): a photographic comparison of recent deglaciation. Erdkunde, 46, 58–64.

Stute, M., Forster, M., Frischkorn, H., Serejo, A., Clark, J. F., Schlosser, P., Broecker, W. S., and Bonani, G. 1995. Cooling of tropical Brazil (5°C) during the last glacial maximum. Science, 269, 379–383.

Thompson, L. G., 2000. Ice core evidence for climate change in the Tropics: implications for our future. Quaternary Science Reviews, 19, 19–35.

Thompson, L. G., Mosley-Thompson, E., Bolzan, J. F., and Koci, B. R. 1985. A 1500 year record of tropical precipitation recorded in ice cores from the Quelccaya Ice Cap, Peru. Science, 229, 971–973.

Thompson, L. G., Mosley-Thompson, E., Dansgaard, W., and Grootes, P. M. 1986. The "Little Ice Age" as recorded in the stratigraphy of the tropical Quelccaya ice cap. Science, 234, 361–364.

Thompson, L. G., Mosley-Thompson, E., Davis, M. E., Bolzan, J., Dai, J., Yao, T., Gundestrup, N., Wu, X., Klein, L., and Xie, Z. 1989. Holocene–Late Pleistocene climatic ice core records from Qinghai-Tibetan Plateau. Science, 246, 474–477.

Thompson, L. G., Mosley-Thompson, E., Davis, M. E., Lin, P-N., Yao, T., Dyurgerov, M., and Dai, J. 1993. "Recent Warming": Ice core evidence from tropical ice cores with emphasis upon Central Asia. Global and Planetary Change, 7, 145–156.

Thompson, L. G., Mosley-Thompson, E., Davis, M. E., Lin, P-N, Henderson, K. A., Cole-Dai, J., Bolzan, J. F., and Liu K-b. 1995. Late Glacial Stage and Holocene tropical ice core records from Huascarán, Peru. Science, 269, 46–50.

Thompson, L. G., Davis, M. E., Mosley-Thompson, E., Sowers, T. A., Henderson, K. A., Zagorodnov, V. S., Lin, P-N., Mikhalenko, V. N., Campen, R. K., Bolzan, J. F., and Cole-Dai, J. A. 1998. 25,000 year tropical climate history from Bolivian Ice Cores. Science, 282, 1858–1864.

Thompson, L. G., Yao T., Mosley-Thompson, E., Davis, M. E., Henderson, K. A., and Lin. P-N. 2000a. A high-resolution millennial record of the South Asian Monsoon from Himalayan ice cores. Science, 289, 1916–1919.

Thompson, L. G., Mosley-Thompson, E., and Henderson, K. A. 2000b. Ice core paleoclimate records in tropical South America since the Last Glacial Maximum. Journal of Quaternary Sciences, 15(4) 377–394.

Weyhenmeyer, C. E., Burns, S. J., Waber, H. N., Aeschbach-Hertig, W., Kipfer, R., Loosli, H. H., Matter, A. 2000. Cool glacial temperatures and changes in moisture source recorded in Oman groundwaters. Science, 287, 842–845.

Yao, T., Thompson, L. G., Mosley-Thompson, E., Zhihong, Y., Xingping, Z., and Lin, P-N. 1996. Climatological significance of $\delta^{18}O$ in north Tibetan ice cores. Journal of Geophysical Research, 101, 29, 531–29, 537.

6 | Century-Scale Variation of Seafloor Temperatures Inferred from Offshore Borehole Geothermal Data

Seiichi Nagihara

Department of Geosciences
Texas Tech University
Lubbock, Texas, U.S.A.

Kelin Wang

Pacific Geoscience Centre
Geological Survey of Canada
Sidney, British Columbia, Canada

ABSTRACT

A large amount of hydrographic data obtained in the last three to four decades indicates that temperatures in the deep ocean have been changing globally. However, because of the scarcity of older
data, it is difficult to trace the ocean thermal history farther back in time. In this study, we examine
the possibility of using subseafloor borehole temperature data to estimate the history of the
bottom-water temperature (BWT) in the last two to three centuries. The thermal signal associated
with BWT fluctuation slowly propagates into the subseafloor rock formation, perturbing the otherwise steady-state temperature field. It is possible to extract this signal and reconstruct the BWT history by inverting the borehole temperature measurements. We make such an attempt using data
obtained from a borehole drilled on 669-m-deep seafloor at Ocean Drilling Program Site 1006 in the

Straits of Florida. The observed temperature-depth profile in the depth range of 26 through 349 m below seafloor shows significant curvature in the upper 100 m. The BWT history reconstructed from this profile indicates that the long-term average BWT in the early eighteenth century was about 1°C lower than the present value. It decreased to a minimum at about the turn of the century, and then gradually increased to the present value. The pattern of the inferred BWT variation is similar to that of the surface air temperature at Key West, Florida, and the global surface air temperature average.

INTRODUCTION

During the past three to four decades, the thermal structure of oceans has undergone major changes. These changes have occurred in the deep ocean as well as in the near surface to intermediate water. Evidence for such changes is most abundant in the North Atlantic Ocean (Levitus, 1989; Levitus et al., 1995; Roemmich and Wunsch, 1984). In general, the deep water (roughly 800 m to 2500 m) of the North Atlantic subtropical gyre **(Figure 1)** is warming, while that of the subarctic gyre is cooling (Levitus and Antonov, 1995). For example, the average temperature along the 24°N parallel increased by 0.3°C at 1100 m depth from 1957 through 1992 (Parrilla et al., 1994). It has been debated whether or not this type of decadal-scale redistribution of heat is related to the greenhouse warming of the earth's atmosphere (Parker, 1996; Watts and Morantine, 1991).

Ocean temperatures may change over longer times, but there are not enough hydrographic data to examine such a possibility. It has been only ten years since the worldwide research programs on ocean climate, such as the World Ocean Circulation Experiment (WOCE), were initiated. Even in the North Atlantic Ocean, which has been most heavily sampled, data prior to the 1950s are scarce. Also lacking are semipermanent observatories that record temperature data continuously at fixed locations in the ocean for a long period of time. Only two such stations have been operating in the North Atlantic since the 1950s (Levitus et al., 1995). All other data come from measurements from throwaway devices such as the expendable bathythermograph (XBT) or by lowering STD (salinity-temperature-depth) and CTD (conductivity-temperature-depth) instruments (Snodgrass, 1968). Data from XBTs are often limited in depth.

In this study, we introduce a method that utilizes subseafloor temperature data from offshore boreholes to assess century-scale changes in the bottom-water temperature (BWT). The conductive geothermal heat flow (q) through the subseafloor formation is expressed as:

$$q = K \frac{dT}{dz}$$

where K is the thermal conductivity of the rock, T is the temperature, and z is the subseafloor depth. Here we assume that the radiogenic heat production within the rock and lateral heat flow are negligible. In the absence of thermal perturbations such as pore fluid flow and temporary fluctuation of BWT, q is constant with depth. If the thermal conductivity is uniform, the temperature of the rock increases linearly with depth. If, however, BWT fluctuates with time, the temperature-versus-depth profile would not be linear, because the thermal signal of the bottom-water fluctuation propagates into the rock formation. The shape of the geothermal profile reflects the history of the BWT fluctuation.

Here, we consider three simple examples. In the first example **(Figure 2a)**, BWT suddenly increased by 2°C 120 years ago and remained constant. The present-day geothermal profile deviates positively from the steady-state linear profile at depths mostly shallower than 100 m. In the second example **(Figure 2b)**, BWT has increased gradually in the last 120 years. The positive deviation of

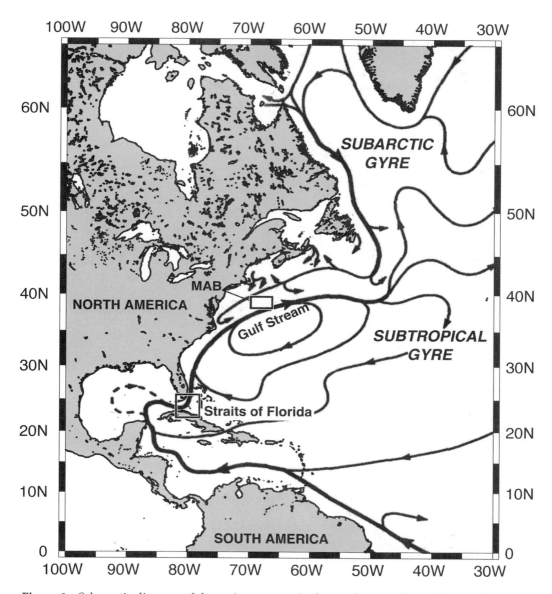

Figure 1 Schematic diagram of the major currents in the northwest Atlantic Ocean (after Loder et al., 1998). The southern rectangular box indicates the coverage of the Straits of Florida map shown in **Figure 3**, and the other indicates the Middle Atlantic Bight area where Drinkwater (1999) compiled the hydrographic data shown in **Figure 8**.

the geothermal profile is similar to that of example (a), but its magnitude is smaller at depths. In the third example **(Figure 2c)**, BWT changes sinusoidally with a 120-year-period. In this case, the geothermal profile deviates negatively at 50 m to 80 m depth, primarily reflecting the cooling event 60 years ago, but deviates positively at shallower depths, reflecting the recent warming trend.

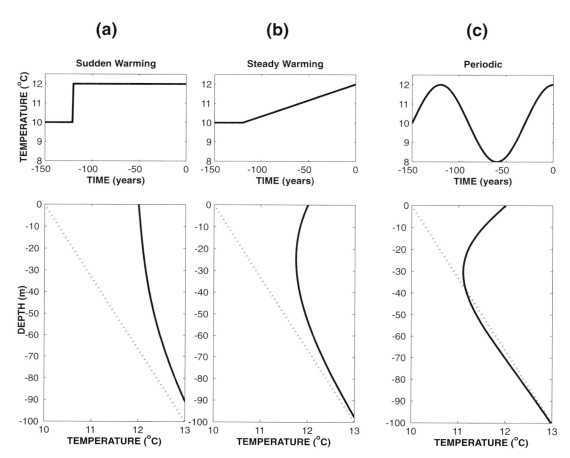

Figure 2 Synthetic examples of geothermal profiles (bottom panels) affected by temporal variation of surface temperature (top panels). The dashed line in each bottom panel shows the steady-
state geotherm. (a) The surface temperature increased instantaneously by 2°C 120 years ago. (b)
The surface temperature linearly increased by 2°C during the last 120 years. (c) The surface temperature changes sinusoidally with a period of 120 years and an amplitude of 2°C.

In example (c), it should be obvious that the thermal signal of an older climatic event is found
at a greater depth than that of a more recent event. Thus, if researchers can obtain detailed and accurate temperature measurements from an offshore borehole, the data may be inverted to reconstruct the local history of BWT. However, because the downward propagation of the thermal
signal is a diffusion process, there is loss of information as the signal travels to greater depths. For
example, it is not generally possible to identify a sharp BWT change such as in example (a) from
borehole data unless the change occurred very recently. Higher-frequency components of the
thermal signal are filtered out at shallow depths. What can be obtained from the borehole temperatures is a smoothed version of the real BWT history with temporal resolution deteriorating
back in time.

Similar methods have been applied to temperature data from boreholes on land to infer the ground-surface temperature (GST) history (Beltrami et al., 1992; Shen and Beck, 1991; Wang, 1992). Deming (1995) summarized such studies previously carried out in North America. It is possible to reconstruct GST history for the past 200 to 300 years, using temperature data from 500- to 700-m-deep boreholes. In many locations in North America, GST histories reconstructed in this manner show a steady warming trend for most of the twentieth century (Wang et al., 1992; Wang et al., 1994).

We believe that this type of study may be of even greater value in the ocean than on land. On land, a considerable amount of surface atmospheric temperature data has accumulated since the mid-nineteenth century (e.g., Hansen and Lebedeff, 1987), but in the ocean, very little information is available, even for the first half of the twentieth century. Researchers involved in the Ocean Drilling Program (ODP) and its predecessor, the Deep Sea Drilling Program (DSDP), have been making high-accuracy measurements in offshore boreholes since the mid-1980s (Horai and Von Herzen, 1985; Hyndman et al., 1987). Most of the ODP/DSDP temperature data were taken from boreholes in very deep seafloor (>2000 m) and did not show indications of significant BWT fluctuation. However, recent measurements made in relatively shallow (<1000 m) seafloor at two localities near the United States yielded geothermal profiles indicative of BWT fluctuation in the last century. The first, from ODP Leg 150 on the continental slope off New Jersey, was reported by Fisher et al. (1999). The present study uses the second data set obtained during ODP Leg 166 in the Straits of Florida **(Figure 3)**, to demonstrate how geothermal data can be used to infer BWT history.

HYDROGRAPHIC REGIME IN THE
VICINITY OF THE STRAITS OF FLORIDA

In the subtropical North Atlantic Ocean, the geostrophic flow is mostly dominated by clockwise rotation. The western boundary current (i.e., the Gulf Stream) originates in the vicinity of the Straits of Florida. Warm water in the Caribbean is transported rapidly northward through the eastern Gulf of Mexico and the Straits of Florida and along the east coast of North America (Loder et al., 1998, **Figure 1**). The main flow of the Gulf Stream **(Figure 3)** enters the Gulf of Mexico through the Yucatán straits, rotates clockwise (Loop Current), and exits through the channel between the Cay Sal Bank and Florida Peninsula (Florida Current). Another branch of flow from the Caribbean through the Old Bahama channel and the Santaren channel continues northward (Leaman et al., 1995).

Site 1006 of ODP Leg 166 is located on nearly flat seafloor in the northern end of the Santaren channel. The water depth is 669 m. The main objective of Leg 166 was to study the sediment-sequence stratigraphy and the pore-fluid chemistry of the western margin of the Great Bahama Bank. Thus, all the other Leg 166 drill sites are located on the margin slope to the east. Only Site 1006 is on a flat seafloor away from the bank and relatively free from complications associated with the downslope sedimentation and the pore fluid migration on the carbonate escarpments (e.g., Kohout et al., 1977). These complications have been discussed in detail elsewhere and will be revisited briefly in a later section when we analyze the temperature data. Site 1006 yielded a greater number of temperature data (18 subbottom depths) than the other stations, and the data are of higher quality.

The thermal structure of the water in the area around Site 1006 is highly complex. This is where the Florida Current meets the flow through the Santaren channel. About 500 temperature data sets (XBT and CTD/STD combined) have been reported previously in this area (24° to 25°N and 79° to 81°W), but most of them reach only ~450 m water depth (National Oceanographic and Atmospheric Administration [NOAA]). In general, the water through the Santaren channel is

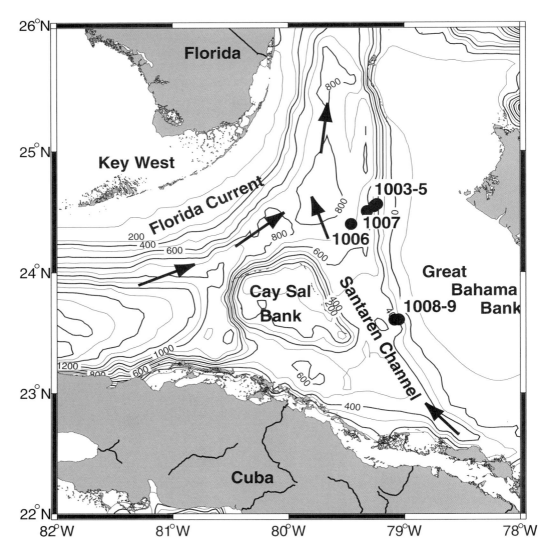

Figure 3 Bathymetric map of the Straits of Florida area. The arrows indicate the directions of currents. Solid circles indicate the locations of the ODP drill sites.

warmer than the Florida Current. The coldest water is found along the Florida shelf edge and the warmest water along the Bahama Bank edge. The temperature difference between the Florida and the Bahama sides is as much as 8°C at 400 m water depth in a relatively short distance of 80 km.

Superimposed on this general trend is the large seasonal variation in the temperature, which affects the entire body of the water in the Straits (greater than 800 m at the deepest point). More than 1°C of seasonal difference has been recorded down to 600 m (Niiler and Richardson, 1973).

Little can be inferred from the available hydrographic data on long-term temperature variations. Unless a series of measurements is made frequently at fixed locations for a decade or two, it is impossible to identify any long-term trend in this area using hydrographic data alone.

BOREHOLE TEMPERATURE MEASUREMENTS

Temperature data are often collected in commercial and scientific drilling operations. The drilling operators almost routinely obtain a continuous downhole temperature log after completing a borehole. However, these data do not necessarily yield the true temperature of the geologic formation, because the temperature around the borehole is greatly perturbed by the drill-fluid circulation. This thermal disturbance can persist for weeks or months, depending on the duration of the drilling operation (e.g., Jaeger, 1961).

There are primarily two ways to obtain more reliable temperature data from offshore boreholes. One is to install a series of temperature sensors and data recorders into the borehole and maintain them for a long period of time until the ambient temperature returns to equilibrium. For example, ODP has established several multipurpose, instrumented boreholes in the Pacific and the Atlantic in the past decade (e.g., Davis and Becker, 1994). The alternative way, less accurate but much more economical, is to insert a long thermal probe into the bottom-hole sediment before drilling perturbation occurs.

ODP has developed two tools for bottom-hole temperature measurements. One is the so-called APC tool, which is used in conjunction with the advanced piston corer, and the other is the water-sampling temperature probe (WSTP) (Fisher and Becker, 1993). The APC is a coring tool used for soft and semiconsolidated sediments. When deployed successfully, it instantaneously penetrates into the bottom-hole sediment and recovers a 9.6-m-long continuous core in a short period of time. The temperature sensor is built into the cutting shoe of the APC. It records the temperature while the APC is held at the bottom of the hole for about 10 minutes. When the sediments are too firm for APC, the WSTP is used. The WSTP is deployed between cores recovered by the extended core barrel (XCB). Its temperature sensor, pushed into the bottom-hole sediments below the previous core, records the temperature for about 10 minutes. For both types of instrumentation, when the temperature sensor is inserted into the sediment, it generates frictional heat. So the tools record the cooling of the sediment as the heat dissipates. The equilibrium temperature of the formation can be theoretically extrapolated from the temperature records (Bullard, 1954; Horai and Von Herzen, 1985).

We used the APC and WSTP at ODP Site 1006 and obtained eighteen temperature measurements at depths 26.1 m below seafloor (mbsf) through 348.9 mbsf (**Figure 4**). We also obtained a total of 220 thermal-conductivity measurements on the core samples from the same hole, using the standard needle probe technique (Von Herzen and Maxwell, 1959). It is clearly shown in the temperature versus depth profile in **Figure 4** that the data points below 100 mbsf fall on a straight line, but those above show gentle curvature. This curvature cannot be explained by the downhole variation of the thermal conductivity (Nagihara and Wang, 2000). The curvature rather suggests that the thermal regime at this site is not steady state or that there is a significant, convective component in the total heat budget. There are three possible mechanisms. The first is temporal fluctuation of BWT. The second is a recent, drastic change in sedimentation rate or erosion. The third is significant migration of pore water through the geologic formation.

In a separate manuscript (Nagihara and Wang, 2000), we discussed this problem in detail and concluded that the first mechanism was the most likely. Here we summarize the discussion. We ruled out the second possibility, because the sedimentation rate at this site (Eberli et al., 1997) was

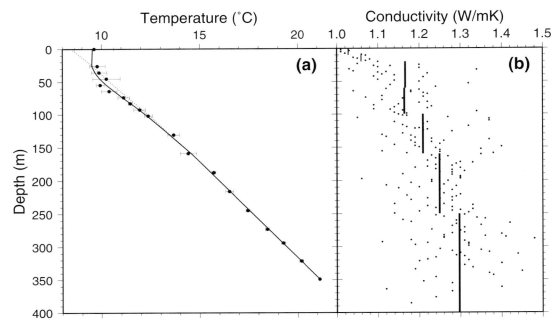

Figure 4   (a) Temperature data obtained at Site 1006 (closed circles). The solid curve shown with the data points corresponds to the bottom-water temperature history shown in **Figure 5**, which is the result of the inversion. The dotted line is the steady-state geotherm, also determined by the inversion. (b) Thermal conductivity data from the cores and the layer structure used in the inversion. Conductivity is expressed in Watts per meter-Kelvin (W/mK).

too small to have a significant thermal effect. Pore-fluid migration is more probable in this relatively porous carbonate platform. In fact, a few of the other Leg 166 drill sites on the upper Bahama margin slope did yield some chemical evidence of fluid flow confined at shallow depths (Eberli et al., 1997; Henderson et al., 1999). The pore fluid chemistry data from Site 1006, especially the Cl^- profile, appears rather uniform down to 20 mbsf, and this can be interpreted as caused by seawater flowing into the formation (Eberli et al., 1997). However, even if there is such influx, it is limited to very shallow depths and cannot cause the anomalous temperatures observed at greater depths. In addition, the constant Cl^- profile may be explained by other mechanisms such as temporal fluctuation in chemical composition of the seawater. There is no porosity variation near 20 mbsf to indicate high permeability of the near-seafloor formation. Thus, we ruled out pore-fluid migration as the primary cause of the curvature in the geothermal profile at Site 1006.

RECONSTRUCTION OF THE BOTTOM-WATER TEMPERATURE HISTORY

Assuming a heat conduction model, we reconstruct the BWT history at Site 1006 by applying the spectral inverse method (Wang, 1992) to the observed geothermal profile. In our inversion, the subbottom sediments are divided into seven layers, each of which has uniform thermal properties

(Figure 4). The thermal diffusivities are estimated from the thermal conductivities assuming a volumetric specific heat capacity of $3.2 \times$ Joules per cubic meter-Kelvin (J/m³K) (Hyndman et al., 1979). As in other inverse problems, the observed geothermal profile **(Figure 4)** may be explained by a number of different scenarios of historical BWT fluctuations within a reasonable margin of error and the solution may become unstable. *A priori* information, or constraints, must be incorporated to minimize the problems of nonuniqueness and instability. Two basic constraints can be applied to the BWT. The first is that BWT changes gradually over time (i.e., smoothness). Excluding high-frequency components in the BWT history alleviates instabilities caused by the noise in the shallow part of the borehole data. The second constraint is that there should be a certain limit to the amplitude of the BWT fluctuation (i.e., boundedness).

In Wang's (1992) inversion method, the BWT history is considered to be a Gaussian stationary stochastic process, and the inversion seeks one of its realizations that satisfies the borehole temperature data and other *a priori* information. The smoothness of the BWT is described by specifying an autocorrelation scale L for the time series. A greater L value yields a smoother BWT history. The boundedness of the BWT is described by a standard deviation (*SD*). This means that BWT at any time is within the *SD* of its mean value with a probability of 68%.

We consider $L = 50$ years and $SD = 1°C$ reasonable values for the inversion of the Site 1006 temperature data. **Figure 5** shows the reconstructed BWT history using these parameters. The model indicates that in the early eighteenth century, the long-term average BWT (~8.2°C) was about 1°C lower than the present value. It slowly cooled down to a minimum at about year 1900, and then increased to the present value of ~9.2°C. Remember that the inversion does not reproduce short-period temperature changes such as seasonal fluctuations. The *a posteriori* geothermal profile, obtained from this BWT history, is shown in **Figure 4a** as a solid line. The observed anomalies in the borehole temperatures are well explained by this BWT fluctuation. The inversion also determines the steady-state geothermal profile, one that is not perturbed by any BWT variations. The steady-state profile is shown in Figure 4a as a dashed line.

Other combinations of L and SD values can also fit the temperature data reasonably well, but they do not show drastic difference in their BWT history estimates. **Figure 6** shows how changing these parameters by a factor of two influences the BWT history solution. In **Figure 6a**, all BWT histories have the same L value (50 years), but they differ in SD: 0.5°C for the dotted line, 1°C for

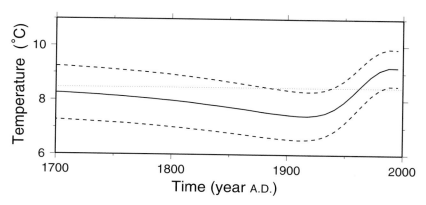

Figure 5 Preferred BWT history reconstructed from the temperature data shown in **Figure 4**. Dashed lines show the one standard deviation error bounds.

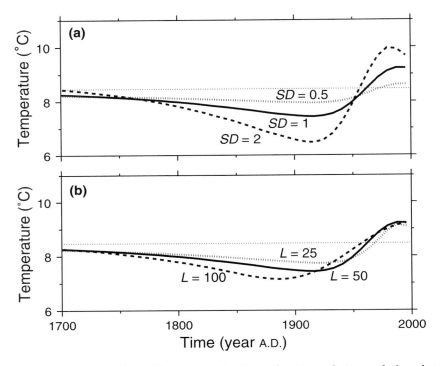

Figure 6 Inversion results using different combinations of autocorrelation scale L and standard deviation SD for the BWT. The solid lines in both (a) and (b) reproduce the preferred solution of **Figure 5**. In the top diagram (a), the three curves have the same L value (50 years) but differ in SD: 0.5°C for the dotted line, 1°C for the solid line, and 2°C for the dashed line. In the bottom diagram (b), the three curves have the same SD value (1°C) but differ in L: 25 years for the dotted line, 50 years for the solid line, and 100 years for the dashed line.

the solid line (preferred solution), and 2°C for the dashed line. The three curves in the bottom diagram (6b) have the same SD value (1°C), but differ in L: 25 years for the dotted line, 50 years for the solid line (preferred solution), and 100 years for the dashed line. Their good fit to the borehole data is illustrated in **Figure 7** by the small differences between the observed and the *a posteriori* subseafloor temperatures, as compared with the error bars of the data. All the BWT history curves in **Figure 6** show the cooling and warming trends similar to those exhibited in our preferred solution of **Figure 5**.

Temperature data are available also from the other Leg 166 drill sites on the Bahama Bank slope, but they are of much smaller numbers (six to nine) and limited in depth (down to 100 to 200 mbsf). Thus, we do not attempt to reconstruct BWT histories for those sites. Their geothermal profiles show some curvature, but the exact shape is difficult to decipher because of the small number of data points. These data have been presented by Nagihara and Wang (2000) and are not shown here. There may have been significant BWT fluctuations at these sites as well, but at some of these sites (1003–1005), pore-water movement may have also impacted the thermal regime.

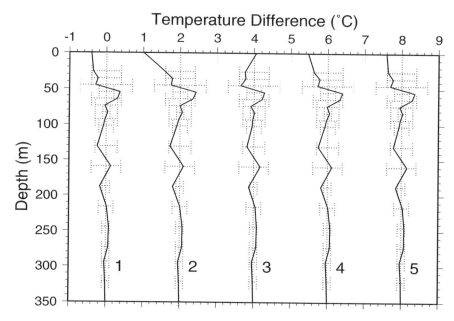

Figure 7 Deviation of the inversion model temperatures from the observed temperature for various combinations of *L* and *SD*. Curve 1 is for the preferred model (*L* = 50 years and *SD* = 1°C). Curve 2 is for *L* = 50 years and *SD* = 0.5°C (dotted line in **Figure 6a**). Curve 3 is for *L* = 50 years and *SD* = 2°C (dashed line in **Figure 6a**). Curve 4 is *L* = 25 years and *SD* = 1°C (dotted line in **Figure 6b**). Curve 5 is *L* = 100 years and *SD* = 1°C (dashed line in **Figure 6b**).

COMPARISON WITH OTHER TEMPERATURE OBSERVATIONS

How does the BWT history at Site 1006 compare with other temperature observations? As mentioned previously, there are no hydrographic temperature records that are long enough to define the century-scale trend. The closest of any long-term temperature measurements in this region is the data set from weather stations in Key West, Florida, which have been archived at the National Climatic Data Center of the National Oceanographic and Atmospheric Administration (NOAA). The weather data in Key West go back to the early nineteenth century. Data obtained after 1850 are considered fairly reliable and have been examined by Hanson and Maul (1993). These authors suggested that the Key West surface air-temperature data serve well as a proxy for sea-surface temperatures. In **Figure 8a**, we show the same data set that they examined. The temperature decreased rapidly in the late nineteenth century, and then gradually increased during the first half of the twentieth century. The general pattern of the temperature variation is similar to the BWT history at Site 1006 **(Figure 8b)**. The warming at Key West appears to have occurred earlier than the BWT, but the exact onset of warming of the BWT is not well constrained. A BWT model solution with earlier warming (*L* = 100 years in **Figure 6**) can also explain the observed geothermal data.

A considerable number of hydrographic data have been collected in the Middle Atlantic Bight, an area off Massachusetts southward through New Jersey, since the 1920s. Lateral variation of water temperature in the area is not as large as in the Straits of Florida. Normally, the warm water of the Gulf Stream occupies the top ~1000 m of this area, but the cold Labrador slope water sometimes invades southward at relatively shallow water depths. Drinkwater (1999) examined the water temperature data (**Figure 1**; 38° to 40°N and 70° to 74°W) at 450-m and 800-m water depths. The data at 450 m show the effect of the Labrador water invasion. The data at 800-m depth, which are reproduced in **Figure 8c**, show a clear trend of warming in the 1920s through the 1970s. The period of warming roughly coincides with those observed at Site 1006 and Key West.

The borehole temperature measurements from ODP Sites 902 and 903 (Fisher et al., 1999) are located on the continental slope off New Jersey. Water depth is 802 m at Site 902 and 453 m at Site 903. These two sites are 5 km apart. Site 902 shows a linear geothermal profile, and the temperatures at Site 903 deviate positively from the linear trend by as much as 1°C to 3°C at 20 mbsf through 60 mbsf. The BWT history reconstructed for Site 903 (Fisher et al., 1999) depicts a BWT increase as large as 5°C in the period 1890 through 1970, followed by a rapid decrease in the last two decades. The timing of the warming is consistent with the other temperature data discussed above, but its magnitude is quite large. Fisher et al. (1999) hypothesize that the subseafloor temperatures at Site 903 may have been influenced by local phenomena such as changes in the depth of the thermocline or across-shelf transport of warm sediment-laden water within shelf/slope channels. Drinkwater (1999) suggests that direct observations are needed to verify whether large, local changes are possible in this setting.

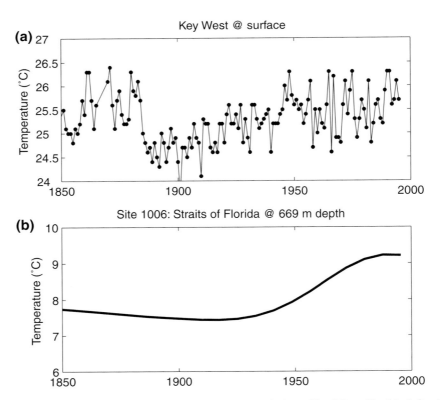

Figure 8 (a) The surface atmospheric temperature records from Key West, Florida (after Hanson and Maul, 1993). (b) BWT history reconstructed from the temperature data at ODP Site 1006.

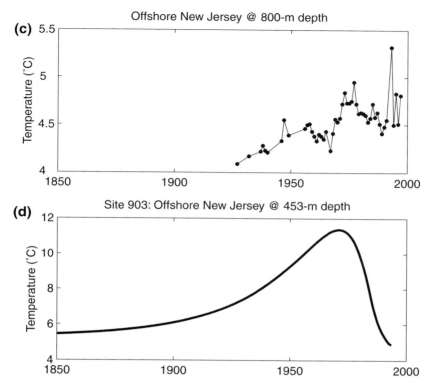

Figure 8 *(continued)* (c) Seawater temperature history at 800-m depth in the Middle Atlantic Bight (after Drinkwater, 1999). (d) BWT history reconstructed from the temperature data at ODP Site 903 (after Fisher et al., 1999).

DISCUSSION AND CONCLUSIONS

The BWT histories estimated at ODP Sites 903 and 1006, the air surface temperature history in Key West, and the seawater temperature history at 800-m depth in the Middle Atlantic Bight all show a period of warming which started in 1900–1920 and lasted at least until 1970–1980. The Key West data and the Site 1006 data also suggest a cooling event in late 1800s. As we mentioned in the beginning of this chapter, there is convincing evidence that the deep water of the North Atlantic subtropical gyre has warmed since the 1960s. All four locations lie within the gyre. In addition, and except for Site 903, the pattern of temperature fluctuation is similar to the global trend of the atmospheric temperature (Hansen and Lebedeff, 1987).

If the correlation between these different observations is not purely coincidental, what is the physical mechanism behind it? Rossby (1959) in his prophetic article suggested that subsurface layers of the ocean might be able to store heat from the atmosphere for a few decades to a few centuries. The deeper water layers are insulated from the atmosphere by well-stratified, warmer near-surface water masses. Thus, if some excess heat is stored in the deep water, it cannot be readily released to the atmosphere. The temperature of the deep layers can change if, for example, there is fluctuation in the production rate of deep water and bottom water in the polar regions (Watts and Morantine, 1991). Changes in surface wind pattern and strength in association with climatic changes may cause such fluctuation.

It is not our intent to discuss in detail global climatic and ocean circulation models, but it seems obvious now that deep ocean water has the potential to be an important part in the global climate. The reconstructed BWT history at Site 1006 may be additional evidence for the warming of the subtropical gyre of the North Atlantic, or at least the Gulf Stream portion of it. It also implies that the warming trend may have started in the beginning of the twentieth century rather than the 1960s, which is the time limit to which the existing hydrographic data can be traced.

Finally, we hope that we have demonstrated the potential usefulness of the BWT history reconstruction by inverting borehole temperature data. Unlike methods that infer temperatures from proxies, the geothermal method directly determines the BWT for the past century or more. However, this method can be used only if the following three conditions are met. First, the borehole temperature data are free from the disturbances associated with drilling operation. Second, sufficient temperature measurements are taken at relatively small depth intervals so that they adequately characterize the curvature of the geothermal profile. Third, the effects of other environmental factors (such as sedimentation, erosion, and pore-fluid migration) are negligible, or they must be accounted for. Information such as biostratigraphy and pore-fluid chemistry data may be necessary to evaluate these effects.

BWT histories reconstructed from offshore borehole temperature data provide valuable constraints to the global climate and ocean circulation models. Future application of this method should help address the following important questions. What are the spatial and temporal scales of BWT variations? How do they relate to ocean circulation and seafloor topography? These questions may be answered if future researchers make a systematic effort to obtain more offshore borehole temperature data. Given the fact that two borehole data sets already exist and that a large number of hydrographic data are available, we suggest that the subtropical North Atlantic is a good place to initiate such an effort.

ACKNOWLEDGMENTS

Kathy Schwehr at the University of Houston assisted Nagihara in compiling the NOAA hydrographic data in the Straits of Florida. Andy Fisher at the University of California, Santa Cruz provided the table of his bottom-water temperature reconstruction at ODP Site 903. Comments by Vaughn Barrie, Jim Lawrence, and three anonymous reviewers were helpful in improving this manuscript. Partial funding for this work was obtained from the United States Scientific Support Program of the Joint Oceanographic Institutions. Geological Survey of Canada contribution number 1999259.

REFERENCES CITED

Beltrami, H., A. M. Jessop, and J.-C. Mareschal, 1992, Ground temperature histories in eastern and central Canada from geothermal measurements: Evidence of climatic change: Paleogeography, Paleoclimatology, Paleoecology, v. 98, p. 167–184.

Bullard, E. C., 1954, The flow of heat through the floor of the Atlantic Ocean: Proceedings from the Royal Society of London, Series A, v. 222, p. 408–429.

Davis, E. E., and K. Becker, 1994, Formation temperatures and pressures in a sedimented rift hydrothermal system: Ten months of CORK observations, Holes 857D and 858G: Proceedings of the Ocean Drilling Program, Scientific Results, v. 139, p. 649–666.

Deming, D., 1995, Climatic warming in North America: Analysis of borehole temperatures: Science, v. 268, p. 1576–1577.

Drinkwater, K. F., 1999, Additional testing needed to confirm warming of water off New Jersey shore: EOS Transactions, American Geophysical Union, v. 80, p. 173–177.

Eberli, G. P., P. K. Swart, M. J. Malone, F. S. Anselmetti, K. Arai, K. H. Bernet, C. Betzler, B. A. Christensen, E. H. De Carlo, P. M. Dejardin, L. Emmanuel, T. D. Frank, G. A. Haddad, A. R. Isern, M. E. Katz, J. A. M. Kenter, P. A. Kramer, D. Kroon, J. A. McKenzie, D. F. McNeill, P. Montgomery, S. Nagihara, C. Pirmez, J. G. Reijmer, T. Sato, N. H. Schovsbo, T. Williams, and J. Wright, 1997, Bahama Transect: Proceedings of the Ocean Drilling Program, Initial Results, v. 166, 850 p.

Fisher, A. T., and K. Becker, 1993, A guide to ODP tools for downhole measurements, College Station, Texas, Ocean Drilling Program Technical Report No. 7, p. 131.

Fisher, A. T., R. P. Von Herzen, P. Blum, B. Hoppie, and K. Wang, 1999, Evidence may indicate recent warming of shallow slope bottom water off New Jersey Shore: EOS Transactions, American Geophysical Union, v. 80, p. 165–173.

Hansen, J., and S. Lebedeff, 1987, Global surface air temperatures: Update through 1987: Geophysical Research Letters, v. 15, p. 323–326.

Hanson, K., and G. A. Maul, 1993, Analysis of temperature, precipitation, and sea-level variability with concentration on Key West, Florida, for evidence of trace-gas-induced climatic change, in G. A. Maul, ed., Climatic Change in the Intra-Americas Seas, London, U.K., United Nations Environmental Program, p. 193–213.

Henderson, G. M., N. C. Slowey, and G. A. Haddad, 1999, Fluid flow through carbonate platforms: Constraints from $^{234}U/^{238}U$ and Cl$^-$ in Bahamas pore-waters: Earth and Planetary Science Letters, v. 169, p. 99–111.

Horai, K., and R. P. Von Herzen, 1985, Measurement of heat flow on leg 86 of the deep sea drilling project: Initial Reports of the Deep Sea Drilling Project, v. 86, p. 759–777.

Hyndman, R. D., E. E. Davis, and J. A. Wright, 1979, The measurement of marine geothermal heat flow by a multipenetration probe with digital acoustic telemetry and in situ thermal conductivity: Marine Geophysical Research, v. 4, p. 181–205.

Hyndman, R. D., M. G. Langseth, and R. P. Von Herzen, 1987, Deep Sea Drilling Project geothermal measurements: A review: Reviews of Geophysics, v. 25, p. 1563–1582.

Jaeger, J. C., 1961, The effect of the drilling fluid on temperatures measured in bore holes: Journal of Geophysical Research, v. 66, p. 563–569.

Kohout, F. A., H. R. Henry, and J. E. Banks, 1977, Hydrogeology related to geothermal conditions of the Floridan plateau, in D. L. Smith, and G. M. Griffin, eds., The Geothermal Nature of the Floridan Plateau, Tallahassee, Florida, Department of Natural Resources, State of Florida, p. 9–41.

Leaman, K. D., P. S. Vertes, L. P. Atkinson, T. N. Lee, P. Hamilton, and E. Waddell, 1995, Transport, potential vorticity, and current/temperature structure across Northwest Providence and Santaren channels and the Florida current of Cay Sal Bank: Journal of Geophysical Research, v. 100, p. 8561–8569.

Levitus, S., 1989, Interpentadal variability of temperature and salinity at intermediate depths of the North Atlantic Ocean, 1970–1974 versus 1955–1959: Journal of Geophysical Research, v. 94, p. 6091–6131.

Levitus, S., and J. Antonov, 1995, Observational evidence of interannual to decadal-scale variability of the subsurface temperature-salinity structure of the world ocean: Climatic Change, v. 31, p. 495–514.

Levitus, S., J. I. Antonov, and T. P. Boyer, 1994, Interannual variability of temperature at a depth of 125 meters in the North Atlantic Ocean: Nature, v. 266, p. 96–99.

Levitus, S., J. I. Antonov, Z. Zengxi, H. Dooley, V. Tsereschenkov, K. Selemenov, and A. F. Michaels, 1995, Observational evidence of decadal-scale valiability of the North Atlantic Ocean, in D. G. Martinson, K. Bryan, M. Ghil, M. M. Hall, T. R. Karl, E. S. Sarachik, S. Sorooshian, and L. D. Talley, eds., Natural Climate Variability on Decade-to-Century Time Scale, Washington, D.C., National Research Council, p. 318–324.

Loder, J. W., W. C. Boicourt, and J. H. Simpson, 1998, Western Ocean Boundary Shelves, in A. R. Robinson and K. H. Brink, eds., The Global Coastal Ocean: Regional Studies and Syntheses: The Sea, New York, John Wiley & Sons, Inc., p. 3–27.

Nagihara, S., and K. Wang, 2000, Geothermal regime of the western margin of the Great Bahama bank: Proceedings of the Ocean Drilling Program, Scientific Results, v. 166, p. 113–120.

Niiler, P. P., and W. Richardson, 1973, Seasonal variability of the Florida current: Journal of Marine Research, v. 31, p. 144–167.

National Oceanographic and Atmospheric Administration, 1994, World Ocean Atlas 1994, Washington, D.C., National Oceanographic and Atmospheric Administration.

Parker, B. B., 1996, Sea level change as indicator of climatic and global change, *in* R. G. Pirie, ed., Oceanography: Contemporary Readings in Ocean Sciences, New York, Oxford University Press, p. 115–129.

Parrilla, G., A. Lavin, H. Bryden, M. Garcia, and R. Milard, 1994, Rising temperatures in the subtropical North Atlantic Ocean over the past 35 years: Nature, v. 369, p. 48–51.

Roemmich, D., and C. Wunsch, 1984, Apparent changes in the climatic state of the deep North Atlantic Ocean: Nature, v. 307, p. 447–450.

Rossby, C. L., 1959, Current problems in meteorology, *in* B. Bolin, ed., The atmosphere and the sea in motion: Scientific contributions to the Rossby memorial volume, New York, Rockefeller Institute Press, p. 9–50.

Shen, P. Y., and A. E. Beck, 1991, Least squares inversion of borehole temperature measurements in functional space: Journal of Geophysical Research, v. 96, p. 19965–19979.

Snodgrass, J. M., 1968, Instrumentation and communications, *in* J. F. Brahtz, ed., Ocean Engineering, New York, John Wiley & Sons, p. 393–477.

Von Herzen, R. P., and A. E. Maxwell, 1959, The measurement of thermal conductivity of deep-sea sediments by a needle probe method: Journal of Geophysical Research, v. 64, p. 1557–1563.

Wang, K., 1992, Estimation of ground surface temperatures from borehole temperature data: Journal of Geophysical Research, v. 97, p. 2095–2106.

Wang, K., T. J. Lewis, D. S. Belton, and P. Y. Shen, 1994, Differences in recent ground surface warming in eastern and western Canada: Evidence from borehole temperatures: Geophysical Research Letters, v. 21, p. 2689–2692.

Wang, K., T. J. Lewis, and A. M. Jessop, 1992, Climatic changes in central and eastern Canada inferred from deep borehole temperature data: Paleogeography, Paleoclimatology, Paleoecology, v. 98, p. 129–141.

Watts, R. G., and M. C. Morantine, 1991, Is the greenhouse gas-climate signal hiding in the deep ocean?: Climatic Change, v. 18, p. 3–4.

Hughes, G. B., and C. W. Thayer, Sclerosponges: Potential high-resolution recorders of marine paleotemperatures, 2001, *in* L. C. Gerhard, W. E. Harrison,
and B. M. Hanson, eds., Geological perspectives of
global climate change, p. 137–151.

7 | Sclerosponges: Potential High-Resolution Recorders of Marine Paleotemperatures

Gary B. Hughes

Raytheon Infrared Operations
Goleta, California, U.S.A.

Charles W. Thayer

Department of Earth and Environmental Science
University of Pennsylvania
Philadelphia, Pennsylvania, U.S.A.

ABSTRACT

Sclerosponges have great potential as seawater temperature recorders. These animals precipitate their skeletons in carbon and oxygen isotopic equilibrium with the surrounding seawater (Druffel and Benavides, 1986). Their skeletons also display chemical properties that vary directly with changes in environmental conditions. Lack of photosynthetic symbionts allows sclerosponges to live below the photic zone, providing the potential to investigate past marine conditions beyond the range of corals. Individual sponges live for several centuries, preserving archives of pre- and postindustrial seawater variations within single specimens (Hartman and Reiswig, 1980). Cross-correlation of successively older specimens could yield up to 2000 years of marine history. Extracting environmental information can be accomplished by determining elemental characteristics preserved in skeletal growth bands. A method is presented here that utilizes energy dispersive spectroscopy (EDS) to provide inexpensive assessment of magnesium (Mg): calcium (Ca) and chlorine (Cl): calcium (Ca) ratios at high spatial resolution, yielding environmental data with correspondingly high temporal resolution. The relationship between environmental conditions and skeletal characteristics is defined by a spectral transfer function, which can then be applied to skeletal carbonate data from ancient sponges to reconstruct past environmental conditions. Accurate reconstruction of seawater temperature and salinity variations is demonstrated here at sub-monthly resolution. The technique's efficiency is ideal for documenting long, high-resolution records of marine paleoenvironments.

SCLEROSPONGES AS POTENTIAL ENVIRONMENTAL RECORDERS

Instrument records of climatological conditions generally represent data gathered for short time spans over limited geographical extent. This is particularly true for the marine environment, in which continuous instrument records spanning more than a few tens of years are rare. The short time span of most instrumental records critically limits the conclusions that can be drawn from analyzing them. The limited spatial extent of instrumental recordings also raises issues with applying even narrow conclusions to larger geographical areas. Investigations focusing on regional and global climate patterns are hampered by this paucity of data. Thus, it is of great interest to pursue possible sources of paleoenvironmental data that complement and extend currently available instrument data and that offer the possibility of investigating longer-term (interdecadal to millennial) climate patterns at regional or global scales.

Biogeochemical records, such as marine sediments, represent rich potential archives of long-time-span paleoenvironmental data. Bottom sediments record important aspects of broad trends in marine environmental parameters such as seawater temperature and salinity. However, data derived from these sources are insufficient for many studies, due to coarse temporal resolution. Turbulence, mixing, and bioturbation tend to inhibit high-frequency environmental variability from being reflected in bottom-sediment characteristics. Annual or better resolution in sediment-derived environmental parameters is uncommon. Studies that rely on information about seasonal fluctuations in climate must search for other archives.

Fortunately, the sea offers some intriguing possibilities in the form of biogenic sedimentary structures such as coral reefs. The animals responsible for constructing coral reefs are influenced by environmental parameters, and studies have shown that environmental influence is discernible in the calcareous skeletal material left behind by the organisms. Similar to corals, sclerosponges (phylum Porifera) also deposit calcareous skeletons as they grow, and they offer some advantages over corals as paleoenvironmental archives (Swart et al., 1998). This study focuses on one potential archive of paleoenvironmental data, sclerosponges. One method of extracting high-resolution data from sclerosponges is presented here that shows the potential of these animals as environmental recorders.

Both corals and sclerosponges precipitate calcium-carbonate ($CaCO_3$) from seawater to construct exoskeletons. As atmospheric carbon dioxide (CO_2) dissolves into the oceans, bicarbonate and carbonate are formed, according to the equilibrium reactions

$$CO_2 + H_2O \leftrightarrow H^+ + HCO_3^- \leftrightarrow 2H^+ + CO_3^{--}$$

Dissolved calcium ions in seawater then combine with carbonate ions to form calcium carbonate, providing the raw material for sclerosponge skeletons:

$$CO_3^{--} + Ca^{++} \rightarrow CaCO_3$$

In this way, dissolved atmospheric CO_2 represents the source material for carbon atoms in coral and sclerosponge skeletal carbonate; specifically, the carbonate carbon atoms are the same carbon atoms that came into the ocean as atmospheric carbon dioxide. As sclerosponges grow, successive growth bands are laid down on top of older layers of the skeleton. At each layer, the precipitated carbonate is drawn from the surrounding seawater. It is the hope of paleoenvironmentalists that some characteristics of the ambient seawater affect skeletal carbonate formation and that these characteristics are immutable after the carbonate has been deposited. That is, some lasting characteristic of the precipitated carbonate must vary directly with environmental conditions for these animals to represent a viable paleoenvironmental archive.

Modern analytical techniques have yielded results demonstrating that several skeletal characteristics of corals and other carbonate-secreting organisms do indeed vary directly with ambient seawater temperature. Mitsuguchi et al. (1996) report successful interpretation of Mg:Ca ratios in coral skeletal carbonate as an accurate paleothermometer. Seawater paleotemperatures have been investigated using Mg:Ca ratios in fossil ostracodes (Dwyer et al., 1995) and mussels (Klein, Lohman, and Thayer, 1996a), which both secrete carbonate exoskeletons. Other studies with corals indicate that trace elements such as strontium incorporated in skeletal carbonate during formation provide definitive environmental proxies (Beck, et al., 1992). Successful interpretation of isotopic data ($\delta^{18}O$, $\delta^{13}C$) as environmental indicators was reported in bivalves by Klein et al. (1996b) and in corals by Dunbar and Wellington (1981).

In addition to corals and other carbonate-secreting organisms, sclerosponges have also attracted attention as potential recorders of past marine conditions. It has been shown that isotopic composition ($\delta^{18}O$, $\delta^{13}C$) of sclerosponge skeletal carbonate is related to ambient seawater temperature (Druffel and Benavides, 1986; Swart et al., 1998, and references therein). Furthermore, the sensitivity of this recording mechanism is sufficient to allow investigation into marine and climatic variability. Sclerosponge skeletons collected in Jamaica and the Bahamas have recorded decreases in the isotopic signature of atmospheric carbon dioxide over the past century, a phenomenon called the carbon-13 Suess effect (Druffel and Benavides, 1986; Quay et al., 1992; Bohm et al., 1996; Swart et al., 1998).

Along with stable isotope analysis, trace-element chemistry shows promise as an environmental recorder in sclerosponge skeletons. In this study, the sponge *Acanthocaetetes wellsi* **(Figure 1)** is used. Hartman (1983) and Hartman and Reiswig (1980) describe their natural history. An explicit method is presented here for determining paleotemperature by measuring Mg:Ca ratios of skeletal carbonate in growth bands of *A. wellsi*. Furthermore, the same method is applied to the reconstruction of paleosalinity using Cl:Ca ratios. Although the measurement of chloride in skeletal carbonate has not yet gained wide acceptance as a salinity indicator, the data presented here hint at an intriguing connection. The method presented here to reconstruct past salinity variations could be exploited if future, more detailed studies validate the use of Cl:Ca ratios as a salinity indicator. Using an electron microprobe, Mg:Ca and Cl:Ca ratios are quickly measured at high spatial resolution in skeletal cross sections. Derived environmental data have correspondingly high temporal resolution. Biweekly records are obtained at measurement resolutions as coarse as 50µm in *A. wellsi*. Data from extant sponges are calibrated with modern instrument records. In the method presented here, the modern relationship between elemental data and environmental conditions is defined by a system transfer function. The transfer function can then be applied to data from ancient sponges to reconstruct past environmental conditions.

The technique's efficiency is ideal for documenting a long, high-resolution record of these marine parameters. Furthermore, individual sponges typically live for several centuries. Living and dead specimens of *A. wellsi* are common from mean tide to at least 200 m depth throughout the western tropical Pacific. So sclerosponges offer the potential for both high-resolution and long time-span records over a large geographical region. These characteristics fit well with the requirements of many studies investigating long-term climate variability, such as El Niño/Southern Oscillation (ENSO).

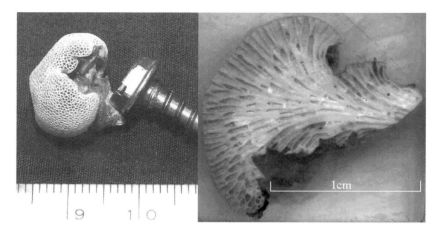

Figure 1 *Acanthocaetetes wellsi* from Palau, harvested in July 1998. Sclerosponges do not depend on symbiotic algae as corals do. As a result, sclerosponges are outcompeted by corals in the photic zone. Sclerosponges are thus restricted to areas of reduced light where most corals cannot grow because of their dependence on algae. Sclerosponges are common in caves and along deep undersea cliffs beyond the penetration of light. This specimen began growing on top of a marker screw shortly after the screw was emplaced in July 1989 (approximate location of the screw is shown). Scale in centimeters. During this time, the sponge grew 1.21 cm, an average of 1.34 mm/yr. The specimen was cross-sectioned for elemental analysis.

ENVIRONMENTAL RECONSTRUCTIONS

Procedure for Analyzing Skeletal Carbonate in *Acanthocaetetes wellsi*

The objective of skeletal carbonate analysis is to obtain time-series data of isotopic and/or elemental ratios recorded across the growth bands. These skeletal characteristics presumably contain markers of environmental conditions that existed when each microlayer formed. That is, each layer deposited at a given point in time should contain approximately constant isotopic and elemental ratios determined by the environmental conditions present at that time. Achieving environmental proxies with a desirably high temporal resolution requires extraction of isotopic and elemental data with correspondingly high spatial resolution at known points along the growth axis of the skeleton. Precise determination of growth-layer boundaries is essential for determining an accurate time-series path over the cross-sectional area. Results presented here show that fast and inexpensive laboratory procedures can be exploited to attain this objective.

A specimen of *A. wellsi* was collected from Palau (Micronesia). This specimen began growing on top of a marker-plate screw shortly after the screw was emplaced in July 1989. The sponge was harvested in July 1998 **(Figure 1)**. During this time period, the specimen grew 1.21 cm, an average of 1.34 mm/yr. The specimen was cross-sectioned, and energy dispersive spectroscopy (EDS) was used to measure elemental ratios at discrete locations across growth bands. At each point, the atomic percentage of sodium (Na), Mg, Cl, and Ca was recorded **(Figure 2)**. EDS gives semiquantitative atomic percentages, often with large associated errors. However, calibration studies with samples of known chemical composition confirm that *ratios* of EDS atomic percentages accurately and repeatably reflect atomic *proportions* in the sample. Repeatability studies of tooth carbonate on

QUANTEX EDS STATISTICS

ELEMENT	WEIGHT PERCENT	ATOMIC PERCENT*
Na	11.49	17.38
Mg	7.04	10.08
Cl	16.34	16.03
Ca	65.12	56.51
TOTAL	99.99	

*NOTE: ATOMIC PERCENT is normalized to 100

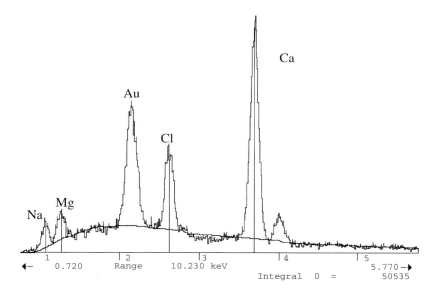

Figure 2   Energy dispersive spectroscopy (EDS) is used to perform elemental analysis of skeletal carbonate. Atomic percentages of each element are derived from the measured energy spectrum and are semiquantitative. However, studies with carbonate samples of known composition verify that *ratios* of semiquantitative atomic percentages are accurate estimates of *elemental proportions* in the skeletal carbonate. On the machine where the current study was performed, EDS-derived elemental ratios are consistently within 3% relative error when both elements being measured comprise >1% of the substrate. Calcium (Ca) and magnesium (Mg) are components of skeletal carbonate. Sodium (Na) and chlorine (Cl) are believed to be incorporated into interstitial spaces of the carbonate crystal lattice, possibly as NaCl. The gold (Au) peak is introduced into the spectrum when the sample is prepared for EDS analysis. Gold is sputtered onto the surface of the cross section to allow excess charge to bleed away from the surface.

the machine where the current study was performed show that EDS-derived elemental ratios have a standard error of the mean (2 sigma) of 3% when both elements being measured comprise >1% of the substrate. It is this feature of EDS analysis that allows quantitative Mg:Ca and Cl:Ca ratios to be determined.

Electron-microprobe techniques allow data to be recorded at high spatial resolution. To translate this high spatial resolution into high temporal resolution, an accurate microspatial map of growth-band boundaries is produced. To accomplish this, the specimen is cross-sectioned and mounted in epoxy resin. High-resolution digital photos of the cross section are produced. Three datum points on the surface are chosen that are recognizable in both the digital image and the electron-microprobe image. The (x, y) locations of the microprobe stage at each datum point are recorded, and a datum plane is established from these values.

The EDS spectrum is recorded at successive points along continuous skeletal ridges exposed in the cross-sectional surface. EDS settings for this study were 10 kilovolts (kv) accelerating voltage, 0° beam angle, and 120-second exposure, at ~100× magnification. The atomic percentage values returned by the EDS software give the estimated percentages (xx.xx%) of each element in the substrate. For a Mg value of mm.mm% and a Ca value of cc.cc%, the Mg:Ca ratios reported here are computed by (mm.mm/cc.cc) × 100, giving cmol Mg/mol Ca. Each Mg:Ca ratio measurement also has an associated location within the datum plane, determined from the microprobe stage readings. Spatial data analysis, such as described by Dettman and Lohmann (1995), is used to create a map of growth-band microtopography over the digital image **(Figure 3)**. Each of the lines in **Figure 3** represents an estimated contour of constant Mg:Ca ratio as calculated by spatial interpolation from EDS measurements. Once the microtopography is established, a time-series path of Mg:Ca ratios can be established. Elemental ratios are then estimated at equally spaced points along the time-series path, spatially interpolated from surrounding known data values **(Figure 4)**. Further details of data extraction and analysis are presented in Hughes (1999). For the Palau

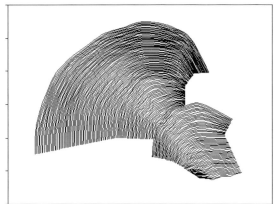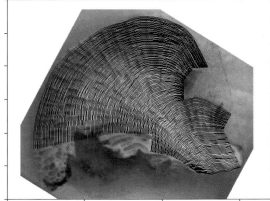

Figure 3 Growth-band contours derived from energy dispersive spectroscopy (EDS) data and spatial data analysis techniques. The key to generating high temporal-resolution environmental reconstructions is obtaining high spatial-resolution elemental data and determining precise boundaries of growth-band layers. Each of the contours represents a line of constant Mg:Ca ratio as estimated by spatial interpolation from EDS measurements. A time-series path is always perpendicular to the growth bands and may follow a convoluted path through the cross section.

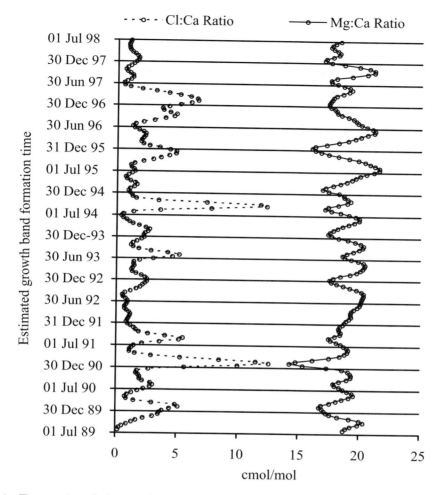

Figure 4 Time series of elemental ratios as determined from energy dispersive spectroscopy (EDS). Spatial locations of the measurements are converted to carbonate formation times by assuming a constant, linear growth rate and by estimating the age of youngest and oldest carbonate. This specimen presumably began growing shortly after July 1989, on top of a marker-plate screw. Data represent one sample every 16.28 days over the period July 1989 through July 1998, when the specimen was harvested. The skeleton left behind by a growing sclerosponge animal is considered here as a biologically precipitated sedimentary structure. The conventional presentation of time-series data for sedimentary structures places the youngest portion of the column at the top, with age of the sedimentary component increasing downward along the vertical axis. The dependent variables (Mg:Ca and Cl:Ca ratios) are still considered to be functions of the independent time variable.

specimen, EDS data were recorded at approximately 50 micron (μm) grid spacing. Using this method, 200 data values were produced over a 1.0 cm time-series path spanning approximately 8 years, 10 months, and 29 days (3256 days). Assuming linear growth of one band directly above the previous band, this translates to one measurement every 16.28 days across the growth bands.

Syndepositional Environmental Data and Correlation with Skeletal Characteristics

The objective here is to compile a high-resolution temperature history from the elemental data. To accomplish this, ambient seawater parameters from the area where the sponge was growing are required. Data for this study were obtained from the Leetmaa Pacific Ocean Analysis (1998, see reference). This analysis uses various sources of environmental data to estimate monthly average seawater temperature and salinity in grid cells of 1° latitude by 1.5° longitude, and at various depths between 5 and 3126.5 m. A direct comparison between the environmental data and the elemental ratios can be made by using linear regression. The regression line then gives a crude mapping from Mg:Ca or Cl:Ca ratios to temperature or salinity, respectively. But variability in the regression will likely occur due to differences between local conditions at the islands of Palau and the grid average, especially for short-lived temperature or salinity spikes caused by upwelling, wind-driven mixing, or rainfall. It is also possible that some amount of difference will arise due to errors in the estimation of growth-band formation times that result in inaccurate mapping of the elemental ratios to the time axis. The growth bands contain calcium carbonate precipitated over a range of times, because a certain amount of "backfilling" occurs on the inner calicle walls, adding small errors to calibration of the time axis. Precise microspatial mapping of growth-band boundaries mitigates the magnitude of this error. In the case of the salinity curve, the relationship between skeletal Cl:Ca ratios and ambient conditions may be more complex than a simple linear dependence, because the Cl is believed to be incorporated into the interstitial spaces of the carbonate crystal. It remains to be determined why Cl:Ca ratios should vary directly with ambient seawater salinity. Data in **Figure 4** suggest some connection. The elemental ratios would likely explain more environmental variability if local ambient conditions were used in the regression.

Spectral Transfer Function Approach

The linear regression approach is not necessarily the best tool to use for establishing a relationship between elemental ratios and environmental conditions. Since environmental conditions are cyclical, the concept of frequency content of the time-series measurements provides a more appropriate means of establishing relationships between elemental ratios and environmental conditions. Spectral techniques can provide powerful tools for investigating correlation between two cyclical variables such as Mg:Ca ratios and seawater temperature. The inaccuracies of regression models can be overcome by exploiting the concept of a spectral *system transfer function*. In linear regression, the model takes each value of the measured variable at a fixed point in the time domain and estimates the value of the predicted variable at the *same point in time*. However, viewing the time series data sets as quasi-periodic functions over finite domains, each signal can alternatively be described by its frequency content. A system transfer function is a model that operates on each frequency of the measured variable signal and estimates how the predicted variable signal will behave at the *same frequency* (see Weaver, 1983).

The system transfer function method for establishing a relationship between elemental ratios and environmental conditions has several advantages for the task at hand. Small, random inaccuracies in the time estimates of the measured variables are effectively smoothed when transforming to the frequency domain, so skeletal growth-rate variations do not pose significant problems as in linear regression. Also, individual frequencies in the measured variable can be amplified or attenuated by the transfer function, allowing complex relationships between the measured and predicted variables. The transfer function can be estimated from basic principles of how Ca, Mg, and

Cl are incorporated into the skeletal carbonate lattice and interstitial spaces, and from how the sampling technique alters the actual signal. Time lags in the response of skeletal carbonate properties to environmental fluctuations can be easily accounted for. Longer time-series data sets provide increasingly narrower constraints on the shape of the transfer function.

System transfer functions have been successfully applied to issues in mixing of abyssal sediments, where depositional characteristics are sought from measurements made on the sediments after mixing has altered the original signal (Schiffelbein, 1984; Goreau, 1980). This concept can also be used to establish the relationship between sclerosponge skeleton elemental ratios and ambient seawater conditions. If instrument records were available from *in situ* monitoring of temperature and salinity at the precise location where the sponge was growing, the transfer function would be constructed by taking the ratio, at each frequency, of power in the temperature data to power in the elemental data. In this case, when the transfer function is applied to the elemental data, the temperature curve would be reproduced exactly over the entire time span of measured data. Thus, the transfer function captures the precise relationship between ambient temperature and skeletal Mg:Ca ratios. Elemental data could then be measured in older sponges for which there is no available temperature information. The transfer function determined from the modern sponge could be applied to the older data to reconstruct ambient seawater temperatures.

For the data extracted from the Palau sclerosponge specimen, the equivalent sampling rate was one measurement every 16.28 days. The highest Fourier frequency (f_c) for these data is approximately 11.22 cycles per year, a period of 0.089 years (~1 month, 2 days). The data were collected over a time span equivalent to 8 years, 10 months, and 29 days (3256 days). This gives the lowest Fourier frequency as 1 cycle in 3256 days, or approximately 0.112 cycles per year, a period of 8 years, 10 months, and 29 days (~107 months). Spectral composition of the Leetmaa temperature and salinity series over the period January 1980 through December 1998 are shown in **Figure 5**. The annual cycle dominates the temperature spectrum. Some less significant cycles appear at longer and shorter periods. Longer period variability is likely due to ENSO, which displays broad-spectrum power at periods between 3 and 7 years (36 to 84 months). The large peak at a period of 6 months is the first harmonic of the annual cycle and represents deviations from an annual sinusoid. The fundamental cause of the 6-month peak is currently a matter of speculation. Spectral composition of the salinity series shows a small annual signal. The most significant salinity cycles are between periods of 3 to 7 years (36 to 84 months), likely driven by ENSO.

For this study, temperature and salinity data were used from the Leetmaa analysis grid. For the Leetmaa data and the Mg:Ca ratios measured in the Palau *A. wellsi* specimen, a system transfer function is shown in **Figure 6**. In this case, the shape of the transfer function is determined from conceptual rules designed to allow local variability to pass through the transfer, while at the same time capturing the overall cyclical elements seen in the grid. Frequencies in the Mg:Ca signal are scaled by the transfer function values shown in the graph. Periods shorter than 2 months are completely filtered, because they are near the Nyquist cutoff frequency. High-frequency fluctuations, represented by periods between 2 and about 3.5 months, can be left unchanged or filtered by the transfer function without significantly affecting the reconstructed signal (the latter choice would be somewhat smoother). Periods longer than 3.5 months are either filtered, attenuated, amplified, or left unchanged by the transfer function, depending on the empirical relationship between the two variables. In the transfer function shown in **Figure 6**, periods at 9 months are amplified because the absolute amplitude of the temperature signal is larger than the amplitude of the Mg:Ca signal at this frequency. This is analogous to the slope of a regression line, but the transfer function approach can use a different "regression slope" at each frequency,

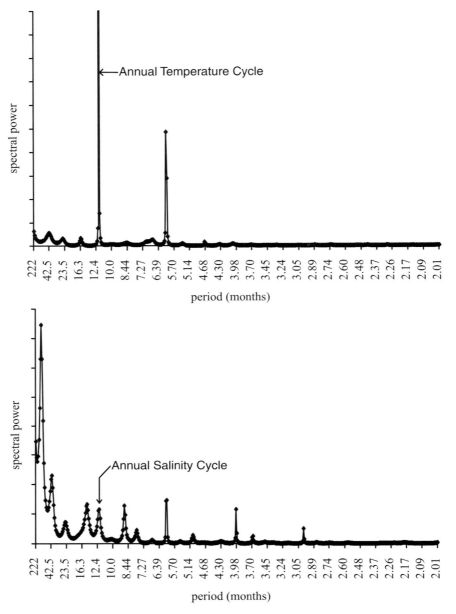

Figure 5 Spectral composition of the Leetmaa Pacific Ocean Analysis (1998) temperature (top) and salinity series for the 1°lat by 1.5°long grid cell centered at 7°N, 134.25°E over the period January 1980 through December 1998, computed by Maximum Entropy Method. The islands of Palau are situated within this grid cell at 7.5°N, 134.5°E. The Leetmaa reconstruction gives average conditions expected within the entire grid cell. Local conditions for sclerosponges growing in Palau may differ from the grid average due to local runoff and different mixing conditions. The annual cycle dominates the temperature spectrum, with some insignificant cycles at longer periods. The large peak in the temperature spectrum at six months is the first harmonic of the annual cycle and represents the majority of deviations from an annual sinusoid. The fundamental cause of the six-month peak is currently a matter of speculation. The salinity spectrum displays a small annual signal, but the most significant cycles are at periods of three to seven years (36 to 84 months), likely driven by ENSO.

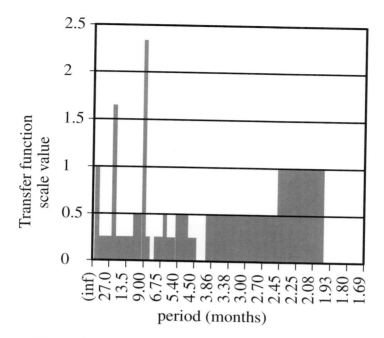

Figure 6 An amplitude-only system transfer function for reconstructing ambient seawater temperature from skeletal carbonate Mg:Ca ratios based on the Palau specimen of *A. wellsi*. The system transfer function is determined from the Fourier spectra of (1) time-series elemental ratios in the cross section and (2) corresponding time-series of ambient conditions. The transfer function captures the spectral relationship between the two variables, and can be used to derive temperature if Mg:Ca ratios are known. The transfer function developed from modern data can thus be used to reconstruct paleotemperatures based on elemental data in ancient sponges.

resulting in better reconstructions. The reconstructed curves for temperature and salinity, determined by spectral system transfer functions, are shown in **Figure 7**. Again, the elemental ratios would likely be closer to environmental values if local ambient conditions were used in the analysis.

CONCLUSIONS AND OUTLOOK FOR GLOBAL ENVIRONMENTAL STUDIES

The system transfer function is developed with modern instrument records of temperature and salinity, and with skeletal-carbonate elemental ratios from modern sclerosponges. As with any environmental proxy, the assumption is that the response of modern specimens to environmental fluctuations is similar to the response of ancient specimens. If this assumption holds, then the system transfer function developed with modern records is also applicable to ancient specimens. A long modern instrument record would serve to constrain the system transfer function. Then, by applying the system transfer function to skeletal-carbonate elemental ratios from ancient specimens, the history of temperature and salinity at the location where each specimen grew is estimated. Both living and dead specimens of *A. wellsi* are common in caves and cavities throughout the tropical Pacific. Radiometric age dating of sclerosponge skeletal carbonate has been accomplished with high

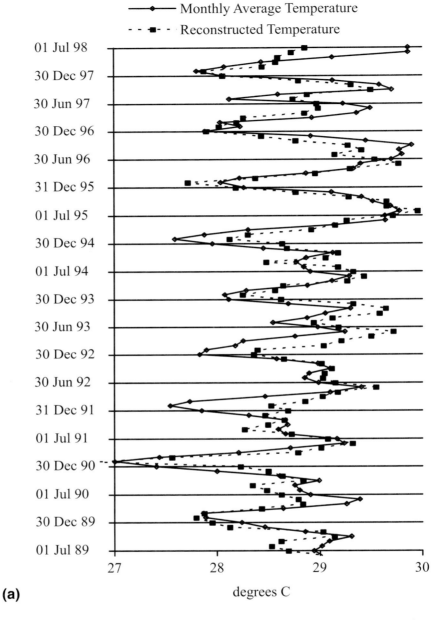

(a)

degrees C

Figure 7 Temperature (a) and salinity (b) signals reconstructed from skeletal carbonate elemental ratios using amplitude-only empirical system transfer functions. The reconstructed curves are compared with estimated monthly average values at 5-m depth in the 1°lat–1.5°long grid cell centered at 7°N, 134.25°E (data from Leetmaa Pacific Ocean Analysis, 1998). Departures between estimated and reconstructed curves likely reflect differences between local conditions where the sclerosponge was growing and the grid average. In any case, the reconstruction confirms that the goal of monthly-average resolution is attainable with these animals. Individual sponges live for several centuries, and as a species range over an extended area throughout the tropical Pacific. Therefore, accurate, high-resolution Pacific Ocean seawater temperature and salinity reconstructions over the past 2000 years are possible by using these archives.

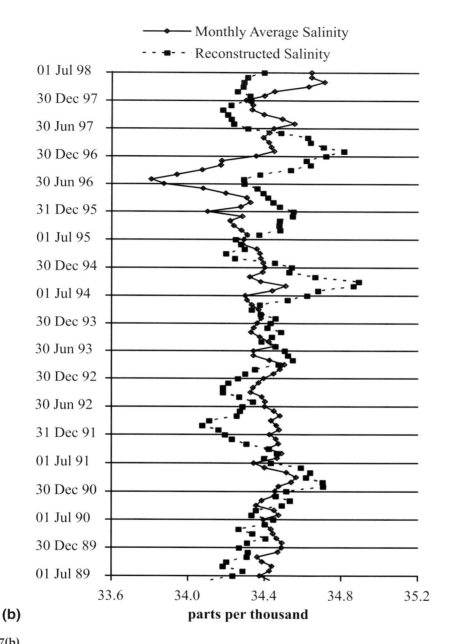

(b)

Figure 7(b)

accuracy (Benavides and Druffel, 1986; Rubenstone et al., 1996), and these methods can be used to constrain the time axis for ancient sponges. Cross-correlation of successively older specimens could yield very long records, possibly up to 2000 years or more.

The potential scope of information that could be gleaned from these animals is extremely tantalizing. Sclerosponges range over most of the tropical Pacific, and are found from the surface to at least 200 m depth. Individual sponges live for several centuries, and specimens that have been dead for many years can be found clinging to cave walls and buried in cave sediments, perhaps representing several thousand years of history. Sclerosponges apparently preserve faithful and sensitive records of paleotemperature and possibly paleosalinity levels at the site where they grew. Efficient and inexpensive laboratory techniques described in this study can be used to reconstruct ambient conditions with submonthly resolution by measuring elemental data in skeletal carbonate growth bands. For ENSO research and other global climate studies, this potential is both rich and unprecedented. By collecting and analyzing many sclerosponges, both living and ancient, a marine climatological history of the tropical Pacific spanning the Medieval Warm Phase could be documented in great detail. This history would represent a critical aspect of studies seeking to understand global change and how ENSO cycles might be affected by ongoing global warming. Such extensive hindsight would undoubtedly lead to a deeper understanding of the fundamental nature of ENSO and other climatological phenomena.

ACKNOWLEDGMENTS

Financial support for isotopic and elemental analysis was provided by Geological Society of America Research Grant No. 6054-97. The Mellon Foundation and the University of Pennsylvania Research Foundation provided financial support for field research in Palau and for isotopic and elemental analysis. Assistance with all aspects of field research from Robin Aiello, Great Adventures Company, Cairns, Australia, is gratefully acknowledged. Assistance with collection and interpretation of isotopic and elemental data from K. C. Lohmann and Lora Wingate, both of the University of Michigan Stable Isotope Laboratory, and from Jerry Harrison, University of Pennsylvania School of Dentistry, is also gratefully acknowledged.

REFERENCES CITED

Beck, J. W., R. L. Edwards, E. Ito, F. W. Taylor, J. Recy, F. Rougerie, P. Joannot, and C. Henin, 1992. Sea-surface temperature from coral skeletal strontium/calcium ratios. Science, vol. 257, 31 Jul 1992, p. 644.

Benavides, L. M., and E. R. M. Druffel, 1986. Sclerosponge growth rate as determined by ^{210}Pb and Δ^{14}C chronologies. Coral Reefs, vol. 4, p. 221–224.

Bohm, F., M. M. Joachimski, H. Lehnert, G. Morgenroth, W. Kretschmer, J. Vacelet, and W. C. Dullo, 1996. Carbon isotope records from extant Caribbean and South Pacific sponges: Evolution of δ^{13}C in surface water DIC. Earth and Planetary Science Letters, vol. 139, p. 291–303.

Dettman, D. L., and K. C. Lohmann, 1995. Microsampling carbonates for stable-isotope and minor element analysis: physical separation of samples on a 20 micrometer scale: Journal of Sedimentary Research A, vol. 65, no. 3, p. 566.

Druffel, E. R. M., and , L. M. Benavides, 1986. Input of Excess CO_2 to the surface ocean based on ^{13}C/^{12}C ratios in a banded Jamaican sclerosponge. Nature, vol. 321, p. 58–61.

Dunbar, R. B. and G. M. Wellington, 1981. Stable isotopes in a branching coral monitor seasonal temperature variation. Nature, vol. 293, p. 453–455.

Dwyer, G. S., T. M. Cronin, P. A. Baker, M. E. Raymo, J. S. Buzas, and T. Correge, 1995. North Atlantic temperature change during late Pliocene and late Quaternary climate cycles. Science, vol. 270, 1347.

Goreau, T. J., 1980. Frequency sensitivity of the deep-sea climatic record. Nature, vol. 287, 16 October 1980, p. 620–622.

Hartman, W. D., 1983. Modern and ancient Sclerospongiae. In Sponges and spongiomorphs: Notes for a short course organized by J. K. Rigby and C. W. Stearn, T. W. Broadhead, ed., University of Tennessee, Department of Geological Sciences, Studies in Geology 7, ISBN-0-910249-06-7.

Hartman, W. D., and H. M. Reiswig, 1980. Morphological and ecological studies of sponges in Pacific reef caves. National Geographic Society Research Reports, vol. 12, p. 339–346.

Hughes, G. B., 1999. Atmospheric carbon dioxide and land-surface air temperatures in geologic and modern-instrument records: Identifying patterns and formulating alternative hypotheses through spectral analysis. Ph.D. dissertation, University of Pennsylvania, Philadelphia, 1999.

Klein, R. T., K. C. Lohman, and C. W. Thayer, 1996a. Sr/Ca and $^{13}C/^{12}C$ ratios in skeletal calcite of *Mytilus trossulus:* Proxies of metabolic rate, salinity and carbon isotopic composition of seawater. Geochimica et Cosmochimica Acta 60(21):4207–4221.

Klein, R. T., K. C. Lohman, and C. W. Thayer, 1996b. Bivalve skeletons record sea-surface temperature and salinity via Mg:Ca and $^{18}O/^{16}O$ ratios. Geology, vol. 24, p. 415–418.

Leetmaa Pacific Ocean Analysis data obtained from the NOAA-CIRES Climate Diagnostics Center, Boulder, Colorado, from their Website at http://www.cdc.noaa.gov/, accessed in November 1998.

Mitsuguchi, T., E. Matsumoto, O. Abe, T. Uchida, and P. J. Isdale, 1996. Mg:Ca thermometry in coral skeletons. Science, vol. 274, 8 November 1996, p. 961.

Quay, P. D., B. Tilbrook, and C. S. Wong, 1992. Oceanic uptake of fossil fuel CO2: Carbon-13 evidence. Science, vol. 256, p. 74–79.

Rubenstone, J. L., M. D. Moore, C. D. Charles, and R. G. Fairbanks, 1996. High precision Th/U TIMS dating of young sclerosponges from the equatorial Pacific Ocean, EOS, supplement to vol. 77, no. 46, F384.

Schiffelbein, P., 1984. Effect of benthic mixing on the information content of deep-sea stratigraphical signals. Nature, vol. 311, 18 October 1984, p. 651–653.

Swart, P. K., J. L. Rubenstone, C. Charles, and J. Reitner, ed., 1998. Sclerosponges: A new proxy indicator of climate. NOAA Climate and Global Change Program, Special Report No. 12, Report from the workshop on the use of Sclerosponges as proxy indicators of climate, 22–24 March 1998, Miami, Florida.

Weaver, H. J., 1983. Applications of discrete and continuous Fourier analysis. John Wiley & Sons, New York, 1983, 375 p.

Ashworth, A. C., Perspectives on Quaternary beetles and climate change, 2001, *in* L. C. Gerhard, W. E. Harrison, and B. M. Hanson, eds., Geological perspectives of global climate change, p. 153–168.

8 | Perspectives on Quaternary Beetles and Climate Change

Allan C. Ashworth

Department of Geosciences
North Dakota State University
Fargo, North Dakota, U.S.A.

ABSTRACT

The response of beetles to climate change during the Quaternary Period is reviewed for the purpose of evaluating their future response to global warming. Beetles responded to Quaternary climatic changes mostly by dispersal, which ultimately led to large-scale changes in geographic distribution. Fragmentation and isolation of populations associated with climate change did not result in either higher rates of speciation or extinction, although local extinctions occurred when dispersal routes were blocked by barriers. Studies from archaeological and late Holocene sites indicate that the fragmentation of the natural landscape by human activities had as great an impact on the local diversity of beetle populations as did climate change. Habitat reduction and fragmentation continue today and are making species increasingly vulnerable to extinction. The major difference between the future and past responses of beetles to climate change is that extinction rates are expected to be much higher, independent of whether the causes of climate change are natural or anthropogenic. The question of determining whether global warming has natural or anthropogenic causes is important because of the ethical implications of extinction.

INTRODUCTION

J. B. S. Haldane, the legendary British evolutionary biologist, is credited with one of the most famous quips in biology. When asked what he could infer about the Creator from his biological studies, he is said to have replied, "An inordinate fondness for beetles." Whether or not Haldane

actually made this irreverent response is open to question (Gould, 1995). What is not in question i
the diversity of beetles implied in Haldane's witticism. In Haldane's day, estimates for the num
bers of beetle species were in the tens of thousands. Currently, the estimates range between 1.2
and 7.5 million species (Hammond, 1992; Erwin, 1997; Stork, 1997). Beetles are the most divers
group of organisms on earth, accounting for about 20% of all species (Erwin, 1997).

Coleoptera (sheathed-wings), or beetles, are an order within the class Insecta that occurs in
most freshwater and terrestrial habitats, including those of remote islands. Beetles are ubiquitous
occurring in the wettest rain forests and in the most arid deserts. There are species with physiolc
gies adapted to survive periodic below-freezing temperatures of high arctic summers. There ar
even species adapted to the littoral zone, where they survive daily tidal inundation by burrowin
in sand. Antarctica is the only continent today that does not have an indigenous beetle fauna. A
few species have been reported from the Antarctic Peninsula, but all of them were probably intrc
duced by humans. Beetles were present in the interior of Antarctica, however, until the Neogene
They became extinct with the growth of the polar ice sheet and the climate change to ultracold and
ultraxeric conditions (Ashworth et al., 1997).

Most beetles are small organisms, ranging in length from 0.25 mm to several centimeters. Av
erage length is estimated to be in the range of 4–5 mm (Crowson, 1981). Small size has ensured tha
their activities mostly go unnoticed. The sheer numbers of species, however, make them importar
members of all terrestrial and freshwater ecosystems. Because of their abundance, beetles are im
portant food items for numerous species of reptiles, birds, small mammals, and fish. Beetles ar
also important agricultural and forestry pests, with numerous species being injurious to crop:
trees, and stored products (Patterson et al., 1999). Their most important beneficial roles are as po
linators and as recyclers of nutrients, activities which ensure the health of ecosystems.

Several of the world's leading biologists believe an extinction crisis faces the earth that has th
potential to match or exceed the mass extinctions at the close of the Permian and Cretaceous Pe
riods. The biggest problem facing beetles is the destruction of natural habitat by human activitie:
It is a sad commentary that for undescribed species, the chance of extinction is greater than th
chance of being described (Stork, 1997). The threat of extinction is most acute in tropical region
but problems exist for beetle species in all densely populated areas.

Record keeping is so inadequate that even in more affluent nations, it is difficult to assess th
number of species that have become extinct during the last century. In the United Kingdom, 64 c
about 4000 species of beetles have not been collected for more than 100 years and are assumed t
be locally extinct (extirpated). In addition, there are perhaps the same number not categorized a
extinct but which have not been found for several decades (Hyman and Parsons, 1992, 1994). Th
locally extinct species are of various ecological types, but the majority are either wetland or fores
floor species. Populations of some of these species survive in isolated patches of old-growth fores
in Europe, but several others have not been collected in many years and are suspected of being e:
tinct throughout their entire geographic ranges (P. M. Hammond, British Museum of Natural Hi
tory, London, personal communication, 2000).

In North America, there are insufficient data to make a similar assessment of the status of the er
tire beetle fauna. Data that are available for the Cicindelidae, or tiger beetles,[1] suggest that 19% c
tiger-beetle species are in a vulnerable to critically imperiled condition (Stein et al., 2000). The tige
beetles are just one of more than 100 families of beetles in North America, and their vulnerability i

---

[1]The tiger beetles are placed in the tribe Cicindelini, family Carabidae, by some taxonomists.

extinction is probably typical. Most beetle diversity is now concentrated in isolated patches. Even in the absence of climate change, natural habitats will shrink and beetles will become extinct as the world's human population, now at more than six billion, grows at a rate of about one billion people a decade.

How will climate change impact an already stressed biota? Lovejoy (1997) wrote that human activity has created an obstacle course for the dispersal of biodiversity and that a climate change, even a natural one, could precipitate one of the greatest biotic crises of all time. Is the situation really as bad as Lovejoy paints it or is he indulging in hyperbole? How can we assess the response of beetles to future climate change? Insect collections provide information that the geographic ranges of some species have changed during the last century. Separating cause and effect from this type of data, however, has proved to be very difficult (Ashworth, 1996, 1997). Experimental physiological data have also been used to predict the future response of species (Patterson et al., 1999). These data, however, are available for only a few beetle species that are of economic significance. In this paper, the response of beetles to Quaternary climatic changes and to human activities is examined for the purpose of speculating on how beetles will respond to future climate change.

QUATERNARY BEETLE FOSSILS

Quaternary beetle fossils consist mostly of disarticulated exoskeletons; setae and scales are frequently preserved, as are pigmented and structural colors **(Figure 1)**. Fossils of sclerotized internal structures, such as the male genitalia, also occur. For practical purposes, however, the parts most

5 mm

Figure 1 A 14,000 cal yr B.P. (12,000 yr B.P.) fossil beetle assemblage from Norwood, Minnesota (Ashworth et al., 1981). The fossils consist of a head, several pronota, and elytra of 12 species of beetles in the families Carabidae, Cicindelidae, Dytiscidae, Heteroceridae, Chrysomelidae, and Curculionidae. Specimens with both structural (metallic appearance) and pigmented coloration are present in the petri dish.

studied by paleoentomologists are heads, pronota (thoraces), and elytra (wing cases). Chitin is a nitrogenous polysaccharide that is insoluble in water, dilute acids, and alkalis. It is stable in anoxic sedimentary environments but is broken down by bacteria during prolonged exposure to the atmosphere. Fossils occur in unconsolidated sediments, mostly from shallow lacustrine, paludal, and fluvial environments. Elias (1994) described the kerosene flotation method used to isolate the fossils, their preparation for study, and their identification. Fossil assemblages typically consist of hundreds of identifiable pieces. The age of fossil assemblages is based on carbon 14 (^{14}C) dating, either directly by accelerator mass spectrometry (AMS dating) or indirectly by dating associated plant fossils. The dates in this paper are reported as calibrated ages (cal yr B.P. [calendar years before the present]) based on a calibration of radiocarbon ages (Stuiver et al., 1998). Original radiocarbon dates are reported in parentheses (yr B.P.) following the calibrated dates.

THE FOSSIL RECORD AND THE RESPONSE OF BEETLES TO CLIMATE CHANGE

Speciation

Beetles are the most species-rich group of organisms on earth, and intuitively, it would seem that they must evolve rapidly. Climatic changes, especially those resulting in the growth and melting of ice sheets, cause the fragmentation and isolation of biological populations. The conditions for reproductive isolation and allopatric speciation would seem to be optimal at times of climate change. Several hundred species of beetles have now been reported from a large number of studies of Pleistocene and Holocene fossil assemblages. References for these studies are listed in Quaternary Bibliography—Insects (QBIB), a comprehensive electronic bibliography (Buckland et al., 1997). Several new species were described in the older literature, but most of these have now been reassigned to extant species following taxonomic revisions. The Quaternary fossil record is unequivocally one of stasis, not speciation **(Figure 2)**.

The most remarkable example of stasis is from a fossil assemblage located near Kap København at latitude 82° 30′ N, Peary Land, northernmost Greenland. Presently, only two species of beetles survive the harsh polar desert conditions of Peary Land. The Kap København Formation was deposited about 2 million years ago (late Pliocene) during a time when the climate in the Arctic was significantly warmer than at present. Preservation of chitin is excellent and 140 species of beetles have been identified from the deposits (Böcher, 1995). Böcher considered only two or three of the species to be extinct. The life histories of most arctic beetle species are either unknown or poorly documented. Even so, it is a reasonable assumption that the generation length of most of the Kap København species is about a year. This means that the genes of the living descendants of Kap København species have been shuffled about two million times during reproduction, with no observable morphological changes. Paradoxically, this remarkable record of stasis has been maintained during a time of repeated glaciations.

Beetles are ectotherms and are dependent on environmental temperatures during all phases of their life cycles. Life-cycle processes affected by climate include life-span duration, diapause, dispersal, mortality, and genetic adaptation (Patterson et al., 1999). Factors such as growth rates and timing of diapause vary within the geographic ranges of species (Butterfield, 1996; Butterfield and Coulson, 1997). For example, in North America and Europe, the date of emergence of adults in northern populations is typically later in the summer than for southern populations. Intraspecific adaptations at the population level, to variations in climate and the amount of daylight, are ubiquitous. Subspecies resulting from these differences are considered to be intermediates in the for-

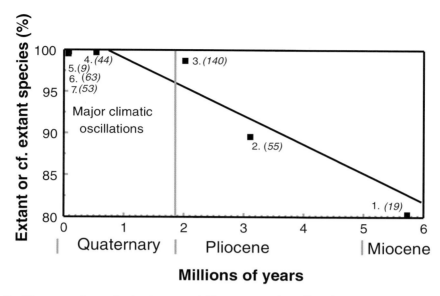

Figure 2 The percentage of extant or near (cf.) extant species of beetles in pre–Late Wisconsinan fossil assemblages (numbers in parentheses refer to numbers of species). 1. Lava Camp, Alaska (Matthews, 1977); 2. Beaufort Formation sites (Matthews, 1977; Fyles et al., 1994); 3. Kap Køben-havn Formation, Greenland (Böcher, 1995); 4. Cape Deceit, Alaska (Matthews, 1974); 5. Gastropod Silts, Minnesota, South Dakota (Garry et al., 1997); 6. Olympic Peninsula, Washington state (Cong and Ashworth, 1996); 7. Titusville, Pennsylvania (Cong et al., 1996).

mation of new species. If even a small percentage of these subspecies resulted in new species, then speciation in beetles should be commonplace. Why then have no new species of beetles been re-ported in the Quaternary fossil record? Is it possible that new species go undetected? Certainly, the fossil record represents only a small percentage of the fauna. Also, preservation favors certain eco-logical types, especially those of mesic habitats. Even so, a large number of species have been de-scribed as fossils, and underrepresentation does not explain why some percentage of those should not be new species. Some speciation may also go undetected because of the inability of paleoento-mologists to separate new species from morphologically similar ancestral species. In summary, it is possible that a low level of speciation goes undetected, but in the high latitudes of the Northern Hemisphere, fossil evidence suggests that there has been little speciation in the beetles during much of the Quaternary Period **(Figure 2)**.

Coope (1978) suggested that stasis might paradoxically be linked to the instability of environ-ments during the Quaternary Period. He suggested that the migrations of species during times of climate change prevented populations from becoming genetically isolated long enough to evolve new species. Elias (1991) also used this mechanism to explain the absence of speciation in fossil assemblages from the Rocky Mountain region. Coope's hypothesis would imply that genetic vari-ation among disjunct populations of living species of beetles should be relatively low and mostly of recent origin. This was not observed, however, in a study of the mitochondrial DNA of the North American ground-beetle species *Amara alpina* Paykull (Reiss et al., 1999). *A. alpina* occurs in tundra and alpine habitats in four widely separated regions: Beringia, the Hudson Bay region, the northern and central Rocky Mountains, and the northern Appalachian Mountains. The genetic evidence indicates a large amount of divergence among Appalachian and Rocky Mountain populations, and indicates that these populations have been reproductively isolated for more than one glacial cycle, and possibly several (Reiss et al.,1999).

Vogler and DeSalle (1993) also detected a large amount of mitochondrial DNA variation in the tiger-beetle species, *Cicindela dorsalis* Say, which has a geographic range in eastern North America. The species inhabits coastal sand dunes from New England to the Gulf Region. Parts of the species' range in New England and New York state were glaciated, and the species must have colonized those areas during the Holocene. Throughout the remainder of the geographic range, eustatic changes in sea level associated with the growth and melting of ice sheets must have resulted in shifts in populations as dunes migrated in response to marine regression and transgression. Vogler and DeSalle identified 17 haplotypes, some representing ancient divergences, others more recent. The species is represented by four subspecies, two in the Gulf Region west of the Florida Panhandle, and two to the east on the Atlantic Coast. The subspecies are defined on the basis of external morphological characteristics. What is especially interesting is that haplotypes are shared between subspecies and generally the correlation between morphology and mitochondrial DNA is not especially strong. At least the major division in subspecies and haplotypes can be traced to the isolation of populations caused by the emergence of the Florida Panhandle during the Pliocene. Once again, this is strong evidence for long-term stasis at the species level.

Bennett (1997) offered an explanation of stasis as part of a hierarchy of evolutionary processes operating during the Quaternary Period. Natural selection, which operates on timescales of thousands of years, results in the accumulation of microevolutionary changes within species. In Bennett's hypothesis, climate changes driven by rhythmic orbital oscillations on 20,000- to 100,000-year frequencies (Milankovitch cycles) continually disrupt the accumulation of microevolutionary changes. The result is stasis of the type associated with the punctuated equilibria hypothesis (Eldredge and Gould, 1972). Eventually, populations become reproductively isolated and allopatric speciation occurs. The amount of time this takes, however, is measured in hundreds of thousands, not tens of thousands, of years.

Most of the evidence for stasis in beetle species is based on studies of fossils from locations in Europe, North America, and South America, which were either glaciated or on the margins of ice sheets. Stasis observations, however, are not confined to glaciated regions. Species identified from the La Brea asphalt deposits in California and packrat (*Neotoma*) middens from the deserts of the American Southwest and northern Mexico also indicate that stasis, not speciation, was the response of beetles to climate change (Elias and Van Devender, 1990; Elias,1992; Miller, 1997). Paleontological evidence supports stasis, but some caveats are necessary before I can conclude that speciation was globally insignificant during the Quaternary Period. The first is that no fossil assemblages have been examined from deposits in the species-rich tropical rain forests. The second is that no fossil assemblages have been examined from regions of high endemism, such as the Caucasus Mountains. It is possible that stasis is the norm but that there could also be speciation "hot spots."

Extinction

Similar conditions to those that favor allopatric speciation also favor extinction. Certainly, the end of the Pleistocene Epoch was marked by a major extinction in North America of more than 40 species of large mammals, including mammoths (Martin and Klein, 1984). There were also several extinctions of large mammals in Europe (Yalden, 1999). Different hypotheses involving climatic and vegetational changes and overkill by Paleo-Indian hunters have been proposed to explain the extinctions (Martin and Klein, 1984). The extinctions of large mammals in both North America and Europe were not accompanied by correspondingly large extinctions in the beetle fauna.

Only two species of dung beetles, *Copris pristinus* Pierce and *Onthophagous everestae* Pierce from the Rancho La Brea asphalt deposits, are reported as Quaternary extinctions (Miller et al., 1981). Climate change, the extinction of large herbivores, and the reduction in dung were all considered to be factors contributing to the extinctions. Miller (1997) considered it a possibility, however, that these species might still be living. He has reason to be cautious. In Europe, d'Orchymont (1927) described a pair of fossil elytra from glacial deposits in Germany as the extinct species *Helophorus wandereri* d'Orchymont. Subsequently, the species described as being extinct was discovered to be part of the extant fauna of Siberia, where it had been described as *H. obscurellus* Poppius. This synonymy is a reminder that beetle extinction is rare in the Quaternary fossil record.

Migration

Most beetles are highly mobile animals. Large numbers of species fly. Some are clumsy fliers and the distances they travel are small. Others, however, are strong fliers capable of traveling tens or hundreds of kilometers in strong winds. The long-term response of beetle populations to climate change is referred to as migration. Migration used in this sense, however, is different from the seasonal movement from one area to another, but represents the long-term shift in geographic distributions resulting from dispersal and survival. The result has also been referred to as species-tracking climate change, but at the level of species, the process is more stochastic than deterministic **(Figure 3)**.

There are many examples of Quaternary climatic changes causing major shifts in geographic ranges of beetle species. In Europe, the ground beetle *Diacheila arctica* Gyllenhal inhabited the British Isles during the last glaciation. The species now has a distribution in northern Scandinavia and adjacent parts of Russia **(Figure 4)**. At the onset of the last glaciation, individuals of *D. arctica* living in arctic regions began to migrate southward. Some populations survived and colonized the British Isles from mainland Europe. Low sea level during the last glaciation exposed a corridor of

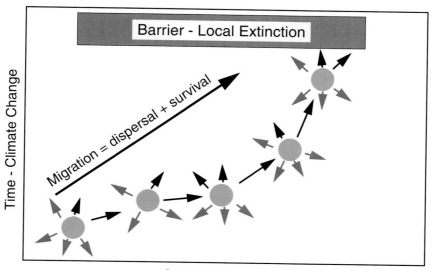

Figure 3 Schematic diagram representing the relationship among dispersal, survival, and migration. As individuals disperse, some survive to reproduce the next generation (black arrows) and others die out (gray arrows). The geographic range shifts as the climate changes.

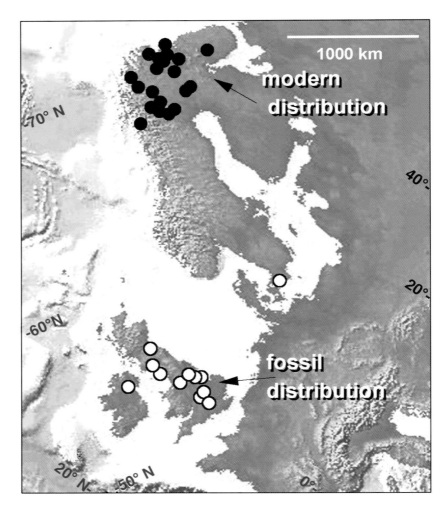

Figure 4 The modern (black dots) and late-glacial fossil distribution (white dots) of the subarctic ground beetle species *Diacheila arctica* (data for modern and fossil distributions from Buckland et al., 1997). The map is modified from Sterner (1996).

land in the vicinity of the English Channel. Rapid warming at about 15,600 cal yr B.P. (13,000 yr B.P.) caused the local extinction of the species from lowland areas in the British Isles **(Figure 5)**. The return to a cold climate during the Younger Dryas Stade resulted in the reappearance of the species in lowland areas. Whether or not it had become "extinct" in the British Isles and then recolonized from mainland Europe is unknown. Small populations could have survived in refugia in upland areas and colonized the lowlands as the climate cooled. Populations of the species inhabited the British Isles during the 1200 years of the Younger Dryas Stade, but in the rapid warming at 11,500 cal yr B.P. (10,000 yr B.P.) they became locally extinct.

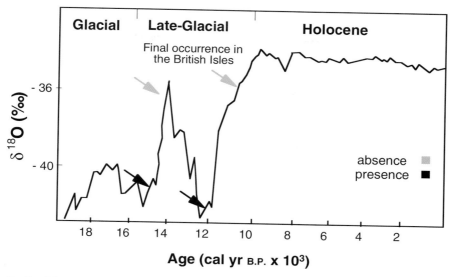

Figure 5 Rapid response of cold-adapted beetles to late-glacial climatic changes in the British Isles. Arctic species were extirpated when the climate changed from cold to warm (gray arrows). They reappeared when the climate cooled (dark arrows). Climatic warming at about 11,500 cal yr B.P. (10,000 yr B.P.), at the end of the Younger Dryas Stade, caused *Diacheila arctica* and other sub-arctic beetle species to become locally "extinct" in the British Isles. Curve for $\delta^{18}O$ is based on the Greenland Ice Core Project (GRIP).

In eastern and central North America, arctic beetle species replaced boreal forest species along the southern margin of the Laurentide ice sheet at about 25,500 cal yr B.P. (21,500 yr B.P.) **(Figure 6)**. Colonization was probably from populations that dispersed southward in front of the growing Laurentide ice sheet and from populations that dispersed westward and eastward from montane refugia in the Appalachian and Rocky Mountains, respectively. This "glacial" fauna, dominated by tundra species, inhabited the ice margin until about 17,400 cal yr B.P. (14,500 yr B.P.) (Morgan, 1987; Schwert, 1992; Ashworth, 1996). Mean July temperature along the ice margin is estimated to have been 10–12°C lower than at present. Warming, in combination with the ice sheet acting as a barrier to northward dispersal, caused the extirpation of tundra species in the midlatitudes at about 17,400 cal yr B.P. (14,500 yr B.P.) (Schwert and Ashworth, 1988). The ice marginal fauna, until 14,700 cal yr B.P. (12,500 yr B.P.), was characteristic of open terrain but warmer climatic conditions than had existed earlier. By 14,700 cal yr B.P. (12,500 yr B.P.), this fauna was replaced by a boreal forest fauna.

In southern South America, several fossil beetle assemblages have been studied from lowland sites (Hoganson and Ashworth, 1992; Hoganson et al. 1989; Ashworth and Hoganson, 1993). From about 28,000 cal yr B.P. (24,000 yr B.P.) until between 18,000 and 17,000 cal yr B.P. (15,000 to 14,000 yr B.P.) the beetle fauna consisted of only a small number of species, mostly from open, wet habitats **(Figure 7)**. Species of an earlier forest fauna had been extirpated as the climate cooled. The survivors of the extirpation were joined by species from habitats at or above treeline in the Cordillera de la Costa and the Cordillera de los Andes. These species were able to survive in the lowlands

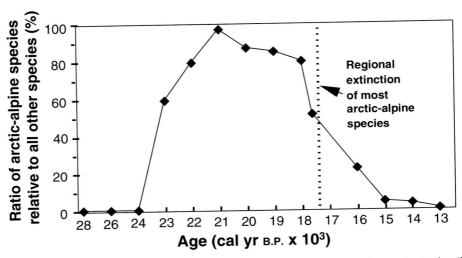

Figure 6 The appearance and disappearance of subarctic and arctic beetle species in fossil sites located along the southern margin of the Laurentide ice sheet from Iowa to New York state. The more cold-adapted species became "extinct" along the margin of the ice sheet at about 17,400 cal yr B.P. (14,500 yr B.P.) as the climate warmed (Morgan, 1987; Schwert and Ashworth, 1988; Schwert, 1992; Ashworth and Willenbring, 1998).

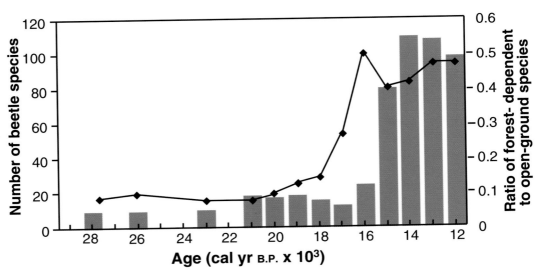

Figure 7 Rapid replacement of open-ground species by forest species as the climate warmed at the end of the last glaciation in southern Chile between 18,000 to 17,000 cal yr B.P. (15,000 to 14,000 yr B.P.) (Hoganson et al., 1989; Hoganson and Ashworth, 1992; A. C. Ashworth, unpublished data).

because of colder climatic conditions. Based on the occurrence of high-elevation species at low elevations, the mean January temperature is estimated to have been 4–5°C lower than present. The glacial fauna is very different from the species-rich forest fauna that started to replace it at about 17,000 cal yr B.P. (14,000 yr B.P.) (Ashworth and Hoganson, 1993). The change in the beetle fauna is first marked by an increase in the ratio of forest to open-ground species and later by an increase in the total number of species **(Figure 7)**. The last fossil occurrence of the most characteristic beetle of the glacial fauna, the weevil *Germainiellus dentipennis* Germain, occurs at about 15,450 cal yr B.P. (12,800 yr B.P.). By 15,100 cal yr B.P. (12,500 yr B.P.), several species of the forest fauna had recolonized the deglaciated terrain. The fauna at this time is inferred to have had a diversity similar to that of the present day.

Human Activities

The response of beetles to human activity has been inferred from the study of beetles in archaeological sites, especially from the British Isles (Buckland and Dinnin, 1993; Osborne, 1997; Whitehouse, 1997; Dinnin and Sadler, 1999). Forest clearance by Neolithic to Bronze Age peoples during the period from about 5700 to 2800 cal yr B.P. (5000 to 2700 yr B.P.) resulted in the extirpation of 35 species. A further six species of beetles were extirpated between the Roman and Medieval periods (Dinnin and Sadler, 1999). The species most affected were those of old-growth forests, especially those associated with dead pine logs (Whitehouse, 1997). Aquatic species were also reduced in numbers. The decline in woodland and aquatic species was accompanied by an increased representation of grassland and dung-feeding species that exploited the newly created pastures. A small amount of climatic change may have played a role in the extirpation of species, but habitat fragmentation resulting from human activities is believed to have been largely responsible.

Europeans colonized Iowa during the mid-nineteenth century. The forest patches were cut down and the prairie was plowed. Ground-beetle species declined and dung beetles increased, as they had after the Neolithic forest clearance in the British Isles. The most spectacular change, however, was in the aquatic environment. This is documented by fossiliferous sediments deposited in the floodplain of Roberts Creek. Before colonization, the stream was rich in elmid beetles, indicating clear running water. After colonization, elmids essentially disappeared, indicating turbid water, presumably caused by increased runoff (Baker et al., 1993). The perturbations of the beetle fauna by farming practices in mid-Holocene Britain and those in nineteenth-century North America have striking parallels. In both places, the changes to the beetle fauna are as great as those resulting from climatic change.

THE RESPONSE OF BEETLES TO GLOBAL WARMING

Meteorological records indicate that the earth has warmed during the last century. The present warming trend started in 1970, with several of the warmest years being recorded during the last decade of the century. The oceans have also warmed. A temperature increase during the last 40 years has been detected in the Pacific, Indian, and Atlantic Oceans to a depth of 3000 m, with the largest amount of warming, 0.3°C, in the upper 300 m (Levitus et al., 2000).

Effects attributed to global warming have also been reported from both polar regions. On the Antarctic Peninsula, a temperature increase of 2.5°C during the last 40 years is associated with the rapid disintegration of the northern Larsen and Wilkins Ice Shelves and the formation of huge icebergs (Drinkwater, 1999). On March 23, 2000, the National Ice Center reported the detection of what may be the largest iceberg (B-15) ever detected. The iceberg, recently calved from the Ross Ice

Shelf, is 292 km long and 37 km wide. In the Arctic region, warming has resulted in the reductio
of sea ice, and that in turn is linked to a population decline in polar bears (Stirling, 2000).

Mann et al. (1998) analyzed high-resolution proxy climate records for the last six centurie
Their data clearly show an increase in temperature during the last century. They reported that var
ations in solar radiation and volcanic aerosols were the main causes of climatic variation before th
last century. During the last century, however, they attributed at least part of the climatic forcing
the buildup of greenhouse gases.

Beetles have survived times of past climatic change because of their small size, population d
namics, and ability to disperse. In the future, beetles will undoubtedly respond to climate change b
dispersal, but it is less certain whether this will ensure their survival. The geographic ranges of som
species will expand and others will contract. Undoubtedly, beetles will adapt genetically to diffe
ences in diurnal and seasonal temperatures and to day length. The relationship between genet
variation and morphological variation at the subspecific level is poorly understood, but it seem
probable that genetic isolation on the scale of decades and centuries will not lead to speciation.

One of the surprises of the Quaternary fossil beetle record is that species extinction is not assoc
ated with climate change. On the face of it, this would appear to be good news for the future. The u
expectedly high numbers of species that were extirpated from the British Isles between the Neolith
and Medieval periods, however, is a clue that the rate of extinction may be significantly higher in th
future than it was in the past. The reason is that natural habitats have become increasingly fra
mented by human activities. Many species currently have very restricted geographic ranges and a
more vulnerable now to extinction than they were during most of the Quaternary Period.

Global warming potentially makes a bad situation even worse. For many species, dispers
will mean abandoning the security of patches of natural habitat for disturbed areas where they w
probably be more vulnerable to predators, to disease, and to being poisoned by pesticide residu
in soils and by chemicals ingested from genetically altered crops. Some individuals will make it
the next patch of natural habitat, but the analogy that humans have set up an obstacle course
dispersal is a good one (Lovejoy, 1997). The chances of dispersing beetles finding habitat in whi
they can survive have been reduced by human activities, and so extinction is more probable than
was in the past **(Figure 8)**.

Increasingly, evidence suggests that the global temperature increase of the past century is re
and that it may be caused by human activities (Mann et al.,1998). With respect to this highly co
tentious and politicized issue, Overpeck et al. (2000) framed a series of questions, namely:

- How much warming has occurred due to anthropogenic increases in atmospheric trace g
  levels?
- How much warming will occur in the future?
- How fast will the climate warm?
- What other kinds of climatic change will be associated with future warming?

Answers to these questions relating to the magnitude and speed of future climate change, a
to questions relating to regional effects, are needed to refine predictions concerning insects a
health, pest management, and conservation.

Organisms will respond to global warming whether it is of natural or anthropogenic origin.
them, the scientific questions of cause are irrelevant. For human beings it is a different issue, and
much a philosophical as a scientific question. If, at some point, we determine beyond reasonal
doubt that our activities are contributing to global warming, then we will have placed ourselv
firmly on the horns of a moral dilemma. To continue on our current course will place us in the u
tenable situation of knowingly being the agents of extinction.

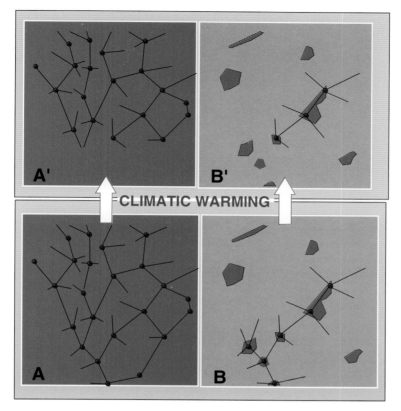

Landscape before human colonization

Landscape after human colonization

Figure 8 Schematic diagram representing a comparison in the effect of climatic warming on a beetle species before (A–A') and after (B–B') landscape modification by human activities. On the precolonization landscape, some extirpation of populations occurs as the climate warms (A'). On the postcolonization landscape (B), the geographic distribution of the same species has been reduced because of fragmentation of the landscape. As the climate warms, the species not only is extirpated from the part of its range that it would have been under natural conditions, but it is also extirpated from other parts of its range because the distances between habitat patches are beyond its dispersal capabilities (B'). The reduction in populations (from 19 to three populations) makes the species more vulnerable to extinction.

REFERENCES CITED

Ashworth, A. C., 1996, The response of arctic Carabidae (Coleoptera) to climate change based on the fossil record of the Quaternary Period: Annales Zoologici Fennici, v. 33, p. 125–131.

Ashworth A. C., 1997, The response of beetles to Quaternary climatic changes, *in* B. Huntley, W. Cramer, A. V. Morgan, H. C. Prentice, and J. R. M. Allen, eds., Past and future rapid environmental changes: The spatial and evolutionary response of terrestrial biota: NATO ASI Series1: Global Environmental Change, Berlin, Springer, v. 47, p. 119–127.

Ashworth, A. C., and J. W. Hoganson, 1993, The magnitude and rapidity of the climate change marking the end of the Pleistocene in the midlatitudes of South America: Palaeogeography, palaeoclimatology, palaeoecology, v. 101, p. 263–270.

Ashworth, A. C., and J. K. Willenbring, 1998, Fossil beetles and climate change at the Sixmile Creek site, Ithaca, New York: American Paleontologist, v. 6, p. 2–3.

Ashworth, A. C., D. M. Harwood, P. N. Webb, and M. G. C. Mabin, 1997, A weevil from the heart of Antarctica, in A. C. Ashworth, P. C. Buckland and J. P. Sadler, eds., Studies in Quaternary Entomology—an inordinate fondness for insects: Quaternary Proceedings, v. 5, p. 15–22.

Ashworth, A. C., D. P. Schwert, W. A. Watts, and H. E. Wright, Jr., 1981, Plant and insect fossils at Norwood in southcentral Minnesota: A record of Late Glacial succession: Quaternary Research, v. 16, p. 66–79.

Baker, R. G., D. P. Schwert, E. A Bettis, III, and C. A. Chumbley, 1993, Impact of EuroAmerican settlement on a riparian landscape in northeast Iowa, midwestern USA: An integrated approach based on historical evidence, floodplain sediments, fossil pollen, plant macrofossils and insects: The Holocene, v. 3, p. 314–323.

Bennett, K. D., 1997, Evolution and Ecology—the pace of life: Cambridge, U.K., Cambridge University Press, 241 p.

Böcher, J., 1995, Palaeoentomology of the Kap København Formation, a Plio-Pleistocene sequence in Peary Land, North Greenland: Meddelelser om Grønland: Geoscience, v. 33, p. 1–82.

Buckland, P. C., and M. H. Dinnin, 1993, Holocene woodlands, the insect evidence, in K. J. Kirby and C. M. Drake, eds., Dead wood matters: The ecology and conservation of saproxylic invertebrates in Britain: English Nature Science Series No. 7, Peterborough, U.K., English Nature, p. 6–20.

Buckland, P. C., G. R. Coope, and J. P. Sadler, 1997 (latest revision), QBIB, <http://www.umu.se/envarchlab/BUGS/QBIB/QBIBFRAM.HTM>, accessed March 24, 2000.

Buckland, P. I., Y. Don, and P. C. Buckland, 1997. BUGS Database Management SystemVersion 4. North Atlantic Biocultural Organization, Hunter College, New York.

Butterfield, J., 1996, Carabid life-cycle strategies and climate change: A study on an altitude transect: Ecological Entomology, v. 21, p. 9–16.

Butterfield, J. E. L., and J. C. Coulson, 1997, Terrestrial invertebrates and climate change: physiological and life cycle adaptations, in B. Huntley, W. Cramer, A. V. Morgan, H. C. Prentice, and J. R. M. Allen, eds., Past and future rapid environmental changes: The spatial and evolutionary response of terrestrial biota: NATO ASI Series1: Global Environmental Change, Berlin, Springer, v. 47, p. 401–412.

Cong, S., and A. C. Ashworth, 1996, Palaeoenvironmental interpretation of middle and late Wisconsinan fossil coleopteran assemblages from western Olympic Peninsula, Washington: Journal of Quaternary Science, v. 11, p. 345–356.

Cong, S., A. C. Ashworth, D. P. Schwert, and S. Totten, 1996, Fossil beetle evidence for a short warm interval near 40,000 yr B.P. at Titusville, Pennsylvania: Quaternary Research, v. 45, p. 216–225.

Coope, G. R., 1978, Constancy of insect species versus inconstancy of Quaternary environments, in L. A. Mound and N. Waloff, eds., Diversity of Insect Faunas (Symposia of the Royal Entomological Society of London, 9), Oxford, Blackwell, p. 76–187.

Crowson, R. A., 1981, The biology of the Coleoptera: London, Academic Press, 802 p.

Dinnin, M. H., and J. P. Sadler, 1999, 10,000 years of change: the Holocene entomofauna of the British Isles, in K. J. Edwards and J. P. Sadler, eds., Holocene environments of prehistoric Britain: Quaternary Proceedings No. 7, Chichester, U.K., John Wiley and Sons Ltd., p. 545–562.

Drinkwater, M. R., 1999, Accelerated summer sea ice melt and retreat of Antarctic ice. shelves under regional warming trend <http://polar.jpl.nasa.gov/>, accessed March 23, 2000.

Eldredge, N., and S. J. Gould, 1972, Punctuated equilibria: An alternative to phyletic gradualism, in T. J. M. Schopf, ed., Models in Paleobiology: San Francisco, Freeman, p. 82–115.

Elias, S. A., 1991, Insects and climate change: Fossil evidence from the Rocky Mountains: Bioscience, v. 41, p. 552–559.

Elias, S. A., 1992, Late Quaternary zoogeography of the Chihuahuan Desert insect fauna, based on fossil records from packrat middens: Journal of Biogeography, v. 19, p. 285–297.

Elias, S. A., 1994, Quaternary insects and their environments: Washington, D. C., Smithsonian Institution Press, 284 p.

Elias, S. A., and T. R. Van Devender, 1990, Fossil insect evidence for late Quaternary climatic change in the Big Bend region, Chihuahuan Desert, Texas: Quaternary Research, v. 34, p. 249–261.

Erwin, T. L., 1997, Biodiversity at its utmost: Tropical forest beetles, in M. L. Reaka-Kudla, D. E. Wilson, and E. O. Wilson, eds., Biodiversity II: Washington, D.C., Joseph Henry Press, p. 27–40.

Fyles, J. G., L. V. Hills, J. V. Matthews Jr., R. Barendregt, J. Baker, E. Irving, and H. Jetté, 1994, Ballast Brook and Beaufort Formations (Late Tertiary) on northern Banks Island, Arctic Canada: Quaternary International, v. 22/23, p. 141–171.

Garry, C. E., R. W. Baker, J. P. Gilbertson, and J. K. Huber, 1997, Fossil insects and pollen from late Illinoian sediments in Midcontinental North America, in A. C. Ashworth, P. C. Buckland, and J. P. Sadler, eds., Studies in Quaternary Entomology—an inordinate fondness for insects: Quaternary Proceedings, v. 5, p. 113–124.

Gould, S. J., 1995, A special fondness for beetles, in Dinosaurs in a haystack: New York, Harmony Books, p. 377–387.

Hammond, P. E., 1992, Species inventory, in B. Groombridge, ed., Global biodiversity, status of the earth's living resources: London, Chapman and Hall, p. 17–39.

Hoganson, J. W., and A. C. Ashworth, 1992, Fossil beetle evidence for climatic change 18,000–10,000 Years B.P. in South Central Chile: Quaternary Research, v. 37, p. 101–116.

Hoganson, J. W., M. Gunderson, and A. C. Ashworth, 1989, Fossil beetle analysis, in T. D. Dillehay, ed., Monte Verde, A Late Pleistocene Settlement in Chile, 1. Palaeoenvironment and Site Context: Washington, D.C., Smithsonian Institute Press, p. 215–226.

Hyman, P. S., and M. S. Parsons, 1992, A review of the scarce and threatened Coleoptera of Great Britain, Part 1: U.K. Nature Conservation No. 3, Peterborough, U.K., Nature Conservation Committee, 484 p.

Hyman, P. S., and M. S. Parsons, 1994, A review of the scarce and threatened Coleoptera of Great Britain, Part 2: U.K. Nature Conservation No. 12, Peterborough, U.K., Nature Conservation Committee, 248 p.

Levitus, S. J., I. Antonov., T. P. Boyer, and C. Stephens, 2000, Warming of the world ocean: Science, v. 287, p. 2225–2229.

Lovejoy, T. E., 1997, Biodiversity. What is it?, in M. L. Reaka-Kudla, D. E. Wilson, and E. O. Wilson, eds., Biodiversity II: Washington, D.C., Joseph Henry Press, p. 7–14.

Mann, M. E., R. S. Bradley, and M. K. Hughes, 1998, Globalscale Temperature Patterns and Climate Forcing over the Past Six Centuries: Nature, v. 392, p. 779–787.

Martin, P. S., and R. G. Klein, 1984, Quaternary extinctions: Tucson, the University of Arizona Press, 892 p.

Matthews, J. V., Jr., 1974, Quaternary Environments at Cape Deceit (Seward Peninsula, Alaska): Evolution of a Tundra Ecosystem: Bulletin of the Geological Society America, v. 85, p. 1353–1384.

Matthews, J. V. Jr., 1977, Tertiary Coleoptera fossils from the North American Arctic: Coleopterists Bulletin, v. 31, p. 297–308.

Miller, S. E., 1997, Late Quaternary insects of Rancho La Brea, California, U.S.A: in A. C. Ashworth, P. C. Buckland, and J. P. Sadler, eds., Studies in Quaternary Entomology—an inordinate fondness for insects: Quaternary Proceedings, v. 5, p. 185–191.

Miller, S. E., R. D. Gordon, and H. F. Howden, 1981, Reevaluation of Pleistocene scarab beetles from Rancho La Brea, California (Coleoptera, Scarabaeidae): Proceedings of the Entomological Society of Washington, v. 83, p. 625–630.

Morgan, A. V., 1987, Late Wisconsin and early Holocene paleoenvironments of eastcentral North America based on assemblages of fossil Coleoptera, in W. F. Ruddiman and H. E. Wright Jr., eds., North America and adjacent oceans during the last deglaciation: The Geology of North America K3. Boulder, Geological Society of America, p. 353–370.

National Ice Center, 2000, Icebergs B15, B16 Calve off Ross Ice Shelf, <http://www.natice.noaa.gov/>, accessed March 23, 2000.

d'Orchymont, A., 1927, Über zwei neue diluviale Helophoren-Arten: Sitzungsberichte und Abhandlungen der Naturwissenschaftlichen Gesellschaft, Isis, v.103, p. 100–104.

Osborne, P. J., 1997, Insects, man, and climate in the British Holocene, *in* A. C. Ashworth, P. C. Buckland, and J. P. Sadler, eds., Studies in Quaternary Entomology—an Inordinate Fondness for Insects: Quaternary Proceedings, v. 5, p. 193–198.

Overpeck, J. T., C. Woodhouse, R. S. Webb, and D. Anderson, 2000, A paleo perspective on global warming, <http://www.ngdc.noaa.gov/paleo/globalwarming/home.html>, accessed March 23, 2000.

Patterson, D. T., J. K. Westbrook, R. J. V. Joyce, P. D. Lindgren, and J. Rogasik, 1999, Weeds, insects, and diseases: Climatic Change, v. 43, p. 711–727.

Reiss, R. A., A. C. Ashworth, and D. P. Schwert, 1999, Molecular genetic divergence for the post-Pleistocene divergence of populations of the arctic-alpine ground beetle *Amara alpina* (Paykull) (Coleoptera:Carabidae): Journal of Biogeography, v. 26, p. 785–794.

Schwert, D. P, 1992, Faunal transitions in response to an ice age: The late Wisconsinan record of Coleoptera in the northcentral United States: Coleopterists Bulletin, v. 46, p. 68–94.

Schwert, D. P., and A. C. Ashworth, 1988, Late Quaternary history of the northern beetle fauna of North America: A synthesis of fossil and distributional evidence: Memoirs of the Entomological Society of Canada, v. 144, p. 93–107.

Stein, B. A., L. S. Kutner, and J. S. Adams, eds., 2000, Precious heritage—the status of biodiversity in the United States: Oxford, Oxford University Press, 416 p.

Sterner, R., 1996, Map of Europe, <http://fermi.jhuapl.edu/temp/el_big.jpg>, accessed March 23, 2000.

Stirling, I., 2000, Running out of ice? Polar bears need plenty of it: Natural History, v. 109, p. 92.

Stork, N. E., 1997, Measuring global biodiversity and its decline, *in* M. L. Reaka-Kudla, D. E. Wilson, and E. O. Wilson, eds., Biodiversity II, Washington, D. C., Joseph Henry Press, p. 41–68.

Stuiver, M., P. J. Reimer, E. Bard, J. W. Beck, G. S. Burr, K. A. Hughen, B. Kromer, F. G. McCormac, J. v.d. Plicht and M. Spurk, 1998, INTCAL98 radiocarbon age correlation, 24,000–0 cal B.P.: Radiocarbon, v. 40, p. 1041–1083.

Vogler, A. P., and R. DeSalle, 1993, Phylogeographic patterns in coastal North American tiger beetles, *Cicindela dorsalis*, inferred from mitochondrial DNA sequences: Evolution, v. 47, p. 1192–1202.

Whitehouse, N. J., 1997, Insect faunas associated with *Pinus sylvestris* L. from the mid-Holocene of the Humberhead Levels, Yorkshire, U.K., *in* A. C. Ashworth, P. C. Buckland, and J. P. Sadler, eds., Studies in Quaternary Entomology—an inordinate fondness for insects: Quaternary Proceedings, v. 5, p. 293–303.

Yalden, D., 1999, The history of British mammals: London, T & A D Poyser Ltd., 305 p.

Kürschner, W. M., F. Wagner, D. L. Dilcher, and H. Visscher, Using fossil leaves for the reconstruction of Cenozoic paleoatmospheric CO$_2$ concentrations, 2001, *in* L. C. Gerhard, W. E. Harrison, and B. M. Hanson, eds., Geological perspectives of global climate change, p. 169–189.

9 | Using Fossil Leaves for the Reconstruction of Cenozoic Paleoatmospheric CO$_2$ Concentrations

Wolfram M. Kürschner
Friederike Wagner
Laboratory of Palaeobotany and Palynology
Utrecht University
Utrecht, The Netherlands

David L. Dilcher
Florida Museum of Natural History
University of Florida
Gainesville, Florida, U.S.A.

Henk Visscher
Laboratory of Palaeobotany and Palynology
Utrecht University
Utrecht, The Netherlands

ABSTRACT

In the present contribution, we address the relationship between climate and atmospheric carbon-dioxide (CO$_2$) concentration on different timescales, from long-term trends through the Cenozoic to short-term variations in the recent past. The inverse relationship between stomatal frequency of angiosperm leaves and the CO$_2$ concentration of the ambient air is used as a robust method for quantifying paleoatmospheric CO$_2$ levels. Short-term, century-scaled CO$_2$ fluctuations are reflected in the stomatal frequency pattern of early Holocene birch leaves. Changes in paleoatmospheric CO$_2$

correlate with major environmental and climatic changes, indicated in the terrestrial palynological record and by $\delta^{18}O$ fluctuations in polar ice. Further evidence for significant perturbations in the global carbon cycle during the early Holocene is revealed by concomitant changes in atmospheric radiocarbon (^{14}C) content. Warm climatic phases during the Cenozoic represent a particularly challenging test of our understanding of stomatal frequency response to past CO_2 concentrations. The principal question is whether an enhanced greenhouse effect was responsible for these periods of increased global temperature. The data available so far indicate that during the late Neogene, when the temperature was significantly increased for the last time in the geological history, the paleoatmospheric CO_2 concentration was close to the present level of about 360 parts per million volume (ppmv). During the peak warmth of the early middle Eocene, however, paleoatmospheric CO_2 concentration was significantly elevated, to about 500 ppmv.

INTRODUCTION

The concentration of the infrared active trace gas CO_2 is rapidly increasing in the atmosphere due to human activities. The rise is expected to raise the earth's mean surface temperature by 2–5°C in this century. To help to understand future climate change, earth scientists are undertaking major efforts to reveal (1) to what extent atmospheric CO_2 concentrations have varied naturally in the past, and (2) what the relationship is between CO_2 concentration and temperature. Records of CO_2 concentration covering a longer period than those that are available from instrumental monitoring are essential for assessment of anthropogenic impact versus the manifold natural causes for climate change. The instrumental record of atmospheric CO_2 covers only the last six decades. Consequently, a number of methods have been developed to reconstruct the history of atmospheric CO_2 in the geological past, each having its own pros and cons. In the Quaternary, analysis of air trapped in polar ice reveals direct evidence for the sensitivity of atmospheric CO_2 concentration for large-scale glacial–interglacial climatic cycles over the last 420,000 years (Petit et al., 1999). Paleoatmospheric CO_2 concentrations cannot be directly measured with the same precision before this time, but are inferred from proxy signals derived mainly from carbon isotope analysis of paleosols (e.g., Cerling 1991, 1992; Ekart et al., 2000), peat (White et al., 1994), or marine organic matter (e.g., Popp et al., 1989; Freeman and Hayes, 1992; Pagani et al., 1999a, b); from boron isotope analysis of planktonic foraminifera (Pearson and Palmer, 1999); and by modeling the long-term global carbon budget (Berner, 1991, 1994).

In addition to isotope analysis, stomatal frequency analysis on fossil leaf remains has emerged as a powerful biological/plant physiological proxy. An inverse relationship between atmospheric CO_2 and stomatal frequency on leaves of woody plants was initially pointed out by Woodward (1987), who analyzed herbarium leaf material collected over the past 200 years. Concomitant growth experiments with tree seedlings in controlled environmental chambers under preindustrial CO_2 levels confirmed the observations on the historical leaf material (Woodward and Bazzaz 1988). A near-annual stomatal frequency analysis of a 40-year record of leaves from a solitary growing birch tree has convincingly confirmed that deciduous trees can be equipped with a plastic phenotype, capable of a lifetime adjustment of stomatal frequency to anthropogenic increase of CO_2 (Wagner et al., 1996). Likewise, natural archives such as peat and lake deposits containing a prolonged record of fossil leaves are highly promising and currently under investigation to extend the record of atmospheric CO_2 with near-instrumental accuracy over the last millennium.

In the past decade, stomatal frequency data from fossil leaves have been used mainly in two ways. In the first approach, studies that compare long-term stomatal frequency changes in fossil leaves of extant tree species with polar ice-core records have revealed a significant correlation with glacial-interglacial CO_2 dynamics (e.g., Beerling 1993; van de Water et al., 1994). Similarly, the carbon model of Berner (1994) has been used for a comparison between long-term trends in stomata frequency data on extinct fossil plant taxa and changes in paleoatmospheric CO_2 (McElwain and Chaloner, 1995; McElwain 1998; McElwain et al., 1999a). In the second approach, stomatal frequency data from extant taxa are used for independent quantitative reconstructions of paleoatmospheric CO_2 content (van der Burgh et al., 1993; Beerling et al., 1995; Rundgren and Beerling, 1999; Wagner et al., 1999a, b). This is possible by comparison of the fossil data with a modern calibration set. Calibration curves are inferred from historical herbarium material, a series of growth experiments at different CO_2 levels (typically, 280–700 ppmv), natural CO_2 springs, or field studies using altitudinal transects.

Despite the excellent results achieved by using ice-core records, this foremost tool used for reconstruction of paleoatmospheric CO_2 levels is restricted to a rather short interval of geological history, i.e., the late Quaternary. Secular variations in paleoatmospheric CO_2 through the Phanerozoic have been successfully tracked by means of carbon isotope analysis of paleosols, but this technique is limited by a low time resolution and a decreased accuracy at CO_2 concentrations below 1000 ppmv. In contrast to the two former approaches, stomatal frequency analysis holds the potential to unveil the history of atmospheric CO_2 from long-term trends through major parts of the Phanerozoic to higher-order fluctuations of the recent past. The objectives of our contribution are a review of the independent paleobotanical evidence from extant and extinct angiosperms for Cenozoic paleoatmospheric CO_2 variations on different timescales. Emphasis is given to (1) a century-scaled record based on data from the early Holocene, and (2) a documentation of long-term trends of paleoatmospheric CO_2 on a timescale $>10^6$ years throughout the late Neogene. In the latter part of the paper, we present a new approach, in which we compare stomatal counts on Eocene leaf remains of an extinct species of the genus *Gordonia* (L.) Ellis with the stomatal frequency response of the nearest living relative.

THE RELATIONSHIP BETWEEN STOMATAL FREQUENCY AND ATMOSPHERIC CO_2

Carbon dioxide not only plays an important role in the earth's climate as an atmospheric constituent absorbing longwave radiation, but also is the substrate for plant photosynthesis. Physiological responses of land plants to changes in the supply of basic resources (solar radiation, CO_2, water, nutrients) require morphological, anatomical, and/or biochemical adjustments. If these adjustments can be recognized in fossil plant remains, their temporal analysis may provide a long-term record of the natural response of plants to past environmental changes. Thus, after careful calibration to a series of known atmospheric CO_2 levels, the past physiological response of land plants can be used as a sensitive measure of paleoatmospheric CO_2 levels. This paleobotanical proxy is based on the inverse relationship between the stomatal frequency of leaves of woody plants and the CO_2 concentration of the ambient air. Stomata consist of an opening surrounded by a pair of cells, the guard cells, located in the epidermal tissue of leaves of most higher land plants. Their primary purpose is the regulation of gas exchange between the ambient air and the intercellular air spaces within the leaf. Stomatal frequency is conventionally expressed in terms of stomatal density [n/mm^2] and/or stomatal index; the stomatal index (SI) [%]= [stomatal density/ (stomatal density + epidermal cell density)] × 100 (Salisbury, 1927). In the past decade, the inverse relationship between stomatal densities and/or indices in leaves of woody plants and atmospheric

CO_2 concentration has been demonstrated repeatedly by analysis of herbarium material collected over the past 200 years and subfossil leaves derived from natural archives and by growing seedlings under preindustrial CO_2 levels (e.g., Woodward, 1987; Peñuelas and Matamala, 1990; Kürschner et al., 1996; Wagner et al., 1996, 1999a). It is evident from growth experiments that photosynthetic rates in C_3 plants rapidly decline under reduced CO_2 concentration (< 300 ppmv) (e.g., Allen et al., 1991; Baker and Allen, 1993; Tissue et al., 1995). Therefore, the stomatal frequency response represents phenotypical acclimation that ensures C_3 plants a high maximum stomatal conductance and assimilation rate at unfavorably low atmospheric CO_2 (Woodward and Bazzaz, 1988; Wagner et al., 1996; Kürschner et al., 1998). Evidence for stomatal frequency response to elevated CO_2 above the present level is more equivocal. Growth experiments at elevated CO_2 show only a small reduction (e.g., Woodward and Bazzaz, 1988; Berryman et al., 1994; Rey and Jarvis, 1997; Kürschner et al., 1998) or even no change at all (e.g., Estiarte et al., 1994). These studies suggest that the phenotypic plasticity of leaf morphological/anatomical adjustments is species specific and limited to a certain CO_2 regime. The nonlinear nature of the stomatal frequency response reduces the sensivity of this proxy to elevated CO_2 concentrations in the geological past. In short-term growth experiments, the stomatal frequency response ceases at elevated CO_2 because only the phenotypical acclimation of one individual plant is triggered. However, a remarkably reduced number of stomata (Fernández et al., 1998) or a reduced size in the guard cells of the stomata (Miglietta and Raschi, 1993) have been found in plants growing under extremely high CO_2 near natural CO_2 springs. Either type of anatomical change results in a reduction of the stomatal conductance to CO_2 diffusion. These observations indicate that plant populations not only show short-term stomatal frequency acclimation at suboptimal CO_2 conditions but also an adaptation in response to supraoptimal CO_2 supply that may, however, occur only over many generations. The latter adaptation may be explained as a result of a shortage in other components of the photosynthetic carbon-reduction cycle (for example, Rubisco limitation of photosynthesis) at elevated atmospheric CO_2 concentration (e.g., Lambers et al., 1998).

Uncertainties with respect to the interpretation of stomatal frequency data in terms of paleoatmospheric CO_2 levels are notably related to two aspects: (1) the intrinsic variation of stomatal parameters and (2) the effects of growth parameters other than atmospheric CO_2 (see also reviews by Beerling, 1999; Poole and Kürschner, 1999). With regard to the first problem, it should be realized that intrinsic variation of stomatal frequency within and between leaves of an individual tree species is generally large (e.g., Poole et al., 1996; Kürschner, 1997). However, leaf areas with extremely low or extremely high frequencies are relatively rare and do not seriously hamper the replication of significant unidirectional trends of mean stomatal densities or indices for specific leaf areas, preferably of sun morphotypes, that are selected for comparative analysis (Wagner et al., 1996; Kürschner, 1997). Moreover, measurements of extreme values can be avoided by concentrating sampling effort on the midlamina region of the leaves. Smaller leaf fragments may be used if the variability has been assessed in more complete specimens. If marginal pieces are used, usually the apical or basal leaf region is excluded from the sample set. Growth parameters other than CO_2, notably light, humidity, and temperature, can influence stomatal density or stomatal index. Light-induced leaf anatomical alterations found in sun and shade morphotypes are by far the most prominent and involve significant variation in stomatal density. It has been shown in numerous studies that the stomatal index is relatively stable under varying irradiance and water supply because changes in the rate of leaf expansion and final leaf size do not affect the ratio between stomata and the epidermal cells (e.g., Salisbury, 1927; Poole et al., 1996; Kürschner, 1997; Wagner, 1998).

SHORT-TERM FLUCTUATIONS IN ATMOSPHERIC CO_2 (<1000 YEARS)

Current understanding of the role of the concentration of atmospheric CO_2 on Quaternary climatic fluctuations has become largely dependent on high-latitude glaciochemical records. A consistently positive correlation between overall trends in CO_2 level and inferred temperature during the last 420,000 years has confirmed that variation in greenhouse gas concentrations is a principal factor in long-term glacial-interglacial climate evolution (Petit et al., 1999). Furthermore, CO_2 data seem to correlate with millennium-scaled temperature changes during glacial periods (Stauffer et al., 1998). A correlation of short-term paleoatmospheric CO_2 fluctuation with climatic variation during the interglacials and interstadials, such as the Holocene, is less obvious, although a few studies suggest that the "preindustrial level" of 280 ppmv has not been stable over the last 1000 years. Century to multidecadal fluctuations in the order of 10 ppmv around the Little Ice Age (LIA) have been measured in Antarctic ice cores (Etheridge et al., 1996; Indermühle et al., 1999).

There is, however, ever-growing evidence from marine (Bond et al., 1997; Barber et al., 1999; Bianchi and McCave, 1999) and terrestrial (Björk et al., 1996; von Grafenstein et al., 1999; Willemse and Törnqvist, 1999; Verschuren et al., 2000) proxy records and modeling experiments (Campbell, et al. 1998) suggesting that the Holocene has been punctuated regularly by century-scaled climatic deteriorations, in which the LIA is considered to be the youngest event (Bond et al., 1997; Broecker, 2000). Up to now, evidence for an involvement of atmospheric CO_2 in well-defined events other than the LIA has not been revealed in ice-core records.

Because trapping of air is not an instantaneous process, each ice-core air sample represents a mean value over a period of time ranging from a few decades (30–50 years at Byrd; Neftel et al., 1988; Indermühle et al., 1999) to more than a hundred years (140 years at Taylor Dome, Indermühle et al., 1999). Since the rates of snow accumulation are low, most of the Holocene ice-core records from Antarctica do not have high temporal resolution. Consequently, rapid shifts in the atmospheric CO_2 concentration may become blurred as a result of this smoothing. In Greenland ice, records from Dye 3 and Greeland Ice Sheet Project II (GISP2) cores have repeatedly indicated short-term CO_2 fluctuations, including periods with CO_2 concentrations well above the preindustrial level of 280 ppmv, that may be associated with the onset of the Holocene (Oeschger et al., 1984; Stauffer et al., 1985; Smith et al., 1997). However, these high early Holocene CO_2 concentrations are generally considered to be the result of postdepositional alteration (Anklin et al., 1995).

Alternatively, early Holocene atmospheric CO_2 concentration has been reconstructed by stomatal frequency analysis of fossil birch leaves buried in peat (Wagner et al., 1999a,b). The leaf material studied originated from a peat section that was temporarily exposed at the Borchert archaeological site near Denekamp in the northeastern part of the Netherlands (52.23°N, 7.00°E). A comprehensive paleoecological record based on palynomorphs and plant macrofossils of the section and its paleoclimatic interpretation has been documented in van Geel et al. (1981). Six available ^{14}C dates indicate that the interval studied covers approximately the first 700 years of the Holocene **(Figure 1)**. With 16 horizons sampled, the resolution of 40–50 years is high enough to detect short-term CO_2 variations which may occur in conjunction with known century-scaled climatic events. The lithology of the section studied and average stomatal index values per horizon are summarized in **Figure 1**. The peat section studied by Wagner et al. (1999a) contained *Betula pendula* L. and *Betula pubescens* L. leaves. Because both species show an essentially similar stomatal frequency response (Kürschner et al., 1997; Wagner et al., 1999a; Wagner et al., 2000), the mixed leaf assemblage was treated as one category. The rate of historical CO_2 responsiveness of tree birches **(Figure 2)** was used to infer a paleoatmospheric CO_2 curve from the early Holocene birch leaf record. The oldest

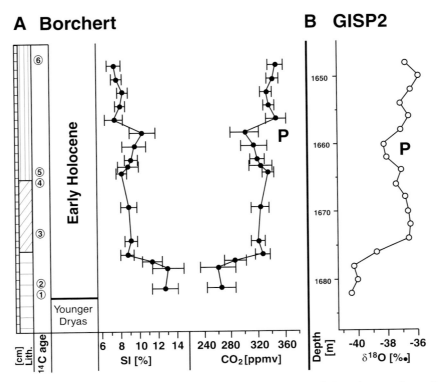

Figure 1 (A) Mean stomatal index values for *Betula pendula* and *B. pubescens* from the early Holocene part of the De Borchert section and reconstructed paleoatmospheric CO_2 levels and reconstructed paleoatmospheric CO_2 concentrations (after Wagner et al., 1999a). The scale of the section is in centimeters. Three lithological (Lith.) units can be recognized: a basal gyttja (=), succeeded by *Drepanocladus* peat (/ /), which is subsequently overlain by *Sphagnum* peat (| |). Six conventional ^{14}C dates (in years before present) are available (indicated by circled numbers): (1) 10,070±90; (2) 9,930±45; (3) 9,685±90; (4) 9,770±90; (5) 9,730±50; (6) 9,380±80. Quantification of atmospheric CO_2 concentrations according to training set in **Figure 2**. "P" indicates the reconstructed position of the Preboreal Oscillation. (B) $\delta^{18}O$ profile for the Younger Dryas–Holocene transition in the Greenland GISP2 ice core after Grootes et al. (1993). "P" denotes the $\delta^{18}O$-inferred cooling of the Preboreal Oscillation, starting at about 11,300 calendar years before present.

leaf remains in the section indicate the lowest CO_2 concentrations, with values of 265 ppmv, followed by a drastic rise to 335 ppmv in the Preboreal. After this first maximum, CO_2 concentrations continuously decline to a minimum of 300 ppmv, followed by significant increase to 350 ppmv. In the uppermost part of the interval studied, the CO_2 concentrations stabilize with values between 335 and 350 ppmv.

The initial rapid decline of the SI suggests a rise in the paleoatmospheric CO_2 concentration by 65 ppmv within less than a century. This increase is contemporaneous with major environmental changes reflected in the lithology and the palynological record. The basal gyttja formation is followed by a rapid hydroseral succession of the formerly open-water site, resulting in a local vegetation dominated by the bryophyte *Drepanocladus* Schwaeg. Regional woodland expansion is reflected by the rising *Betula* pollen curve and the occurrence of *Betula* macrofossils (Wagner et al., 1999a). The

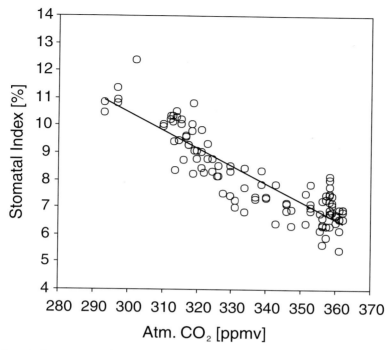

Figure 2 Historical response of mean stomatal index for *Betula pendula* and *Betula pubescens* to global atmospheric CO$_2$ increase in the period 1896 and 1995 (after Wagner et al., 1999a).

sharp rise in CO$_2$ concentration and the environmental changes correspond to the general climatic amelioration at the onset of the Holocene. The exact beginning of this CO$_2$ increase could not be determined because no leaf remains were recovered from the Younger Dryas–Holocene transition. It is intriguing that the oldest part-leaf-based CO$_2$ curve corresponds exactly with the Antarctic ice-core records (Byrd in Neftel et al., 1988; Staffelbach et al., 1991; Taylor Dome in Indermühle et al., 1999), all indicating 260 ppmv. The general timing of the rise also agrees well with the CO$_2$ record of the Antarctic Byrd ice core (Neftel et al., 1988; Staffelbach et al., 1991), where the Younger Dryas–Holocene transition is defined by a sudden CO$_2$ increase from 260 ppmv to 280 ppmv. In the leaf-based CO$_2$ record, however, the magnitude of the rise is significantly higher, resulting in CO$_2$ concentrations well above 300 ppmv. Additional evidence for this magnitude of change in atmospheric CO$_2$ at the Younger Dryas–Holocene transition has recently been provided by stomatal frequency analysis on *Dryas integrifolia* and *Picea glauca* remains from sites in New Brunswick, Atlantic Canada (McElwain et al., 1999b). Accordingly, a similar sharp CO$_2$ rise has been inferred from δ^{13}C analysis of sedges and mosses from a peat core in South America (Figge and White, 1995).

We emphasize that the paleoatmospheric CO$_2$ curve based on stomatal frequency analysis is a real time record because the phenotypic adaptation to changes in the ambient CO$_2$ environment appears immediately during the time of leaf formation (see also discussion above). Consequently, the chronology and time resolution achieved for the CO$_2$ curve directly match the stratigraphy of the leaf-bearing sediments. By contrast, although the age of the ice can be obtained by counting annual layers and a chronological record of its physical properties (δ^{18}O, δH, etc.) can be easily established, the age of the enclosed atmospheric CO_2 is more difficult to estimate and can differ significantly from the age of the ice. Despite the contradictions between leaf-based and ice-based CO_2 records, there is a clear match between Wagner's (1999a) paleoatmospheric CO_2 curve and the sharp positive shift $\delta^{18}O$ which reflects the onset of the Holocene warming in high resolution records from Greenland ice **(Figure 1)**. A few centuries after the onset of the Holocene, a $\delta^{18}O$ minimum in Greenland ice reflects a short-term cooling. A short-lived climate deterioration starting at about 300 calendar years after the termination of the Younger Dryas is also reported from numerous marine and terrestrial records throughout Europe (e.g., Behre, 1966; Karpuz and Jansen, 1992; Hansen and Knudson, 1995; Björk et al., 1997). Although exact dating is hampered by the occurrence of $\Delta^{14}C$ plateaus during the early Holocene, recent multidisciplinary studies suggest that all events reported reflect the Preboreal Oscillation, a century-scaled cooling pulse that occurred shortly after the onset of the Holocene (Björk et al., 1996; Björk et al., 1997). A relationship between the Preboreal Oscillation and atmospheric CO_2 dynamics has been supposed by these authors on the basis of $\Delta^{14}C$ records but could not be confirmed by the ice-core data.

Since our leaf-based CO_2 curve contradicts the ice-core records (Indermühle et al., 1999; Wagner et al., 1999a, b), the question arises whether there is other evidence for large perturbations in the global carbon cycle during the early Holocene and whether a mechanism could be found that may explain the observed pattern. If major rapid changes in the paleoatmospheric CO_2 concentration took place during the early Holocene, they might be reflected by secular variations in other atmospheric physical properties, such as the content of radiocarbon (^{14}C) in the air. Basically, the amount of ^{14}C is dependent on production by cosmic rays in the atmosphere, their radioactive decay, and the exchange rates between the different reservoirs of the global carbon cycle. Tree rings and plant organic matter derived from laminated lacustrine sediments have been used to establish a chronological record of the ^{14}C content in the atmosphere (Kromer and Becker, 1993; Goslar et al., 1995). Because no indication of major changes in the global carbon cycle during the Holocene has been found to date, the principal patterns in the Holocene $\Delta^{14}C$ curve have been explained by variations in the production rate controlled by the strength of the earth's magnetic field and by solar activity (e.g., Stuiver et al., 1991). The transition between the Last Glacial Maximum and the Holocene, however, is characterized by a highly variable atmospheric $\Delta^{14}C$ content, and variations of the geomagnetic field are too weak to explain the pattern observed (Goslar et al., 1999). Alternatively, Lal and Revelle (1984) suggested changes in the atmospheric CO_2 concentration as a possible mechanism to explain the $\Delta^{14}C$ record. A general rise from a CO_2 level of about 200 ppmv 17,000 years ago to a Holocene level of 280 ppmv could have produced an increase in $\Delta^{14}C$ by 25–35%. The gradual increase in CO_2 during the Late Glacial–Holocene transitional period has been rejected as a reason for the rapid variations of $\Delta^{14}C$ (Goslar et al., 1995). However, this contradiction between the pattern in atmospheric $\Delta^{14}C$ and CO_2 may be the result of the insufficient time resolution of the ice-core CO_2 record. Wagner et al. (1999a) suggested a correlation between the occurrence of ^{14}C plateaus and periods of high atmospheric CO_2 concentrations. **Figure 3** depicts for the first time the changes in atmospheric CO_2 together with changes in atmospheric $\Delta^{14}C$, both inferred from plant organic matter. In the interval studied, two increases in atmospheric CO_2 by 60 ppmv between 11,500 and 11,400 calendar years before the present (cal yr B.P.) and by 40 ppmv at 11,000 and 10,900 calendar years before the present (cal yr B.P.) correspond approximately to a significant decrease in $\Delta^{14}C$ by 40‰ (per thousand) and 30‰, respectively. Although wiggle matching of each data point with new AMS measurements is needed for an exact correlation between the two curves, the patterns observed suggest a significant correlation between atmospheric $\Delta^{14}C$ and CO_2. On the one hand, decreased $\Delta^{14}C$ values indicate a reduction in atmospheric ^{14}C, which may imply periods of more effective uptake of atmospheric ^{14}C by the oceanic reservoir and/or increased outgassing by aged oceanic ^{14}C into

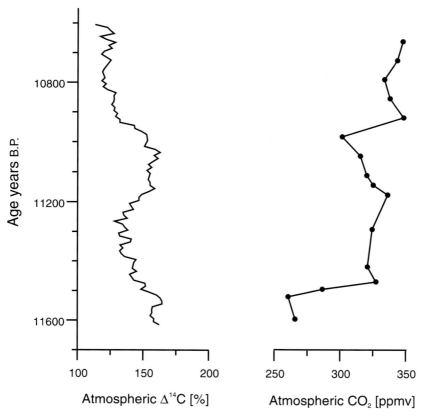

Figure 3 Reconstructed early Holocene atmospheric $\Delta^{14}C$ curve from Stuiver et al. (1998) and atmospheric CO_2 curve from Wagner et al. (1999a), both inferred from plant organic matter.

...he atmospheric reservoir. Outgassing of oceanic carbon may explain the observed periods of high ...tmospheric CO_2 during the early Holocene. On the other hand, between 11,150 and 11,050 B.P. the ...$\Delta^{14}C$ values increase again, which indicates an ocean ventilation minimum. The ventilation of the ...modern deep oceans is driven mostly through sinking of large surface-water masses in the North ...Atlantic, where North Atlantic deep water (NADW) is formed. Changes in the rate of NADW for-...mation have been suggested as an explanation for glacial-interglacial climatic cycles and the ...Younger Dryas cooling episode (see review by Hay, 1993). The weaker interglacial/Holocene cli-...matic cycles may have been driven by alterations in the mode of deepwater formation, too (Bond et ...al., 1997; Broecker, 2000). Late glacial short-term shifts in paleoatmospheric CO_2 may have been the ...result of changes in mode of thermohaline circulation (Figge and White, 1995). Reduced NADW for-...mation would have resulted in a decline of atmospheric CO_2, which is indicated by the CO_2 curve ...during the Preboreal Oscillation. The amount of ice-rafted debris (IRD) in North Atlantic sediments ...has revealed initial evidence for oscillations in the strength of the North Atlantic conveyor during ...the early Holocene (e.g., Björk et al., 1996; Bianchi and McCave, 1999).

    Although a conceptual link between the concentration of radiocarbon and CO_2 in the atmos-...phere and the mode of NADW formation appears to be plausible, the results are in contradiction with the idea of a rather constant CO_2 level of about 280 ppmv throughout the Holocene until the impact of industrial revolution (Indermühle et al., 1999). Before we can draw reliable conclusions about the natural century-scale to multidecadal variability of atmospheric CO_2 concentration, more data are needed which will confirm or reject the concept of short-term CO_2 fluctuations and periodically high CO_2 concentrations during the early Holocene. We leave the question open as to whether atmospheric CO_2 has responded to early Holocene climate changes or whether changes in atmospheric CO_2 have contributed to the climate variations ("the chicken-egg dilemma"). It should be noted, however, that these results imply that not only the magnitude of natural short-term variation in atmospheric CO_2 is higher than previously suggested, but also that all changes in atmospheric CO_2 so far documented occurred together with significant changes in climate and environment.

LONG-TERM FLUCTUATIONS IN ATMOSPHERIC CO_2 (>1×10⁶ YEARS)

The Late Neogene

The warm periods during the Cenozoic have attracted considerable attention as time intervals in the geological distant past during which elevated atmospheric CO_2 concentrations may have contributed significantly to the earth's increased surface temperature. Because of the complexity of the climate system, however, the geological record is not likely to reveal analogs of immediate future climate change (Hay et al., 1997). It nevertheless provides invaluable information about the possible range of long-term perturbations in the climate system and carbon cycle. The late Neogene was probably the last period in which the global climate was significantly warmer than today (Crowley, 1996). Elevated concentration of atmospheric CO_2 (Crowley, 1991) or an increased oceanic heat transport (Dowsett et al., 1992) have been put forward as the principal mechanisms of late Neogene warmth. To test the hypothesis of elevated CO_2 levels, the stomatal frequency pattern was studied in two extant species of the genera *Quercus* (oak) and *Betula* (birch) (van der Burgh et al., 1993; Kürschner et al., 1996; Kürschner, 1996). All fossil leaves were derived from coastal plain deposits exposed in open-cast brown coal mines and clay pits in the Lower Rhine Embayment (Germany and the Netherlands, 50°55'N, 6°50'E). Regional climate history during the Late Cenozoic has been inferred by means of palynological analysis (Zagwijn and Hager, 1987). The taxonomical relationship between the fossil leaf remains and their modern relatives has been established based on leaf morphology and cuticle anatomy (van der Burgh, 1993; Belz and Mosbrugger, 1994). Generally, the stomatal index of fossil oak- and birch-leaf remains was equal to the stomatal index of their modern equivalents during intervals when a warm-temperate to subtropical climate prevailed. During time intervals when a cooler climate prevailed, the stomatal index increased consistently in both taxa. Comparisons with training sets based on historical responsiveness of stomatal frequency of the two taxa suggest that the paleoatmospheric CO_2 concentrations were not significantly higher than the present value of ~ 360 ppmv. Mio-Pliocene CO_2 levels fluctuated between about 280 and 360 ppmv, in phase with major regional climatic variations with higher CO_2 in warmer periods and lower CO_2 levels during cooler intervals **(Figure 4)**.

On a global scale, intervals with reduced CO_2 levels so far observed match with early pulses of substantial glaciation in the Northern Hemisphere, indicated by the occurrence of ice-rafted debris in deep-sea sediments of the Northern Atlantic (Kürschner et al., 1996). This covariation with major climatic changes during the late Neogene strongly implies a causal relationship between fluctuations in paleoatmospheric CO_2 and the temperature regimes, also on a timescale of about 1×10⁶ years. With the temporal resolution achieved, however, the question remains open whether atmospheric CO_2 was responding to late Neogene climate changes or vice versa. It should be

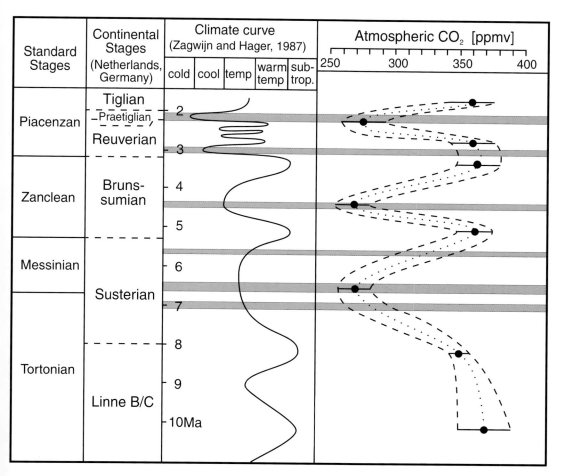

Figure 4 Late Neogene paleoatmospheric CO_2 curve compared with a regional climate curve based on palynological analysis (from Zagwijn and Hager, 1987) after Kürschner et al. (1996). The occurrences of ice-rafted debris in the North Atlantic are indicated by gray shaded intervals (Jansen and Sjøholm, 1991; Larsen et al., 1994).

noted that the late Neogene CO_2 levels based on stomatal frequency analysis are in good agreement with those reconstructed by means of geochemical proxies from both the marine realm Raymo et al., 1996; Pagani et al., 1999a) and the terrestrial realm (Cerling, 1992).

New data predict that a paleoatmospheric CO_2 level similar to and even lower than the present evel prevailed throughout the Miocene (Pagani et al., 1999b). The proposed CO_2 fluctuations do not show covariation with long-term climate change in the Miocene. Particularly, low CO_2 levels (about 80–200 ppmv) existed during the warmest period of the whole Neogene, the climatic optimum between about 17 million and 14.5 million years ago. According to their reconstruction, rising CO_2 evels accompanied global cooling and expansion of the East Antarctic ice sheet about 12.5–14 million years ago. Extension of the stomatal frequency record of fossil angiosperm leaves can provide ndependent paleobotanical evidence for CO_2 levels in this crucial time interval, which is currently highly debated. Preliminary data from studies on late-early Oligocene birch leaves indicate a slightly higher paleoatmospheric CO_2 level of about 400 ppmv (Kürschner, unpublished data).

The Early-Middle Eocene

By consensus, the Paleocene–Eocene transition, together with the early-middle Eocene, represents the warmest and wettest period of the whole Cenozoic (e.g., Barron, 1987; Wing and Greenwood, 1993; Greenwood and Wing, 1995; Wilf et al., 1998). The concept of a warmer climate is also supported by the lack of continental ice (Miller et al., 1987) and low-latitudinal temperature gradients in oceanic surface waters (Zachos et al., 1994). The most common explanations for this peak warmth are an elevated level of atmospheric CO_2 (e.g., Sloan and Barron, 1992; Sloan and Rea, 1995) or an increased oceanic heat transport (Rind and Chandler, 1991). To test the hypothesis that the warmest period of the Cenozoic was related to elevated CO_2, fossil angiosperm leaves have been analyzed for their stomatal frequency pattern. The fossil leaf material was derived from the lower part of the Claiborne Group in the Lamkin Pit (Graves County, Kentucky, 36°54'56"N, 88°41'48"W), which has been assigned to the lower-middle Eocene (Potter and Dilcher, 1980). Palynological studies have enabled the correlation with the European early Tertiary stages. According to Fairchild and Elsik (1969), the lower part of the Claiborne Group correlates approximately with the European Lower Lutetian Stage. The Gulf Coastal Plain sections in the Mississippi Embayment are the only sites outside the Bighorn Basin in the northern Rockies where Paleocene and Eocene megafloras are common. Unfortunately, most of the major megafloral localities occur in stratigraphically isolated clay lenses that are difficult to correlate with each other or with standard marine zonations. However, the early Cenozoic floras of the Gulf coastal plain are among the most diverse and most pristine preserved fossil floras known in the world (Dilcher, 1963, 1971). By contrast, the Bighorn Basin fossil flora is preserved only as impressions and is therefore not suitable for stomatal frequency analysis. Dilcher (1974) studied the taxonomic relationships of particular elements of the Gulf Plain flora by means of their morphological and anatomical features to extant genera.

Whereas extant species among late Neogene floras are still common, their number decreases rapidly with increasing age. *Gordonia lasianthus* (L.) Ellis (Theaceae) was chosen as a suitable taxon to reconstruct the concentration of CO_2 in the middle Eocene atmosphere. Fossil leaves similar to the leaves of the extant genus *Gordonia* are known to have existed in North America and Europe since the early Eocene (Kvaček and Walther, 1984a, b). Grote and Dilcher (1992) established the presence of the extant genus *Gordonia* through the preserved remains of fruits and seeds in the middle Eocene Claiborne Group. These co-occur with leaf remains named *Ternstroemites* Berry, which conform to the leaf morphology and cuticle anatomy of the extant species *Gordonia lasianthus* and, therefore, together with the evidence from the fructifications, can be related reliably to that extant genus. The taxonomic relationship is very important, because the more closely related the fossil and living taxa are to each other, the more reliable the CO_2 correlations. The stomatal frequency is basically a species-specific feature, and profound differences in stomatal density can be found between species of one particular genus (e.g., Metcalfe and Chalk, 1979). There are, however, also examples of similar stomatal frequency values and stomatal frequency responsiveness to changing atmospheric CO_2 among closely related species within one genus, as, for example, within European tree birches, *Betula pendula* and *B. pubescens*, and oaks, *Quercus petraea* and *Q. robur* (e.g., Kürschner et al., 1997; Wagner et al., 1999a; Wagner et al., 2000). It should be noted that our taxonomical approach differs from the concept of the nearest living equivalent method, which has been introduced by McElwain and Chaloner (1995). To achieve a (semi-) quantitative reconstruction of paleoatmospheric CO_2 throughout the Phanerozoic, they compared data from fossil plants with those derived from living taxa that have a comparable ecological setting and/or structural similarity. In our study, the specific identification of the fossil taxon and its living relative is constrained by the morphology of the fossil leaf and co-occurring fruits and seeds as well as by the anatomy of the leaf cuticle.

In contrast to the late Neogene birch, oak, and beech leaves, the mean stomatal index of the middle Eocene *Gordonia* leaf remains is lower by 6% in comparison with that of the modern equivalent **(Figure 5)**. Herbarium leaf material collected between A.D. 1876 and 1919 was analyzed to assess historical responsiveness to anthropogenic CO_2 increase. The stomatal index was increased by 4.5% at approximately 290 ppmv. The changes in the stomatal index were accompanied by

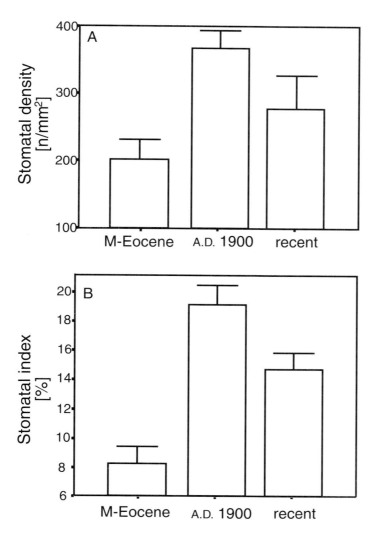

Figure 5 Mean stomatal density (A) and mean stomatal index (B) of recent (1995), historical (A.D. 1876–1919) and middle Eocene *Gordonia* leaves.

prominent changes in the stomatal density **(Figure 5)**. The marked decrease in the stomatal index and stomatal density of the middle Eocene *Gordonia* clearly signals higher paleoatmospheric CO_2 than the present value.

Our data are in good agreement with stomatal frequency measurements on Lutetian Lauraceae leaf remains derived from the London Clay flora (McElwain, 1998). The two genera *Lindera* and *Litsea* show consistently decreased stomatal density and stomatal index values in comparison with their modern equivalents **(Figure 6)**. Consequently, all three genera from geographically different regions indicate that paleoatmospheric CO_2 concentration during the Lutetian was signifi-

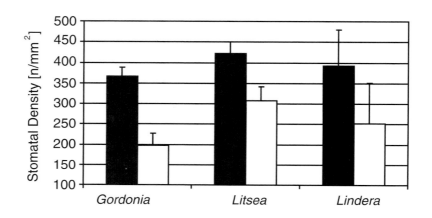

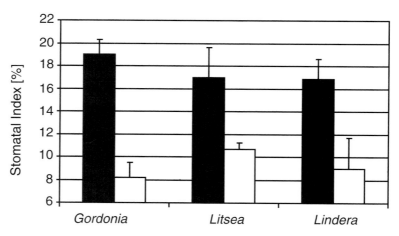

Figure 6 Mean stomatal density and stomatal index of early-middle Eocene (Lutetian) *Gordonia* leaves from the lower part of the Claiborne Group (Kentucky, U.S.A.), and two Lauraceae genera *Litsea** and *Lindera** from the London Clay (U.K.) and their modern relatives (*data from McElwain, 1998).

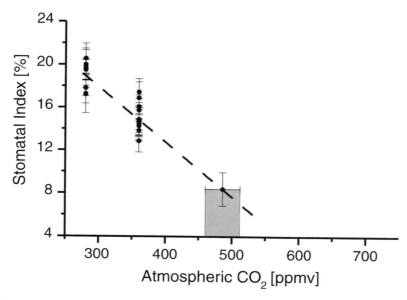

Figure 7 Estimation of middle Eocene paleoatmospheric CO_2 by means of linear extrapolation of the responsiveness of the stomatal index of *Gordonia lasianthus* to changing atmospheric CO_2.

cantly higher than the present-day value. Because of the lack of a calibration of the stomatal frequency response of *Gordonia* to elevated CO_2 by growth experiments, we have chosen a linear extrapolation of the historical stomatal responsiveness to infer a quantitative estimation of the atmospheric CO_2 in the early-middle Eocene.

As a first-order approximation, our data indicate that the paleoatmospheric CO_2 level was between about 450 and 500 ppmv **(Figure 7)**. A semiquantitative interpretation of the stomatal frequency trends observed in the leaf remains from the London Clay yielded a CO_2 level in the same order (1.4–3 times the preindustrial value CO_2; McElwain, 1998). Our reconstruction agrees well with estimates based on $\delta^{13}C$ values of pedogenic carbonate suggesting an Eocene CO_2 level of about 500 ppmv (Sinha and Scott, 1994; Ekart et al., 2000), but is more conservative than the higher CO_2 level of about 700 ppmv that was inferred from $\delta^{13}C$ values of marine organic matter (Freeman and Hayes, 1992). However, the nonlinear nature of stomatal frequency response (as discussed above) would imply that our first-order approximation represents a lower limit for the early-middle Eocene paleoatmospheric CO_2 level rather than the average CO_2 niveau in this period. With more data, a recalibration of the stomatal indices from the fossil *Gordonia* leaves with a logistic model may confirm the higher estimations of about 700 ppmv that are indicated by the marine geochemical proxy record (Freeman and Hayes, 1992).

Dickens et al. (1997) suggest the catastrophic releases of methane from large oceanic gas hydrate reservoirs, which is then oxidized to CO_2 in the atmosphere, as a possible mechanism to explain the short-term warming and the negative carbon isotope excursion near the Paleocene-Eocene transition. Yet several of these hyperthermal events may have occurred throughout the early and early-middle Eocene (Thomas et al., 2000). Stomatal frequency analysis of leaf records crossing this time interval promise a sublime opportunity to estimate the impact of the short-term peaks in CO_2 on the most prominent temperature excursions of the whole Cenozoic.

CONCLUDING REMARKS

It should be apparent from our review that stomatal frequency analysis of fossil angiosperm leaves offers an effective alternative method for tracing secular variations in atmospheric CO_2 concentration levels, at least throughout the Cenozoic. Although our work is preliminary, we are able to show evidence for a relationship between climate and atmospheric CO_2 on different timescales, from short-term century-scaled fluctuations in the early Holocene to long-term trends throughout the Tertiary. We may draw the following conclusions:

- In the early Holocene, century-scaled CO_2 fluctuations occur concomitantly with major environmental and climatic changes indicated by the terrestrial palynological record and $\delta^{18}O$ fluctuations in polar ice, respectively. Perturbations in the global carbon cycle are also inferred from rapid variations in atmospheric $\Delta^{14}C$.
- In the late Neogene, when the earth was considerably warmer than today, the paleoatmospheric CO_2 concentration did not significantly exceed the present level. Lower-order climatic fluctuations (1×10^6 years) are accompanied by paleoatmospheric CO_2 fluctuations between about 280 ppmv and 360 ppmv.
- The peak warmth in the early-middle Eocene was characterized by a significantly higher paleoatmospheric CO_2 level of at least 500 ppmv.

ACKNOWLEDGMENTS

Parts of this study were conducted during a visit of W. M. Kürschner at the Florida Museum of Natural History, University of Florida (Gainesville). The financial support by a travel (SIR) grant from the Netherlands Research Council (NWO) to WMK is greatly appreciated. We thank Kent Perkins at University of Florida Herbarium (FLAS), Gainesville for providing *Gordonia* herbarium leaf material, and Harald Walther and Lutz Kunzmann (Staatliches Museum für Mineralogie und Geologie zu Dresden) for providing *Betulaceae* leaf remains. Michael Krings, Sharon Klavins (University of Kansas, Lawrence), and three anonymous reviewers are acknowledged for their thoughtful comments on an earlier version of this manuscript. This is Netherlands Research School of Sedimentary Geology publication no. 20000602 and Florida Museum of Natural History publication no. 517.

REFERENCES CITED

Allen, L. H., Jr., Bisbal, E. C., Boote, K. J., and Jones, P. H., 1991. Soybean dry matter allocation under subambient and superambient levels of carbon dioxide. Agronomy Journal, 83:875–883.

Anklin, M., Barnola, J. M., Schwander, J., Stauffer, B., and Raynaud, D., 1995. Processes affecting the CO_2 concentrations measured in Greenland ice. Tellus, 47(B):461–470.

Baker, J. T., and Allen, L. H. Jr., 1993. Contrasting crop species response to CO_2 and temperature: Rice, soybean and citrus. Vegetatio, 104/105:239–260.

Barber, D. C., Dyke, A., Hillaire-Marcel, C., Jennings, A. E., Andrews, J. T., Kerwin, M. W., Bilodeau, G., McNeely, R., Southon, J., Morehead, M. D., and Gagnon J.-M., 1999. Forcing of the cold event of 8,200 years ago by catastrophic drainage of Laurentide lakes. Nature, 400:344–348.

Barron, E. J., 1987. Eocene equator-to-pole surface ocean temperatures: A significant climate problem? Paleoceanography, 2:729–739.

Beerling, D. J., 1993. Changes in the stomatal density of *Betula nana* leaves in response to increases in atmospheric carbon dioxide concentration since the Late-Glacial. Special Paper in Paleontology, 49:181–187.

Beerling, D. J., 1999. Stomatal density and index: Theory and application. *In* Jones, T. P., and Rowe, N. P. (eds.): Fossil plants and spores: Modern techniques. Geological Society, London: 251–256.

Beerling, D. J., Birks, H. H., and Woodward., F. I., 1995. Rapid late glacial atmospheric CO_2 changes reconstructed from the leaf stomatal density record of fossil leaves. Journal of Quaternary Science, 10:379–384.

Behre, K. E., 1966. Untersuchungen zur spätglazialen und frühpostglazialen Vegetationsgeschichte Ostfrieslands. Eiszeitalter und Gegenwart, 17:69–84.

Belz, G., and Mosbrugger, V., 1994. Systematic, palaeoecologic, and palaeoclimatic analysis of Miocene and Pliocene leaf floras from the Lower Rhine Embayment (NW-Germany). Palaeontographica (B), 233:19–156.

Berner, R. A., 1994. Geocarb II: A revised model of atmospheric CO_2 over Phanerozoic time. American Journal of Science, 294:56–91.

Berner, R. A., 1991. A model for atmospheric CO_2 over Phanerozoic time. American Journal of Science, 291:339–375.

Berryman, C. A., Eamus, D., and Duff, G. A., 1994. Stomatal responses to a range of variables in two tropical tree species grown with CO_2 enrichment. Journal of Experimental Botany, 45:539–546.

Bianchi, G. G., and McCave, N., 1999. Holocene periodicity in North Atlantic climate and deep-ocean flow south of Iceland. Nature, 397:515–517.

Björk, S., Kromer, B., Johnson, S., Bennike O., Hammerlund, D., Lemdal, G., Possnert, G., Rasmussen, T. L., Wohlfarth, B., Hammer, C. U., and Spurk, M., 1996. Synchronized terrestrial-atmospheric deglacial records around the North Atlantic. Science, 274:1155–1160.

Björk, S., Rundgren, M., Ingólfsson, Ó., and Funder, S., 1997. The Preboreal oscillation around the Nordic Seas: Terrestrial and lacrustine responses. Journal of Quaternary Science, 12(6):455–465.

Bond, G., Showers, W., Cheseby, M., Lotti, R., Almasi, P., Demenocal, P., Priore, P., Cullen, H., Hajdas, I., and Bonani, G., 1997. A pervasive millennial-scale cycle in North Atlantic Holocene and glacial climates. *Science*, 278:1257–1266.

Broecker, W. S., 2000. Was a change in thermohaline circulation responsible for the Little Ice Age? Proceedings of the National Academy of Sciences USA, 97 (4):1339–1342.

Campbell, I. D., Campbell, C., Apps, M. J., Rutter, N. W., and Bush, A. B. G., 1998. Late Holocene ~ 1500 yr climatic periodicities and their implications. Geology, 26:483–487.

Cerling, T. E., 1991. Carbon dioxide in the atmosphere: Evidence from Cenozoic and Mesozoic paleosols. American Journal of Science, 291:377–400.

Cerling, T. E., 1992. Use of carbon isotopes in paleosols as an indicator of the $P(CO_2)$ of the paleoatmosphere. Global Biogeochemical Cycles, 6 (3):307–314.

Crowley, T. J., 1991. Modeling Pliocene warmth. Quaternary Science Review, 10:275–282.

Crowley, T. J., 1996. Pliocene climates: The nature of the problem. Marine Micropaleontology, 27:3–12.

Dickens, G. R., Castillo, M. M., and Walker, J. G. C., 1997. A blast of gas in the latest Paleocene: Simulating first-order effects of massive dissociation of oceanic methane hydrate. Geology, 25:259–262.

Dilcher, D. L., 1963. Cuticular analysis of Eocene leaves of *Octoea obtusifolia*. American Journal of Botany, 50:1–8.

Dilcher, D. L., 1971. A revision of the Eocene Floras of Southeastern North America. The Palaeobotanist, 20:7–18.

Dilcher, D. L., 1974. Approaches to the identification of angiosperm leaf remains. Botanical Review, 40:1–157.

Dowsett, H. J., Cronin, T. M., Poore, R. Z., Thompson, R. S., Whatley, R. C., and Wood, A. M., 1992. Micropaleontological evidence for increased meridional heat transport in the North Atlantic Ocean during the Pliocene. Science, 258:1133–1135.

Ekart, D. D., Cerling, T. E., Montañez, I. P., and Tabor, N. J., 1999. A 400 million year carbon isotope record of pedogenic carbonate: Implications for atmospheric carbon dioxide. American Journal of Science, 299:805–827.

Estiarte, M., Peñuelas, J., Kimball, B. A., Idso, S. D., LaMorte, R. L., Pinter, P. J., Wall, G. W., and Garcia, R. L., 1994. Elevated CO_2 effects on stomatal density of wheat and sour orange trees. Journal of Experimental Botany, 45:1665–1668.

Etheridge, D. M., Steele, L., Langenfelds, R., Francey, R., Barnola, J.-M., and Morgan, V., 1996. Natural and anthropogenic changes in atmospheric CO_2 over the last 1000 years from air in Antarctic ice and firn. Journal of Geophysical Research, 101:4115–4128.

Fairchild, W. W., and Elsik, W. C., 1969. Characteristic palynomorphs of the Lower Tertiary in the Gulf Coast. Palaeontographica (B), 128:81–89.

Fernández, M. D., Pieters, A., Donoso, C., Tezara, W., Azkue, M., Herrera, C., Rengifo, E., and Herrera, A., 1998. Effects of a natural source of very high CO_2 concentration on the leaf gas exchange, xylem water potential and stomatal characteristics of plants of Spathiphyllum cannifolium and Bauhinia multinerva. New Phytologist, 138:689–697.

Figge, R. A., and White, J. W. C., 1995. High-resolution Holocene and late glacial atmospheric CO_2 record: Variability tied to changes in thermohaline circulation. Global Biogeochemical Cycles, 9 (3):391–403.

Freeman, K. H., and Hayes, J. M., 1992. Fractionation of carbon isotopes by phytoplankton and estimates of ancient CO_2 levels. Global Biogeochemical Cycles, 6 (3):185–198.

Goslar, T., Arnold, M., Bard, E., Kuc, T., Pazdur, M. F., Ralska-Jasiewiczowa, M., Różański, Tisnerat, N., Walanus, A., Wicik, B., and Wieckowski, K., 1995. High concentration of atmospheric ^{14}C during the Younger Dryas cold episode. Nature, 377:414–417.

Goslar, T., Wohlfarth, B., Björck, S., Possnert, G., and Björck, J., 1999. Variations of atmospheric ^{14}C concentrations over the Allerød-Younger Dryas transition. Climate Dynamics, 15:29–42.

Greenwood, D. L., and Wing, S., 1995. Eocene continental climates and latitudinal temperature gradients. Geology, 23:1044–1048.

Grootes, P. M., Stuiver, M., White, J. W. C., Johnson, S. J., and Jouzel, 1993. Comparison of oxygen isotope records from the GISP2 and GRIP Greenland ice cores. Nature, 366:552–554.

Grote, P. J., and Dilcher, D. L., 1992. Fruits and seeds of the tribe Gordonieae (Theaceae) from the Eocene of North America. American Journal of Botany, 79:744–753.

Hansen, A., and Knudson, K. L., 1995. Recent foraminiferal distribution in Freemansundet and Early Holocene stratigraphy on Edgøya, Svalbard. Polar Research, 14:215–238.

Hay, W. W., 1993. The role of polar deep water formation in global climate change. Annual Review of Earth and Planetary Science, 21:227–254.

Hay, W. W., Deconto, R. M., and Wold, C. N., 1997. Climate—Is the past the key to the future? Geologische Rundschau, 86:471–491.

Indermühle, A., Stocker, T. F., Joos, F., Fischer, H., Smith, H. J., Wahlen, M., Deck, B., Mastroianni, D., Tschumi, J., Blunier, T., Meyer, R., and Stauffer, B., 1999. Holocene carbon-cycle dynamics based on CO_2 trapped in ice at Taylor Dome, Antarctica. Nature, 398:121–126.

Jansen, E., and Sjøholm, J., 1991. Reconstruction of glaciation over the past 6 Myr from ice-borne deposits in the Norwegian Sea. Nature, 349:600–603.

Karpuz, N. K., and Jansen, E., 1992. A high-resolution diatom record of the last deglaciation from the SE Norwegian Sea: Documentation of rapid climatic changes. Paleocanography, 7:499–520.

Kromer, B., and Becker, B., 1993. German oak and pine ^{14}C calibration, 7200–9439 B.C. Radiocarbon, 35:125–135.

Kürschner, W. M., 1996. Leaf stomata as biosensors of palaeoatmospheric CO_2 levels. Ph.D. thesis, Utrecht University. LPP Contributions Series, 5:153 pp.

Kürschner, W. M., 1997. The anatomical diversity of recent and fossil leaves of the durmast oak (Quercus petraea, Lieblein/Q. pseudocastanea, Goeppert)—Implications for their use as biosensors of palaeoatmospheric CO_2 levels. Review of Palaeobotany and Palynology, 96:1–30.

Kürschner, W. M., van der Burgh, J., Visscher, H., and Dilcher, D. L., 1996. Oak leaves as biosensors of late Neogene and early Pleistocene palaeoatmospheric CO_2 levels. Marine Micropaleontology, 27:299–312.

Kürschner, W. M., Wagner, F., Visscher, E. H., and Visscher, H., 1997. Predicting the stomatal frequency response to a future CO_2 enriched atmosphere: Constraints from historical observations. Geologische Rundschau, 86:512–517.

Kürschner, W. M., Stulen, I., Wagner, F., Kuiper P. J. C., 1998. Comparison of palaeobotanical observations with experimental data on the leaf anatomy of Durmast Oak [Quercus petraea (Fagaceae)] in Response to Environmental Change. Annals of Botany, 81:657–664.

Kvaček, Z., and Walther, H., 1984a. Nachweis tertiärer Theaceen Mitteleuropas nach blatt-epidermalen Unter-suchungen. I. Epidermale Merkmalskomplexe rezenter Theaceae. Feddes Repertorium, 95:209–227.

Kvaček, Z., and Walther, H., 1984b. Nachweis tertiärer Theaceen Mitteleuropas nach blatt-epidermalen Unter-suchungen. II. Bestimmung fossiler Theaceen Sippen. Feddes Repertorium, 95:331–346.

Lal, D., and Revelle, R., 1984. Atmospheric PCO_2 changes recorded in lake sediments. Nature, 308:344–346.

Lambers, H., Stuart Chapin III, F., and Pons, T. L., 1998. Plant physiological ecology. Springer, New York. 540 pp.

Larsen, H. C., Saunders, A. D., Clift, P. D., Beget, J., Wei, W., Spezzaferri, S., ODP Leg 152 Scientific Party, 1994. Seven million years of glaciation on Greenland. Science, 264:952–955.

McElwain, J., 1998. Do fossil plants signal palaeoatmospheric CO_2 concentration in the geological past? Philosophical Transactions of the Royal Society of London, B, 353:83–96.

McElwain, J., and Chaloner, W. G., 1995. Stomatal density and index of fossil plants track atmospheric carbon dioxide in the Palaeozoic. Annals of Botany, 76:389–395.

McElwain, J., Beerling, D. J., and Woodward, F. I., 1999a. Fossil plants and global warming at the Triassic–Jurassic boundary. Science, 285:1386–1390.

McElwain, J. C., Beerling, D. J., and Mayle, F. E., 1999b. Does atmospheric CO_2 lead or lag Younger Dryas temperature change? Poster and abstract, INQUA XV International Congress, 3–11 August 1999, Book of abstracts p.121.

Metcalfe, C. R., and Chalk, L., 1979. Anatomy of Dicotyledons (2. Edition) Vol. I: Systematic anatomy of the leaf and stem. Oxford Univ. Press, Oxford. 276 pp.

Miglietta, F., and Raschi, A., 1993. Studying the effect of elevated CO_2 in the open in a naturally enriched environment in Central Italy. Vegetatio, 104/105:391–400.

Miller, K. G., Janacek, T. R., Katz, M. R., and Keil, D. J., 1987. Abyssal circulation and benthic foraminiferal changes near the Paleocene / Eocene boundary. Paleoceanography, 2:741–761.

Neftel, A., Oeschger, H., Staffelbach, T., and Stauffer, B., 1988. CO_2 record in the Byrd ice core 50,000–5,000 years B.P. Nature, 331:609–611.

Oeschger, H., Beer, J., Siegenthaler, U., Stauffer, B., Dansgaard, W., and Langway, C. C., 1984. Late Glacial climate history from ice cores. In Climate Processes and Climate Sensitivity. Geophysical Monographs, 29:299–306.

Pagani, M., Freeman, K. H., and Arthur, M. A., 1999a. Late Miocene atmospheric CO_2 concentrations and the expansion of C4 grasses. Science, 285:876–879.

Pagani, M., Arthur, M. A., and Freeman, K. H., 1999b. Miocene evolution of atmospheric carbon dioxide. Paleoceanography, 14:273–292.

Pearson, P. N., and Palmer, M. R., 1999. Middle Eocene seawater pH and atmospheric carbon dioxide concentrations. Science, 284:1824–1826.

Peñuelas, J., and Matamala, R., 1990. Changes in N and S leaf content, stomatal density and specific leaf area of 14 plant species during the last three centuries of CO_2 increase. Journal of Experimental Botany, 41:1119–1124.

Petit, J. R., Jouzel, J., Raynaud, D., Barkov, N. I., Barnola, J.-M., Basile, I., Bender, M., Chappellaz, J., Davis, M., Delaygue, G., Delmotte, M., Kotlyakov, V. M., Legrand, M., Lipenkov, V. Y., Lorius, C., Pépin, L., Ritz, C., Saltzman, E., and Stievenard, M., 1999. Climate and atmospheric history of the past 420,000 years from the Vostok ice core, Antarctica. Nature, 399:429–436.

Poole, I., Weyers, J. D. B., Lawson, T., and Raven, J. A., 1996. Variations in stomatal density and index: Implications for palaeoclimatic reconstructions. Plant, Cell and Environment, 19:705–712.

Poole, I., and Kürschner, W. M., 1999. Stomatal density and index: The praxis. In Jones, T. P., and Rowe, N. P. (eds.): Fossil plants and spores modern techniques. Geol. Soc., London: 256–260.

Popp, B. N., Takigiku, R., Hayes, J. M., Louda, J. W., and Baker, E. W., 1989. The post-Palaeozoic chronology of C-13 depletion in primary marine organic matter. American Journal of Science, 289:436–454.

Potter, F., and Dilcher, D. L., 1980. Biostratigraphic analysis of Middle Eocene floras of western Kentucky and Tennessee. In Dilcher, D. L., and Taylor, T. N. (eds.): Biostratigraphy of fossil plants: Successional and paleoecological analysis. Dowdon, Hutchinson and Ross Publishers: 211–225.

Raymo, M. E., Grant, B., Horowitz, M., and Rau, G. H., 1996. Mid-Pliocene warmth: Stronger greenhouse and conveyor. Marine Micropaleontology, 27:312–326.

Rey, A., and Jarvis, P. G., 1997. Growth response of young birch trees (*B. pendula* Roth.) after four and a half years of CO₂ exposure. Annals of Botany, 80:807–816.

Rind, D., and Chandler, M. A., 1991. Increased ocean heat transport and warmer climate. Journal of Geophysical Research, 96:7437–7461.

Rundgren, M., and Beerling, D. J., 1999. A Holocene CO₂ record from the stomatal index of subfossil *Salix herbacea* L. leaves from northern Sweden. The Holocene, 9:509–513.

Salisbury, E. J., 1927. On the causes and ecological significance of stomatal frequency, with special reference to the woodland flora. Philosophical Transactions of the Royal Society of London, B, 216:1–65.

Sinha, A., and Scott, L. D., 1994. New atmospheric PCO₂ estimates from paleosols during the late Paleocene/early Eocene global warming interval. Global and Planetary Change, 9:297–307.

Sloan, L. C., and Barron, E. J., 1992. Eocene climate model results: Quantitative comparison to paleoclimatic evidence. Palaeogeography, Paleoclimatology Palaeoecology, 93:183–202.

Sloan, L. C., and Rea, D. K., 1995. Atmospheric carbon dioxide and early Eocene climate: A general circulation modeling sensitivity study. Palaeogeography, Paleoclimatology Palaeoecology, 119:275–292.

Smith, H. J., Wahlen, M., Mastroianni, D., and Taylor, K. C., 1997. The CO₂ concentration of air trapped in GISP2 ice from the Last Glacial Maximum–Holocene transition. Geophysical Research Letters, 24:1–4.

Staffelbach, T., Stauffer, B., and Sigg, A., 1991. CO₂ measurements from polar ice cores: More data from different sites. Tellus, 43(B):91–96.

Stauffer, B., Neftel, A., Oeschger, H., and Schwander, J., 1985. CO₂ concentration in air extracted from Greenland ice samples. Geophysical Monographs, 33:85–89.

Stauffer, B., Blunier, T., Dällenbach, A., Indermühle, A., Schwander, J., Stocker, T. F., Tschumi, J., Chappelaz, J., Raynaud, D., Hammer, C. U., and Clausen, H. B., 1998. Atmospheric CO₂ concentration and millennial-scale climate change during the last glacial period. Nature, 392:59–62.

Stuiver, M., Braziunas, T. F., Becker, B., Kromer, B., 1991. Climatic, solar, oceanic, and geomagnetic influences on late-glacial and Holocene atmospheric ¹⁴C/¹²C change. Quaternary Research, 35:1–24.

Stuiver, M., Reimer, P. J., Bard, E., Beck, J. W., Burr, G. S., Hughen, K. A., Kromer, B., McCormac, F. G., van der Plicht, J., Spurk, M., 1998. INTCAL98 radiocarbon age calibration, 24,000–0 cal BP. Radiocarbon, 40:1041–1083.

Thomas, E., Zachos, J. C., and Bralower, T. J., 2000. Deep-sea environments on a warm earth: latest Paleocene-early Eocene. *In* Huber, B.T., MacLeod, K. G., and Wing, S. L. (eds.): Warm climates in earth history. Cambridge University Press, Cambridge, pp. 132–160.

Tissue, D. T., Griffin, K. L., Thomas, R. B., and Strain, B. R., 1995. Effects of low and elevated CO₂ on C3 and C4 annuals. II. Photosynthesis and leaf biochemistry. Oecologia, 101:21–28.

van der Burgh, J., Visscher, H., Dilcher, D. L., and Kürschner, W. M., 1993. Paleoatmospheric signatures in fossil leaves. Science, 260:1788–1790.

van der Burgh, J., 1993. Oaks related to *Quercus petraea* from the upper Tertiary of the Lower Rhenish Basin. Palaeontographica (B), 230:195–201.

van de Water, P. K., Leavitt, S. W., and Betancourt, J. L., 1994. Trends in stomatal density and ¹³C/¹²C ratios of *Pinus flexilis* needles during the last Glacial-Interglacial cycle. Science, 264:239–243.

van Geel, B., Bohncke, S. J. P., and Dee, H., 1981. A palaeoecological study of an upper Late Glacial and Holocene sequence from "De Borchert," The Netherlands. Review of Palaeobotany and Palynology, 31:367–448.

Verschuren, D., Laird, K. R., and Cumming, B. F., 2000. Rainfall and drought in equatorial Africa during the past 1,100 years. Nature, 403:410–414.

von Grafenstein, U., Erlenkeuser, H., Brauer, A., Jouzel, J., and Johnsen, J., 1999. A Mid-European isotope-climate record from 15,500 to 5000 years B.P., Science, 284:1654–1657.

Wagner, F., 1998. The influence of environment on the stomatal frequency in *Betula*. Ph.D. Thesis, Utrecht University. Laboratory of Palaeobotany and Palynology Contributions Series, 9:102 pp.

Wagner, F., Below, R., De Klerk, P., Dilcher, D. L., Joosten, H., Kürschner, W. M., and Visscher, H., 1996. A natural experiment on plant acclimation: Lifetime stomatal frequency response of an individual tree to annual atmospheric CO₂ increase. Proceedings of the National Academy of Sciences USA, 93:11705–11708.

Wagner, F., Bohncke, S. J. P., Dilcher, D. L., Kürschner, W. M., van Geel, B., and Visscher, H., 1999a. Century-scale shifts in early Holocene atmospheric CO_2 concentrations. Science, 284:1971–1973.

Wagner, F., Bohncke, S. J. P., Dilcher, D. L., Kürschner, W. M., van Geel, B., Visscher, H., 1999b. Response to Technical Comments by Indermühle et al., Birks et al. on "Century-scale shifts in early Holocene atmospheric CO_2 concentrations." Science, 286, 1815a.

Wagner, F., Neuvonen, S., Kürschner, W. M., and Visscher, H., 2000. The influence of hybridization on epidermal properties of birch species and the consequences for palaeoatmospheric CO_2 reconstructions. Plant Ecology, 148:61–69.

White, J. W. C., Ciais, P., Figge, R. A., Kenny, R., and Markgraf, V., 1994. A high resolution record of atmospheric CO_2 content from carbon isotopes in peat. Nature, 367:153–156.

Wilf, P., Wing, S. L., Greenwood, D. R., Greenwood, C. L., 1998. Using fossil leaves as paleoprecipitation indicators: An Eocene example. Geology, 26:203–206.

Willemse, N. W., and Törnqvist, T. E., 1999. Holocene century-scale temperature variability from West Greenland lake records. Geology, 27:580–584.

Wing, S., and Greenwood, D. L., 1993. Fossils and fossil climate: The case of equable continental interiors in the Eocene. Philosophical Transactions of the Royal Society of London, B, 341:243–252.

Woodward, F. I., 1987. Stomatal numbers are sensitive to increases in CO_2 concentration from pre-industrial levels. Nature, 327:617–618.

Woodward, F. I., and Bazzaz, F., 1988. The response of stomatal density to CO_2 partial pressure. Journal of Experimental Botany, 39:1771–1781.

Zachos, J. C., Stott, L. D., and Lohmann, K. C., 1994. Evolution of early Cenozoic marine temperatures. Paleoceanography, 9:353–387.

Zagwijn, W. H., and Hager, H., 1987. Correlations of continental and marine Neogene deposits in the southeastern Netherlands and the Lower Rhine District. Mededelingen van de Werkgroep voor Tertiair en Kwartair Geologie, 24:59–78.

Part III | Natural Variability and Studies of Past Temperature Changes

Any assessment of potential man-induced modification of climate must be measured against the naturally occurring changes in temperature that have taken place in the absence of human activity. Climate changes involving small to large temperature variations over short to long periods of time have been demonstrated by various techniques. Such studies are critical if we are to fully understand the relative contributions of the broad range of natural drivers.

Bluemle, J. P., J. M. Sabel, and W. Karlén, Rate and magnitude of past global climate changes, 2001, *in* L. C.
Gerhard, W. E. Harrison, and B. M. Hanson, eds.,
Geological perspectives of global climate change,
p. 193–211.

10 | Rate and Magnitude of Past Global Climate Changes

John P. Bluemle
North Dakota Geological Survey
Bismarck, North Dakota, U.S.A.

Joseph M. Sabel
U.S. Coast Guard Civil Engineering Unit Oakland
Oakland, California, U.S.A.

Wibjörn Karlén
Department of Physical Geography, Stockholm University
Stockholm, Sweden

ABSTRACT

Existing data indicate that the earth's climate is probably warming. Politicians and the media typically assume this warming is the result of human activity. This article summarizes previous climate
changes to test the validity of assigning causality to human activity.

Records of glacial advances and retreats indicate relative summer temperature. Lacustrine and
subaerial sediments afford a record of glacier advances and retreats from the Pleistocene to the
present time. Palynology offers a record of species succession in response to climate changes.
Dendrochronology is another indicator of summer temperature. Isotopic paleontology offers a
measurement of temperature at the time of marine sediment deposition, and isotopic evaluation of
continental ice is an indicator of temperature at the time of precipitation. Anthropologic sources
contain significant climate data, such as information about villages overrun by glaciers, open-
ocean iceberg density, or harbors filled with ice. Today, scientists are capable of direct measurement
of climatic conditions.

These sources record continual changes in climate. Broadly, the temperature changed 15° to 20°C from the Paleocene to the Neogene. Perhaps there was as much as another 10°C change in the Pleistocene. Correlative data from North America, Greenland, and Scandinavia indicate many climate changes were truly global in scope. Although it is difficult to develop precise paleothermometry, qualitative evaluations indicate sudden and dramatic changes in climate. Some are perhaps as great as a change from conditions warmer than today to a full glacial climate in as little as 100 years. The converse can be true. Current data indicate a trend of change that is substantially severe but no greater in rate or magnitude, and probably less in both, than many changes that have occurred in the past.

INTRODUCTION

The perceived increases in observed and predicted global temperatures attributed to CO_2 have been called the "greenhouse effect." Discussions of global climate in the popular press, as well as in national and international political debate, assume or conclude that (1) we are in a warming trend; (2) the warming rate is unusually high; and (3) this warming is the result of humans burning fossil fuels emitting CO_2 (it is anthropogenic).

This is important because an increase in the earth's temperature may have profound effects. Some agricultural areas would suffer setbacks while others would benefit. Sea level may rise due to thermal expansion of the seawater and melting glaciers, particularly the Antarctic and Greenland ice caps. Coastal population centers may flood. There could be an increase in icebergs, thus endangering maritime commerce, particularly in Alaska (Carter, 1975). The list of possible effects of climate change is long and uncertain (Lindholm, 1976). If the warming is anthropogenic, there is a logical corollary that it could be reversed.

To evaluate the validity of the three assumptions or conclusions, current climatic conditions and trends must be compared with past events. If the current variations and conditions are within past magnitude and rate limits, the conclusion that there is a measurable anthropogenic effect is suspect. The question thus posed is, "What has past global climatic variability been in magnitude and rate?"

The Pleistocene Epoch was preceded by a long-term, general decline in the earth's temperature, beginning near the end of the Oligocene Epoch (Dorf, 1960; Anderson and Borns, 1997) or possibly earlier, in the Tertiary Period (Margolis et al., 1975). The overall decline in temperature in central Europe from the beginning of the Tertiary Period to the Pleistocene was between 15° and 20°C (Anderson and Borns, 1997) **(Figure 1)**, and the average temperature declined at least another 10°C during the Pleistocene Epoch. Whereas extensive continental glaciation is a phenomenon of the last two or three million years, a large ice sheet was already well established in west Antarctica in late Miocene time, seven million years ago (Denton et al., 1971). Glaciers locally reached the edge of the Antarctic continent approximately five million years before that.

In North Dakota during the Pleistocene, at least six separate major glacial advances and retreats have been identified (Hobbs and Bluemle, 1987). With the exception of the most recent of these, we do not know precisely when the advances took place. Studies of the Holocene record in the United States, Scandinavia, Greenland, and continental Europe provide evidence of a series of smaller, more recent fluctuations.

Each major glacial and interglacial episode generally lasted from 100,000 to 200,000 years and involved a temperature decrease on the order of 10°C during glaciation (Anderson and Borns, 1997), followed by an increase of similar magnitude during the subsequent interglacial period.

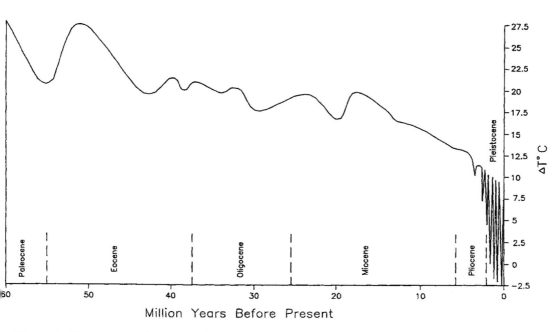

Figure 1 Temperature fluctuation (mean annual temperatures) in central Europe during Tertiary time, the past 60 million years. Except for a peak in early Eocene time, temperature decreased throughout the Tertiary. Beginning in late Pliocene/early Pleistocene time, with the onset of glacial conditions, temperatures fluctuated widely, ranging from full glacial to interglacial conditions. The modern condition is approximately +4 to +5°C. Modified and adapted from Anderson and Borns (1997).

Other sources of data confirm climate fluctuations. Often, the methodologies are used to confirm or calibrate one another. Foraminifera studies on cores from the Gulf of Mexico and oxygen isotope studies of speleothems from France and Iowa (U.S.A.) also suggest periods of rapid climate change that confirm ice-core studies. Some other methodologies include lichenometry, dendrochronology, sedimentology, paleontology, and historical evidence, such as tax and weather records. It is the evaluation of past events through these kinds of records that provides the reference of comparison with current climatic data.

GEOLOGIC RECORD OF CLIMATIC FLUCTUATION

The geological record has several quantitative and qualitative catalogs of paleoclimatic condition. These include records of glacial advance and retreat, sedimentology, palynology, dendrochronology, and isotopic paleontology.

Glaciation Record

Evidence of early glacier fluctuations is found throughout middle and high latitudes. Glacier advances are a result of a positive mass balance between winter precipitation and summer melt over an extended period of time. A 50-year record of glacier mass balance shows that in northern Sweden, summer temperature is the factor that is most important (Holmlund et al., 1996). The maximum size of a glacier will occur after a minimum in summer temperature; that is, glaciers grow during years with cool summers. Therefore, studying the historical record of glacier advances and retreats provides one record of general temperature variations.

A large number of age dates on frontal moraines in Norway and Sweden have been obtained by lichenometry, a technique based on measurements of the diameter of lichens. Dates obtained by this technique are quite precise for Little Ice Age (LIA)[1] moraines (A.D. 1400–1850) but less so for older moraines. The lichen growth rate of northern Sweden is calibrated back only to ~3000 years B.P. (Karlén, 1976). Lichenometry dates on moraines older than 3000 years B.P. therefore are not as well constrained. Nevertheless, they indicate that early and mid-Holocene glacier advances were extensive (Karlén, 1976; Matthews and Shakesby, 1984).

Late Pleistocene Glacier Fluctuations   The geologic record documents growth and retreat of large ice-age glaciers during the Pleistocene Epoch. The last major glaciation, the Wisconsinan in North America (the Würm or Weichselian in Europe), was marked by the growth of several large ice sheets in the Northern Hemisphere and smaller glaciers in alpine and polar areas.

At the time of the Wisconsinan glacial maximum, between 20,000 and 14,000 years B.P., sea level was ~130 m lower than it is today, and glacier ice covered ~27% of the earth's land surface compared with ~10% today (Anderson and Borns, 1997).

The end of the Pleistocene, between 13,000 and 10,000 years B.P., was marked by a series of climatic fluctuations (Coope and Lemdahl, 1995). At 13,000 years B.P., the climate in Europe and North America warmed rapidly. A widespread glacier retreat, accompanied by a rapid rise in sea level, followed the Wisconsinan glacial maximum. By 11,000 years B.P., the climate had changed back to colder conditions, and arctic species once again colonized the British Isles (Allan Ashworth, personal communication, 1998). The arctic species persisted until ~10,000 years B.P., when they were replaced by more temperate species. The difference in mean July temperatures between the peak warm and cold episodes during this time interval is estimated to have been between 5° and 8°C (Lowe et al., 1994).

Holocene Glacier Fluctuations   The Holocene Epoch, the most recent of many interglacial ages, includes recorded human history. Holocene time has been characterized by numerous climatic fluctuations. Many of the fluctuations appear shorter in length and of lesser magnitude than those that marked the glacial and interglacial periods of the Pleistocene Epoch. Whether this difference is real or an artifact of more precise measurement owing to direct observation in historic times is a question deserving more attention.

The magnitude and duration of climate change is significant. Evidence suggests that the magnitude of change is related to the duration of the change. The changes with the greatest magnitude occur during longer period fluctuations (Anderson and Borns, 1997).

The beginning of the Holocene Epoch was marked by a rapid warming event that virtually all paleoclimate studies show, starting ~10,000 years ago. By 8000 years B.P., the earth's climate was relatively warm, roughly 2.5°C higher than temperatures during subsequent cool periods (Lindholm, 1976).

This generally warm climate lasted several thousand years but was punctuated by numerous fluctuations. Studies of ice cores from the Greenland ice cap show numerous small-scale climate fluctuations (Figure 2). One such fluctuation was a rapid drop in temperature at 9500 years B.P. The data indicate that within 100 years, the climate changed from warmer than today (in Greenland) into full glacial severity (Dansgaard et al., 1989). In Sweden, a major cold event reached its maximum ~8200 years B.P. and this cold "spike" is also noted on the Greenland ice cores (see Figure 2). Following ~1500 years of warmer Scandinavian temperatures, cooling events occurred at 6600–6200 B.P., 4800–4300 B.P., 3000–2000 B.P., and the last 500 years (except for the past 100 years).

In North Dakota, the last glacial ice to melt was the debris-covered glacier ice in the Turtle Mountains and Missouri and Prairie Coteau areas. The ice in North Dakota had probably all melted by 8000 years ago. Radiocarbon (^{14}C) dates of ancient tree remnants and pollen records from bogs and lakes in North Dakota and other midcontinent sites indicate that the time of maximum warmth was reached close to 6000 years ago. At that time, the mean world temperature was probably ~2 or 3°C warmer than it is now (Denton and Porter, 1970). By ~7000 years ago, the Scandinavian ice sheet had almost completely disintegrated. By 5000 years B.P., the last remnants of the once huge Laurentide ice sheet had melted away in the Hudson Bay region. Nearly 2000 years of somewhat lower temperatures followed (between 5000 and 3000 years B.P.).

The climate then fluctuated rapidly during the following centuries. A relatively warm climate around 1500 years B.P. is inferred from high tree limit (Karlén and Kuylenstierma, 1996). This was followed by a marked cold climate between 1200 and 1100 years B.P.

Several glacier expansions during this period are well dated. At Engabreen on the coast of northern Norway, several trees were sheared off during a glacier advance. A cold climate between A.D. 800 and 900 (1100–1200 years B.P.) is documented by dendrochronology (Briffa et al., 1992). The following warm period is known as the Medieval Warm Period (MWP), which is known from dates on high pine tree limit and dendrochronology.

A major cold period, the LIA started ~800 years ago (A.D. 1200). In Sweden, this cold period was split by a relatively warmer interval of time between A.D. 1350 and 1550 (Briffa et al., 1992). Many glaciers reached their Holocene maximum during the 1600s and early 1700s.

In Norway and Sweden, changes in the size of glaciers, changes in the elevation of the alpine tree-limit, and variation in the width of tree rings during the Holocene indicate how the Scandinavian summer temperatures have fluctuated considerably. During warm periods, temperatures have been ~2°C warmer than they are today; during cold periods, they were almost as cold as they were during the coldest decades of the previous centuries. Short fluctuations in temperature that are superimposed on the longer-term variations lasted 100–200 years.

A Scandinavian climate chronology, based on glacier and alpine tree-limit fluctuations and dendrochronology, correlates well with studies of ice cores from central Greenland (Briffa et al., 1992; Dansgaard et al., 1993; Cuffy and Clow, 1997). The Scandinavian climate chronology likely represents conditions across a large area. The Scandinavian record can be compared with data on solar irradiation variations, estimated as ^{14}C anomalies obtained from tree rings, and the correlation between climate change and solar radiation variations (Lean et al., 1995; Karlén and Kuylenstierna, 1996). The correlation to solar radiation presents some intriguing potential alternatives to anthropogenic warming.

The earliest recorded direct observations of glaciers in North America were made along the southern coast of Alaska. Count de La Perouse visited Lituya Bay on a voyage of exploration in 1787. Shortly thereafter, in 1794, George Vancouver observed the position of ice in Glacier Bay. At that time, glaciers in southern Alaska had already retreated from their earlier observed positions (Denton and Porter, 1970).

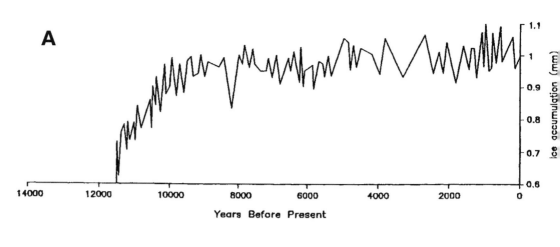

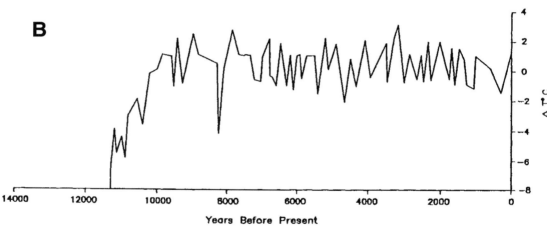

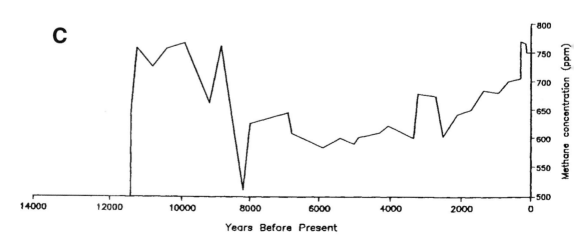

Figure 2 *(facing page)* These curves are derived from the Greenland Ice Sheet Project II (GISP2) cores. (A) Ice-accumulation data. These are probably the most direct measurements of temperature history. Small corrections must be considered for ice flow. Local accumulation is primarily controlled by precipitation minus melt, hence the value of a temperature indicator. (B) The temperature curve is based on $^{18}O/O_{ice}$ ratios. This ratio is a surrogate for temperature by indicating increased preferential evaporation of the lighter isotope in warmer conditions. This process is discussed in more detail in other sections of this article. (C) The methane curve is a direct measurement of a "greenhouse gas" preserved in the ice. Prior to anthropogenic effects, and perhaps after, the authors indicate this is a surrogate for anaerobic conditions associated with wetlands and large-scale weather patterns.

All three of these curves show a remarkable similarity. In addition to documenting the wide and abrupt changes in climate, the data also validate the use of these temperature proxies. The primary source (Alley et al., 1997) also includes Ca and Cl ionic data. Those curves are also very similar. They were not included herein because those lines of evidence are not discussed in this paper. The reader is invited to review Alley et al. (1997) for more detailed information.

Note that there is a strong similarity between the ice data and Sargasso Sea surface temperatures (**Figure 3**).

Detailed regional observations of glaciers in southern Alaska, begun in the late nineteenth century, have documented a general retreat from the Neoglacial maximum in the early eighteenth century. In Glacier Bay, the recession has exceeded 10 km, owing in part to the fact that the glaciers terminate in the sea (Denton and Porter, 1970). Observations of glacier fluctuations in other North American localities provide evidence of general glacier retreat through the first half of the twentieth century. Within the past several decades, glaciers in the Cascade Range and in the Canadian Rocky Mountains readvanced from ~1940–1960 and have tended to retreat since then (Luckman et al., 1997).

Similarities between the Scandinavian and Greenland records are evidence that fluctuations in climate occurred simultaneously across large areas (Denton and Karlén, 1973). The pattern of frequent and rapid changes in climate throughout the Holocene Epoch in these two areas is strong evidence that the warming of the last 100 years is not a unique event (Denton and Karlén, 1973; Bonnefille et al., 1990; Hodell et al., 1991; Mosley-Thompson, 1992; Talma and Vogel, 1992; Vorren et al., 1996; Bond et al., 1997; Overpeck et al., 1997).

The Holocene Epoch period of maximum warmth following the Wisconsinan glaciation and preceding today's relatively cooler climate is commonly referred to as the Hypsithermal Interval. The Hypsithermal Interval was originally thought to be a period of uniformly mild climate and was strictly defined on the basis of pollen-zone boundaries from northwest Europe. It was defined as extending from ~9000 until 2500 years B.P. (Denton and Porter, 1970; Beget, 1983). It is now known that complex low-order changes of climate characterized the Hypsithermal Interval, resulting in several Neoglacial episodes of glacier expansion during that time span. Denton and Porter (1970) introduced the term *neoglaciation* to refer to the interval of rebirth or renewed growth and all subsequent fluctuations of glaciers after the time of maximum Hypsithermal glacier shrinkage. It is likely that maximum shrinkage of glaciers worldwide (alpine and continental) coincided with the period of maximum warmth.

Sedimentologic Record

High-resolution, continuous records of glacier advances have been obtained from proglacial lacustrine sediments (Karlén, 1976, 1981; Nesje and Dahl, 1991; Nesje and Kvamme, 1991; Karlén and

Matthews, 1992; Matthews and Karlén, 1992). When a glacier moves, substrate is eroded and rock flour is released to glacier-fed streams; this rock flour can be dated using ^{14}C techniques to date contained organic materials. Much of this rock flour is deposited in the first few lakes downstream from the glacier. The relative volume of rock flour increases distinctly in the sediments during periods with active glaciers in the drainage areas (Karlén, 1981). It was once believed that the release of rock flour reached a maximum after a glacier retreat had begun. However, glacier retreats observed during this century indicate that the lag is small compared with the dating error. In addition, a few ^{14}C dates have been obtained on organic debris buried beneath glacier moraines in Scandinavia and from peat deposits flooded during glacier advances (Karlén, 1973, 1982; Griffey and Worsley, 1978; Matthews, 1991; Dahl and Nesje, 1994).

Combined with information about tree-limit and lichenometric data, glacier fluctuations indicate that the climate has fluctuated frequently during the Holocene in Scandinavia (Karlén, 1976). One of the most conspicuous events is a cold period ~8200 years B.P., knowledge of which was established in the 1970s by integrating lacustrine sediment data with the other two data sources.

Glacier advances in both Norway and Sweden have been dated to the period between 6700 and 6000 years B.P. Several short, cold events between 6000 and 5000 years B.P., which represent several small advances, are indicated by minor increases in the silt content of lacustrine sediments from the lake Vuolep Allakasjaure (Karlén, 1976, 1981). This lake receives meltwater from a small glacier which, because of its configuration, responds rapidly to even short changes in mass balance.

A major cold event occurred between ~4800 and 4400 years B.P. It is recorded in lacustrine sediment records from a ^{14}C-dated moraine in front of the glacier Nipalsglaciaren in northern Sweden and from dates on peat from Nedrefetene, near Hardangerjokoulen, southern Norway (Dahl and Nesje, 1994). A cold event ~3700 years B.P. is known from Norwegian studies (Nesje et al., 1991) and from the Swedish proglacial lake Vuolep Allakasjaure.

Around 3000 years B.P., another major cooling took place that is documented by lacustrine sediments, ^{14}C dates, and lichenometric dates on moraines (Karlén, 1973; Karlén and Matthews, 1992). Following a short, warmer period, the climate again turned colder ~2000 years B.P. A short, warm event occurred ~1900 years B.P.

Palynologic Record

A general view of the Scandinavian climate was obtained by palynological studies (Birks et al., 1996). However, because of the generally low rate of sedimentation in lakes and peat bogs, this technique revealed mostly large-scale, low-frequency changes in climate. Vorren et al. (1996) found palynological evidence for several Holocene changes in the pine tree line in northern Norway. Studies that have combined glacier fluctuations, tree limit variations, dendrochronology, and palynology have yielded much more detailed information (Karlén, 1982; Karlén and Kuylenstierna, 1996; Vorren et al., 1996).

According to Nesje and Kvamme (1991), forest stands of elm and birch at Sygneskardot, Sunndalen in Norway at ~6000 years B.P. reached a limit that suggests the temperature there was 4°C warmer than today. Pollen records indicate that the Roman Iron Age in northern Europe (extending from A.D. ~50–400) was characterized by a mild climate. The climate remained relatively mild until A.D. ~1200, with the warmest interval apparently occurring between A.D. 800 and 1000, when summer temperatures were a degree or two higher than they are now.

Dendrochronologic Record

Dendrochronology as well as information about the alpine pine tree limit reveal summer temperatures. Summer temperatures seen in an average decade or more correlate quite well with annual temperatures (Manley, 1974). In northern Scandinavia, pine growth is determined largely by summer temperature (Briffa and Schweingruber, 1992; Kullman, 1997). Thus, a reconstruction of temperature history, based on pine tree-ring width, yields a temperature estimate that is closely related to actual summer temperature variations.

Briffa et al. (1992) published a temperature record for Scandinavia based on dendrochronology for the period since A.D. 500 that is accurate to within a few years. It documents a cold climate between A.D. 800 and 900 (1100–1200 years B.P.). The following warm episode is known from dates on high pine tree limit and dendrochronology. This corresponds with the MWP when the Vikings colonized Greenland. The chronology before 1500 B.P. is based on ^{14}C dates that have been calibrated to calendar years according to Stuiver and Braziunas (1995). Generally, the ^{14}C dating of lacustrine sediments can obtain only maximum and minimum dates. That is, the time of a glacier advance is estimated by linear interpolation, a technique that does not allow for precise dating.

Luckman et al. (1997) have reconstructed the record of summer temperatures based on a study of tree rings in Alberta, Canada, at the Columbia Icefield. They found that mean temperatures from the years 1101 until 1900 were 0.7°C below the 1961–1990 reference period and 0.33°C below the 1891–1990 mean of the instrument record. The 1961–1990 reference period was warmer than any equivalent-length period over the last 800 years, presumably a return to more temperate conditions as the area emerges from the LIA.

Isotope Paleontologic Record

An increase of air temperature results in increased evaporation of water. The isotope ^{16}O, a lighter atom of oxygen, is preferentially evaporated. This increases the relative concentration of ^{18}O. Oxygen in seawater is then combined by marine organisms into shell or skeletal structure. The ^{18}O/^{16}O concentration ratio in preserved shell material is thus a proxy measurement of temperature (Boulton et al., 1991).

The oxygen data gathered from sediments from Scotland indicate temperature cycles had less variability, both in frequency and temperature differential, before 2.4 million years B.P. than since then. A significant change in variability occurred ~750,000 years ago. After that time, there was a dramatic increase in climatic variability.

Isotopic data indicate a cooling cycle in Britain culminating ~18,000 years B.P. (Boulton et al., 1991). This closely correlates to the last major midlatitude ice-sheet expansion in Britain. During that time, most of Britain was covered with ice, reaching 1 km in thickness in the Midland Valley of Scotland.

Following the Younger Dryas cooling event (11,000–10,000 B.P.) (Kudrass et al., 1991), a significant warming event is recorded in the isotopic record from Scotland. This dramatic warming coincided with the return of the North Atlantic Drift (the Gulf Stream in the United States). The rate of change may have been 1°C per decade; the data suggest it may have been fully accomplished in a few decades. The isotopic data have been correlated to faunal succession data from correlative sediments (Boulton et al., 1991; Broecker, 1997a, 1997b).

Foraminifera studies on cores from the western Gulf of Mexico (Kennett and Huddlestun, 1972) and oxygen isotopes of speleothems from France (Duplessy et al., 1970; Emiliani, 1971) and Iowa (Dorale et al., 1992) also suggest a period of rapid cooling beginning ~9500 years B.P.

Keigwin (1996) compiled oxygen isotope data from seafloor cores collected in the Sargasso Sea (Figure 3). This area of the Southwestern Atlantic Ocean is well known for calm winds and few currents. The oxygen isotope data were date-calibrated using standard carbon-radiometric techniques. The result is a 3000-year history of ocean temperatures from this portion of the world.

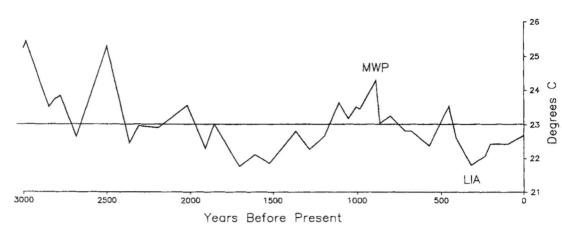

Figure 3 Sargasso Sea surface temperature. The Sargasso Sea surface temperature is derived from $^{18}O/^{16}O$ ratios. These are indicators of evaporation, hence a proxy for sea surface temperature. The Sargasso Sea is a 2,000,000-mi^2 body of water in the North Atlantic Ocean lying roughly between the West Indies and the Azores, at approximately 20–35°N (the "horse latitudes"). It is relatively static through its vertical column so that potential interference from mixing with other water masses and sediment sources is minimal. The isotopic ratios are derived from biotic debris that has vertically precipitated onto the seafloor. Wide and abrupt variations in temperature are indicated. The relative temperature variations of the LIA and MWP are both quite prominently recorded in the data. The horizontal line is the average temperature for this 3000-year period. After Keigwin (1996).

These data show generally declining temperatures from 3000–1500 years B.P. The data support the timing of the MWP between 1300 and 800 B.P. (A.D. 800 and 1500) and the LIA between 300 and 200 B.P. (A.D. 1600 and 1800).

Studies of ice cores recovered during two recent projects in Greenland, an American study, the Greenland Ice Sheet Project II (GISP2), and a European study, the Greenland Ice-Core Project, have added a great amount of detailed information about how the climate has changed during the past 250,000 years (Meese et al., 1994; Alley et al., 1997). The last 14,000 years from GISP2 are graphically represented by three temperature proxy curves in **Figure 2**.

Oxygen isotope ratio studies of the ice cores indicate air temperatures and show how temperatures over Greenland have fluctuated **(Figure 2)**. For example, during the Eemian Interglacial Interval (the Eemian is the European equivalent to the North American Sangamon), which lasted from ~135,000–110,000 years B.P., the climate fluctuated considerably. On two occasions during the Eemian, temperatures dropped from 2°C warmer than present to 5°C colder in less than a few centuries. Temperatures then remained cold for several thousand years. Temperatures during the Eemian rose and fell abruptly many times. In one instance, temperatures dropped 14°C in a decade and 70 years later rose again to their former level (Dansgaard et al., 1993).

Although temperatures during the Holocene have been much more variable than previously believed (as we have shown in this article), they have apparently not fluctuated to the extent they did during the Eemian (Dansgaard et al., 1993; White, 1993). Data from the GISP2 core also verify that temperatures during the Holocene were not as stable as commonly thought (Meese et al., 1994).

ANTHROPOLOGICAL RECORD OF CLIMATIC FLUCTUATION

Evidence of climate changes is not restricted to geologic data. Study of historical documents reveals climatic changes reflected by impacts on economies and life. These records can be qualitative, such as the relative timing of population explosions or famines in prehistoric civilization. The records also can have remarkably quantitative precision, such as land records from a variety of European sources.

This type of information has long been known by anthropologists and historians. The sudden demise of civilizations, known to have occurred on nearly every continent, has long been attributed to famine and prolonged drought—synonyms of climate change. Some of the better known examples of this include the Anasazi of the American Southwest and the Maya of Mexico and Central America. An extreme example of climate effects on human habitation was discovered in 1969. The research submarine *Alvin* located a Clovis-age midden on the Atlantic continental shelf. Now under >43 m of water, this area was submerged by rising sea level due to melting glaciers. The glaciers melted in response to a warmer climate at the beginning of the Holocene (Emery, 1969).

More recently, a period of relatively warm temperatures from A.D. ~700–1200 has become known as the MWP in Europe. Known as the Neo-Atlantic in the Western Hemisphere, this period is well documented in Europe (Lamb, 1982; Grove, 1988). An immediately following period of relative cooling, the LIA (A.D. ~1200–1700) is well documented in history from Europe and North America.

Specific cooling and warming cycles can sometimes be precisely documented. Advances of Norwegian glaciers that affected farmland are documented in tax records as far back as the seventeenth century. At least one Middle Ages glacial advance can be dated in this way (Grove, 1972; Karlén, 1982). The widespread advance of glaciers in western Norway between A.D. 1660 and 1700 is well documented and chronicled by tax records.

The first half of the eighteenth century was also marked by general glacial advance, culminating between A.D. 1740 and 1750. As a result, crops failed to ripen, famine set in, and the death rate in Norway far exceeded the birth rate. Subsequent regional glacier recessions have been interrupted by a number of minor readvances (particularly during the intervals 1807–1812, 1835–1855, 1904–1905, and 1921–1925).

Historical records of the latest glacier fluctuations are available from many other alpine regions in Europe. In the European Alps, the oldest records involve the Great Aletsch Glacier in Switzerland. During the thirteenth century, this glacier advanced over part of an aqueduct used to transport meltwater to a local village. More recent records throughout the Alps chronicle glacier advances (and recessions) over the past few centuries. Villages built in what had been considered safe places were overwhelmed by glaciers in the early seventeenth century. Several of these former villages are still ice covered today. Many distinct advances characterize this period of glacier expansion, major ones culminating in the early 1600s, the 1820s, and the 1850s. A glacial recession, interrupted by a minor readvance between 1912 and 1920, characterized the period from 1850 to ~1960.

Historical records from Iceland indicate that from the beginning of colonization in A.D. 870 until at least A.D. 1200, glaciers were more restricted than they are now. During that time, farms were built in locations that were later overrun, or nearly overrun, by advancing ice early in the eighteenth century. In Iceland, as in Norway and the European Alps, the past three centuries have been generally characterized by widespread glacier expansion that has included several major and minor advances (Denton and Porter, 1970).

Complementing records of glacier activity are other historical references to the climatic changes that accompanied glacier fluctuations. These include records of Viking voyages, the amounts of arctic sea ice near Iceland and southern Greenland, changes in habitats of animals and plants, agricultural data from throughout Europe, and direct meteorological measurements.

During the height of the Roman Empire, from A.D. 50–400, arctic sea ice virtually disappeared near Iceland and southern Greenland. In this period, the Viking voyages of discovery, colonization, and trade in the North Atlantic increased markedly (Denton and Porter, 1970). This correlates with remarkable precision to the palynologic record previously cited.

Climatic cooling characterized the period from A.D. 1200–1400. Arctic sea ice expanded, clogging sailing routes between Iceland and Greenland. This climatic change, together with social changes in Norway, greatly hampered and finally prevented trade between Greenland and the outside world.

The period from A.D. 1400–1550 in Greenland was characterized by a slight warming, which was followed between 1550 and 1880 by cooling, leading to a vast increase in the amount of sea ice near Iceland and southern Greenland. On at least one occasion, the edge of the sea ice is believed to have reached the Faroe Islands, only 250 miles north of Great Britain. It was during this interval that Eskimos reached Scotland by kayak, presumably traveling along the margin of the arctic ice pack (Denton and Porter, 1970).

DIRECT CLIMATE MEASUREMENT

During the nineteenth and twentieth centuries, direct instrumental measurements have been gathered and used to describe the modern climate. Direct-measurement data indicate that a warming trend began in the 1880s and lasted until ~1940, when temperatures began to decline. From 1860 to 1990, the global mean annual surface temperature increased 0.55°C (Parker et al., 1994). During this same 130-year period, the concentration of industrially produced CO_2 gas in the earth's atmosphere increased from 280 to 353 ppmv, leading to the hypothesis that the warmer temperatures signify the climate system's response to anthropogenic influences (Houghton, 1996). However, the warming trend does not pattern the CO_2 trend across this period. Most of the warming took place before most of the increase in CO_2 occurred. Statistical analyses of the climate record since 1860 show significant interannual and interdecadal variability (Lean et al., 1995), suggesting that the cause of the warming is more complex than the influence of increasing greenhouse gases alone. Furthermore, the surface temperature increase of the past 130 years appears to be part of a longer-term warming that commenced in the seventeenth century (Bradley and Jones, 1993), before the industrial epoch. Palaeoclimatic records obtained from ice cores, tree rings, pollen, corals, and glacial events likewise attest to the variability of the immediate preindustrial climate (Lean et al., 1995).

During the 1970s, the ongoing cooling trend then under way was interpreted by some scientists (and by the popular media) as a prelude to a new ice age (Ponte, 1976). In more recent years, direct measurements appear to indicate worldwide temperatures have again been rising.

Similarities between the Scandinavian and Greenland temperature records are evidence that fluctuations in climate occurred simultaneously over large areas (Denton and Karlén, 1973). The pattern of frequent and rapid changes in climate throughout the Holocene Epoch in these two areas is strong evidence that the warming of the last 100 years is not a unique event (Denton and Karlén, 1973; Bonnefille et al., 1990; Hodell et al., 1991; Mosley-Thompson, 1992; Talma and Vogel, 1992; Vorren et al., 1996; Bond et al., 1997; Overpeck et al., 1997).

DISCUSSION

The evidence gathered here indicates that observed climatic changes occur over widespread areas, probably on the global scale. A distinct correlation between climatic events and variations in solar radiation was pointed out by Denton and Karlén (1973) and later by Eddy (1976, 1977). Calder

(1975) generated a chart correlating summer solar energy to climate. He estimated that a plus/minus change of 2% solar output is sufficient to swing from a time of glacier growth to retreat (or vice versa). Wigley and Kelly (1990) found "a significant, but not convincing correlation," using ^{14}C anomalies as an index of solar activity. Karlén and Kuylenstierna (1996) compared the climate chronology of Scandinavia with ^{14}C anomalies and observed an obvious similarity in the timing of major periods of reduced solar radiation and the Scandinavian climate.

The solar forcing of the climate during the last few hundred years has been confirmed by Friis-Christensen and Lassen (1991), Lean et al. (1995), and Svensmark and Friis-Christensen (1997). A model in which solar forcing determines global temperature shows a detailed similarity between modeled and observed temperature (Soon et al., 1996). Observed breakpoints in the temperature trend, such as the low temperature around 1900, the high temperature around 1940, and a decrease between the 1940s and the mid-1970s, are all indicated in models using solar forcing, while models using CO_2 forcing show only a continuous temperature increase since the end of the nineteenth century (Wigley and Kelly, 1990; Soon et al., 1996).

According to a compilation of surface observations, the temperature increase of the last decade is considerable (World Meterological Organization, 1996). However, the fall, winter, and spring warming of Siberia, an area where strong inversions frequently cause extremely low temperatures near the ground, largely determined this global average temperature increase. Even a small change in wind velocity would reduce the frequency of inversions and increase the fall, winter, and spring temperatures considerably. The supposition that the warming of the last few decades is a result of a reduced number of extreme inversions, and not necessarily an indication of global warming, is supported by balloon and satellite temperature observations, which indicate no major change in the temperature of the lower 6000 m of the atmosphere since the late 1950s (Christy and McNider, 1994).

This review of the literature includes several dozen potential temperature curves characterizing periods of time ranging from the very short-term to varying long-term curves. These are generally available to the reader in other sources, and it would serve little purpose to recount many of them here. However, some of them are discussed below.

Compiling an accurate, integrated time/temperature curve is a difficult task. Relative rate and magnitude of change are fairly easy to determine. Precise paleothermometry is somewhat trickier. Additionally, distinguishing between local weather and climate is difficult today, let alone in the geologic record. As an example illustrating this point, see **Figure 4** (water elevation of Devils Lake, North Dakota). This figure integrates the combined effects of precipitation, temperature, and evaporation. The Devils Lake watershed is a 9900 km² closed drainage basin in northeastern North Dakota. A significant high-water stand coincides with the MWP and a low-water stand coincides with the LIA. It is impossible, of course, from these data alone to reach finite conclusions regarding worldwide weather/climate and paleotemperatures and the effect of these factors on this part of North Dakota.

Some of the most pertinent and detailed temperature curves compiled in recent years are those from the Greenland ice-core projects—the GISP2 and Greenland Ice-Core Project curves (see some of the GISP2 curves in **Figure 2**). The ice-accumulation record depicted by these curves provides the best portrayal of climate change across a wide geographic area throughout the Holocene that is currently available.

Figure 5 shows a qualitative temperature curve based on analysis of several different data sets to better illustrate climate variability. The curve, integrating the events described in this article, is a qualitative temperature chronology of the Holocene based on and taking into account all of the factors evaluated here. By using 1950 as a reference temperature, departures from that reference are depicted for the past 13,000 years. The most notable attribute of the curve is the frequent and substantial deviation, plus and minus, from the 1950 (0°C) norm.

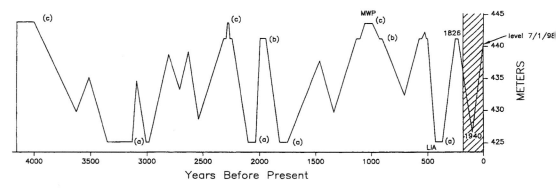

Figure 4 Hydrograph showing how the level of Devils Lake has fluctuated over the past 4000 years. Devils Lake exists in a closed, 9900-km² basin, a subbasin of the Red River of the North in northeastern North Dakota. The lake consists of a series of interconnected, separate, ice-thrust depressions, which become connected at various elevations. If it rises to an elevation of 441 m above sea level, the current (1998) Devils Lake will spill into still another subbasin, the Stump Lake basin (but the system will still remain enclosed, with no outward surface drainage). However, at 445 m, the combined Devils–Stump Lake spills into and becomes part of the Red River of the North drainage to Hudson Bay.

This hydrograph is based on study of cores taken from bottom sediment in the lake, radiocarbon dating of soils directly overlying and overlain by shoreline sediments at progressively higher elevations, and radiocarbon dating and pollen study of soils in the natural outlets from Devils and Stump Lakes. The level of Devils Lake fluctuates constantly, ranging between extremes of being completely dry (level a) to overflowing to Stump Lake (level b) or, at times, to overflowing into the Red River of the North drainage to Hudson Bay (level c). Although the curve shows only lake level, this correlates quite closely (although not exactly) to temperature. The most recent spillage of Devils Lake out of its basin (elevation of 445 m) coincides with the MWP; the most recent time that the lake was completely dry coincides with the LIA.

The changes in the level of Devils Lake are entirely in response to long-term climatic changes. Although the rising lake level of recent years has been attributed by some to anthropogenic causes (agricultural practices, drainage of wetlands, etc.), it is quite obvious that the lake level fluctuated widely and often prior to any human influence. The hachured area represents the time since Europeans settled the area. Modified and updated from Bluemle (1996).

The most important lesson to be learned from the data is simple: climate is in continual flux, the [average annual] temperature is always either rising or falling; and the temperature is never static over a long period of time. The Holocene Epoch, as a whole, has been a remarkably stable period of time with few extremes of either rising or falling temperatures as were common during the Wisconsinan and earlier Pleistocene glacial and interglacial periods. Nevertheless, the Holocene was—and still is—a time of fluctuating climate.

A corollary lesson to the above is as follows: the amount of geologically recent and even historically recent change in the temperature has been considerable. The amount of change (in temperature and other climatic factors) experienced in the past 30 or 50 years has been much less than changes that have previously occurred.

There is a second corollary: all previous changes in the climate have occurred without human influence. It can be expected that irrespective of any anthropogenic influence, global temperatures will rise as the earth continues to emerge from the LIA.

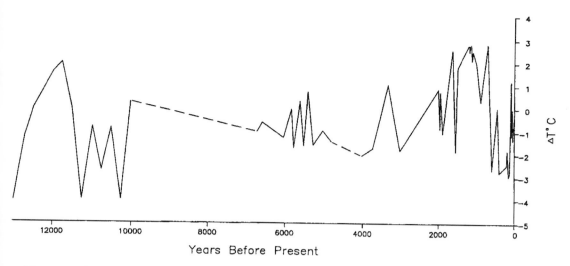

Figure 5 Generalized time/temperature curve showing deviation from current (standardized at 1950) mean annual global temperature. This curve plots the changes that have taken place in temperature over the past 13,000 years. Each point on the curve is documented in the text of this article. The closer frequency of changes in the curve in the years since ~2500 B.P. reflects the availability of more precise data, not increased climatic instability. Intervals with dotted lines lack sufficient data to draw an accurate curve. The curve depicts relative deviation from the mean annual temperature, not absolute temperature change.

SUMMARY

Glacier advance takes place in response to and during times of climatic cooling. Intervening periods of glacial recession, which precede and follow each advance, represent times of climatic warming—often rapid warming. These have been of sufficient magnitude and rate to cause significant changes in glacial extent in many parts of the world. The current (since ~1980) period of warming is just one more example of the repeated short-term variations in annual temperature.

The generally accepted idea of the Holocene Epoch as a time that was characterized by an early warming period followed by overall cooling is oversimplified and possibly seriously flawed. In fact, there is no certainty that there ever was a single identifiable warm period in the earliest Holocene. Instead, the entirety of Holocene climatic history can be more accurately characterized as a sequence of 10 or more global-scale "little ice ages," fairly irregularly spaced, each lasting a few centuries and separated by global warming events.

Grove (1988) speculates that changes in solar irradiance, most probably modified by induced changes in albedo and/or ocean circulation, are the main causes for the fluctuations in climate, increasing and decreasing temperatures. Karlén and Kuylenstierna (1996) reached a similar conclusion. The authors also believe climatic variations may be caused by solar irradiation variations.

Even though there are anthropogenic sources of greenhouse gases such as CO_2, climate changes must be judged against the natural climatic variability that occurs on a comparable timescale. The LIA, the MWP, and similar events are examples of this natural variability. These events correspond to global temperature changes of 1–2°C. The frequency, rate, and magnitude of climate changes during the Holocene do not support the opinion that the climatic changes observed during the last 100 years are unique or even unusual. Recent fluctuations in temperature, both upward and downward, are well within the limits observed in nature.

ACKNOWLEDGMENTS

We acknowledge Dr. John Reid, Professor of Geology at the University of North Dakota, for providing reprints from his personal library of many articles dealing with the issues discussed in this article. Conversations about this topic with Dr. Reid were highly beneficial. Additionally, John Trotter, President of Aquarius II, Inc., provided useful information.

ENDNOTE

[1]Various researchers have stated differing times for the period of the Little Ice Age. The Little Ice Age is a time-transgressive event that was not the same everywhere on earth. Furthermore, studies of different proxy data (lichens, tree rings, sedimentology, paleontology, isotopes, etc.) result in slightly different dates for the cooling that initiated the LIA and the warming that terminated it. I have used each author's own dates when referring to his or her work. The differing dates stated in this paper for the Little Ice Age are not necessarily contradictory.

REFERENCES CITED

Alley, R. B., Mayewski, P. A., Sowers, T., Stuiver, M., Taylor, M., and Clark, P. U. (1997). Holocene climatic instability: A prominent, widespread event 8200 yr ago. Geology, 25, 483–486.

Anderson, B. G., and Borns, H. W. (1997). The ice age world. Oslo: Scandinavian University Press.

Beget, J. C. (1983). Radiocarbon-dated evidence of worldwide early Holocene climate change. Geology, 11, 389–393.

Birks, H. H., Vorren, K. D., and Birks, H. J. B. (1996). Holocene treelines, dendrochronology and palaeoclimate. Palaeoclimatic Research, 20, 1–42.

Bluemle, J. P. (1996). From the state geologist. NDGS Newsletter, 23, 1–2.

Bond, G., Showers, W., Cheseby, M., Lotti, R., Almasi, P., deMenocal, P., Priore, P., Cullen, H. H., and Bonani, G. (1997). A pervasive millennial scale cycle in North Atlantic Holocene and glacial climates. Science, 278, 1557–1566.

Bonnefille, R., Roeland, J. C., and Guiot, J. (1990). Temperature and rainfall estimates for the past 40,000 years in equatorial Africa. Nature, 346, 347–349.

Boulton, G. S., Peacock, J. D., and Sutherland, E. G. (1991). Quaternary. In G. Y. Craig, (Ed.), Geology of Scotland, (pp. 503–543). Bath: Geologic Society of London.

Bradley, R. S., and Jones, P. D. (1993). "Little Ice Age" summer temperature variations: Their nature and relevance to recent global warming trends. Holocene, 3, 367–376.

Briffa, K. R., and Schweingruber, F. H. (1992). Recent dendroclimatic evidence of northern and central European summer temperatures. In R. S. Bradley and P. D. Jones (Eds.), Climate since A.D. 1500 (pp. 366–392). New York: Routledge.

Briffa, K. R., Jones, P. D., Bartholin, T. S., Eckstein, D., Schweingruber, F. H., Karlén, W., Zetterberg, P., and Eronen, M. (1992). Fennoscandian summers from A.D. 500: Temperature changes on short and long time scales. Climatic Dynamics, 7, 111–119.

Broecker, W. S. (1997a). Thermohaline circulation, the Achilles heel of our climate system: Will man-made CO_2 upset the current balance? Science, 278, 1582–1588.

Broecker, W. S. (1997b). Will our ride into the greenhouse future be a smooth one? GSA Today, 7, 1–7.

Calder, N. (1975). The weather machine. New York: Viking Press.

Carter, L. (1975). Icebergs and oil tankers: USGS glaciologists are concerned. Science, 190, 641–643.

Christy, J. R., and McNider, R. T. (1994). Satellite greenhouse signal. Nature, 367, 325.

Coope, G. R., and Lemdahl, G. (1995). Regional differences in the late glacial climate of northern Europe based on coleopteran analysis. Journal of Quaternary Science, 10, 391395.

Cuffy, K. M., and Clow, G. D. (1997). Temperature, accumulation and ice sheet elevation in central Greenland through the last deglacial transition. Journal of Geophysical Research, 102, 26383–26396.

Dahl, S. O., and Nesje, A. (1994). Holocene glacier fluctuations at Hardangerjökulen, central southern Norway: A high resolution composite chronology from lacustrine and terrestrial deposits. Holocene, 4, 269–277.

Dansgaard, W., White, J. W. C., and Johnsen, S. J. (1989). The abrupt termination of the Younger Dryas climate event. Nature, 339, 532–533.

Dansgaard, W., Johnsen, S. J., Clausen, H. B., Dahl-Jensen, D., Gundestrup, N. S., Hammer, C. U., Hvidberg, C. S., Steffensen, J. P., Sveinbjörnsdottir, A. E., Jouzel, J., and Bond, G. (1993). Evidence for general instability of past climate from a 250kyr icecore record. Nature, 364, 218220.

Denton, G. H., and Karlén, W. (1973). Holocene climatic variations—Their pattern and cause. Quaternary Research, 3, 155–205.

Denton, G. H., and Porter, S. C. (1970). Neoglaciation. Scientific American, 222, 101–110.

Denton, G. H., Armstrong, R. L., and Stuiver, M. S. (1971). The Late Cenozoic glacial history of Antarctica. In K. K. Turekian (Ed.), The Late Cenozoic Glacial Ages (pp. 267–306). New Haven, Connecticutt: Yale University Press.

Dorale, J. A., Gonzalez, L. A., Reagon, M. K., Pickett, D. A., Murrell, M. T., and Baker, R. G. (1992). A high-resolution record of Holocene climate change in speleothern calcite from Cold Water Cave, northeast Iowa. Science, 258, 1626–1630.

Dorf, E. (1960). Climatic changes of the past and present. American Scientist, 48, 341–364.

Duplessy, J. C., Labeyrie, D. J., Labou, C., and Hguyen, H. V. (1970). Continental climatic variations between 130,000 and 90,000 years B.P. Nature, 226, 631–633.

Eddy, J. A. (1976). The Maunder minimum. Science, 192, 1189–1202.

Eddy, J. A. (1977). Climate and the changing sun. Climatic Change, 1, 173–190.

Emery, K. O. (1969). The continental shelves. Scientific American, 222, 107–122.

Emiliani, C. (1971). Pleistocene paleotemperatures. Science, 257, 1462.

Friis-Christensen, E., and Lassen, K. (1991). Length of the solar cycle: An indicator of solar activity closely associated with climate. Science, 245, 698–700.

Griffey, N. J., and Worsley, P. (1978). The pattern of Neoglacial glacier variations in the Okstindan region of northern Norway during the last three millennia. Boreas, 7, 1–17.

Grove, J. M. (1972). The incidence of landslides, avalanches, and floods in western Norway during the Little Ice Age. Arctic and Alpine Research, 4, 131–138.

Grove, J. M. (1988). The Little Ice Age. London: Methuen.

Hobbs, H. C., and Bluemle, J. P. (1987). Geology of Ramsey County, North Dakota. Bismarck: North Dakota Geological Survey Bulletin 71, Part I.

Hodell, D. A., Curtis, J. A., Jones, G. A., Higuera-Gundy, A., Brenner, M., Binford, M. W., and Dorsey, K. T. (1991). Reconstruction of Caribbean climate change over the past 10,500 years. Nature, 352, 790–793.

Holmlund, P., Karlén, W., and Grudd, H. (1996). Fifty years of mass balance and glacier front observations at the Tarfala Research Station. Geografiska Annaler, 78A, 105–114.

Houghton, J. T. (1996). Climate change 1995: The science of climate change. Contribution of working group I to the second assessment report of the intergovernment panel on climate change. New York: Cambridge University Press.

Karlén, W. (1973). Holocene glacier and climatic variations, Kebnekaise Mountains, Swedish Lappland. Geografiska Annaler, 55A, 29–63.

Karlén, W. (1976). Lacustrine sediments and treelimit variations as indicators of Holocene climatic fluctuations in Lappland: Northern Sweden. Geografiska Annaler, 58A, 1–34.

Karlén, W. (1981). Lacustrine sediment studies. Geografiska Annaler, 63A, 273–281.

Karlén, W. (1982). Holocene glacier fluctuations in Scandinavia. Strie, 18, 26–34.

Karlén, W., and Kuylenstierna, J. (1996). On solar forcing of Holocene climate: Evidence from Scandinavia. Holocene, 6, 359–365.

Karlén, W., and Matthews, J. A. (1992). Reconstructing Holocene glacier variations from glacial lake sediments: Studies from Östra Peninsula and Jostedalsbreen-Jotunheimen, southern Norway. Geografiska Annaler, 74A, 327–348.

Keigwin, L. D. (1996). The Little Ice Age and Medieval Warm Period in the Sargasso Sea. Science, 274, 1504–1508.

Kennett, J. P., and Huddlestun, P. (1972). Abrupt climatic change at 90,000 yr. B.P.: Faunal evidence from Gulf of Mexico cores. Quaternary Research, 2, 384–395. (Note: Vol. 2, No. 3 is devoted to "The present interglacial: How and when will it end" and includes 24 papers.)

Kudrass, H. R., Erienkeuser, H., Vollbrecht, R., and Weiss, W. (1991). Global nature of the Younger Dryas cooling event inferred from oxygen isotope data from Sulu Sea cores. Nature, 349, 406–408.

Kullman, L. (1997). Tree-limit stress and disturbance—A 25-year survey of geological changes in the Scandes Mountains of Sweden. Geografiska Annaler, 79A, 139–165.

Lamb, H. H. (1982). Climate history and the modern world. London: Methuen.

Lean, J., Beer, J., and Bradley, R. (1995). Reconstruction of solar irradiation since 1610: Implications for climate change. Geophysical Research Letters, 22, 3195–3198.

Lindholm, R. C. (1976). Climatic change—Past, present and future. Journal of Geological Education, 24, 156–163.

Lowe, J. J., Coope, G. R., Keen, D., and Walker, M. J. C. (1994). High resolution stratigraphy of the last glacial-interglacial transition (LGIT) and inferred climatic gradients. In B. M. Funnell, and R. L. F. Kay (Eds.), Paleoclimate of the last Glacial-Interglacial Cycle (pp. 47–52). Cambridge, U.K.: National Environmental Research Council Earth Sciences Directories Special Publication No. 94/2.

Luckman, B. H., Briffa, K. R., Jones, P. D., and Schweingruber, F. H. (1997). Tree-ring based reconstruction of summer temperatures at the Columbia Icefield, Alberta, Canada, A.D. 1073–1983. Holocene, 7, 375–389.

Manley, G. (1974). Central England temperatures: Monthly means, 1659 to 1973. Quaternary Journal of Royal Meteorological Society, 100, 389–405.

Margolis, S. V., Kroopnick, P. O., Goodney, D. E., Dudley, W. C., and Mahoney, M. E. (1975). Oxygen and carbon isotopes from calcareous nanofossils as paleoceanographic indicators, Science, 1889, 555–557.

Matthews, J. A. (1991). The late Neoglacial ("Little Ice Age") glacier maximum in southern Norway: New ^{14}C-dating evidence and climatic implications. Holocene, 1, 219–233.

Matthews, J. A., and Karlén, W. (1992). Asynchronous neo-glaciation and Holocene climatic change reconstructed from Norwegian glaciolacustrine sedimentary sequences. Geology, 20, 991–994.

Matthews, J. A., and Shakesby, R. A. (1984). The states of the "Little Ice Age" in southern Norway: Relative-age dating of Neoglacial moraines with Smidt hammer and lichenometry. Boreas, 13, 333–346.

Meese, D. A., Gow, A. J., Grootes, P., Mayewski, P. A., Ram, M., Stuiver, M., Taylor, K. C., Waddington, E. D., and Zielinski, G. A. (1994). The accumulation record from the GISP2 core as an indicator of climate change throughout the Holocene. Science, 266, 1680–1682.

Mosley-Thompson, E. (1992). Paleoenvironmental conditions in Antarctica since A.D. 1500: Ice core evidence. In R. S. Bradley and P. D, Jones (Eds.), Climate since A.D. 1500 (pp. 572–591). New York: Routledge.

Nesje, A., and Dahl, S. 0. (1991). Holocene glacier variations of Blåisen, Hardangerjökulen, central south Norway. Quaternary Research, 35, 25–40.

Nesje, A., and Kvamme, M. (1991). Holocene glacier and climate variations in western Norway: Evidence for early Holocene glacier demise and multiple Neoglacial events. Geology, 19, 610–612.

Nesje, A., Kvamme, M., Rye, N., and Lövlie, R. (1991). Holocene glacial and climate history of the Jostedalsbreen region, western Norway: Evidence from lake sediments and terrestrial deposits. Quaternary Science Review, 10, 87–114.

Overpeck, J., Hughen, K., Hardy, D., Bradley, R., Case, R., Douglas, M., Finney, B., Gajewski, K., Jacoby, G., Jennings, A., Lamoureux, S., Lasca, A., MacDonald, G., Moore, J., Retelle, J., Smith, S., Wolfe, A., and Zelinski, G. (1997). Arctic environmental changes of the last four centuries. Science, 278, 1251–1255.

Parker, D. E., Jones, P. D., Folland, C. K., and Bevan, A. (1994). Interdecadal changes of surface temperatures since the late nineteenth century. Journal of Geophysical Research, 99, 14373–14399.

Ponte, L. (1976). The cooling. Englewood Cliffs, New Jersey: Prentice Hall, Inc.

Soon, W. H., Posmenties, E. S., and Baliunas, S. L. (1996). Inference of solar irradiance variability from terrestrial temperature changes, 1880–1993: An astronomical application of the sun/climate connection. Cambridge, MA: Harvard-Smithsonian Center for Astrophysics Preprint Series No. 4344.

Stuiver, M. S., and Braziunas, T. F. (1995). The GISP2 $\delta^{18}O$ climate record of the past 16,500 years and the role of the sun, ocean, and volcanoes. Quaternary Research, 44, 341.

Svensmark, H., and Friis-Christensen, E. (1997). Variation of cosmic ray flux and global cloud coverage—A missing link in solarclimate relationships. Journal of Atmospheric and Solar Terrestrial Physics, 59, 1225–1232.

Talma, A. S., and Vogel, J. C. (1992). Late Quaternary paleotemperatures derived from a speleotherm from Cango Caves, Cape Province, South Africa. Quaternary Research, 37, 303–313.

Vorren, K. D., Alm, T., and Mörkved, B. (1996). Holocene pine (Pinus sylvestris L.) and grey alder (Alnus incana Moench.) immigration and areal oscillations in central Troms, northern Norway, and their Palaeoclimatic implications. Palaeoclimatic Research, 20, 271–291.

White, J. W. C. (1993). Don't touch that dial. Nature, 364, 186.

Wigley, T. M. L., and Kelly, P. M. (1990). Holocene climatic change, ^{14}C wiggles and variations in solar irradiance. Philosophical Transactions of the Royal Society of London, 330, 547–560.

World Meteorological Organization. (1996). World Meteorological Organization statement on the status of the global climate in 1995. Geneva: World Meteorological Organization Publ. No. 838.

Davis, J. C., and G. C. Bohling, The search for patterns in
ice-core temperature curves, 2001, *in* L. C. Gerhard, W. E. Harrison, and B. M. Hanson, eds., Geological perspectives of global climate change,
p. 213–229.

11 | The Search for Patterns in Ice-Core Temperature Curves

John C. Davis
Geoffrey C. Bohling
Kansas Geological Survey
Lawrence, Kansas, U.S.A.

ABSTRACT

Predictions of global climate change are based on large computer-simulation models that are "history-matched" to weather records compiled from the early nineteenth century onward. Climate-change model forecasts would be more convincing if they were based on the natural records of the Holocene ($\approx$10,000 years) and were capable of simulating climate characteristics of this epoch. Temperature records estimated from $\delta^{18}O$ measurements on ice cores from the Greenland ice cap and the Antarctic could be used to develop models based on geochronological data rather than historically brief weather records.

The 20-year average record of $\delta^{18}O$ values from the Greenland Ice Sheet Project 2 (GISP2) ice core exhibits a long-term trend of declining temperatures over most of the Holocene, except during the last 100 years, when temperatures have increased—a change widely blamed on carbon-dioxide (CO_2) emissions from fossil fuels. However, the range in temperatures since the start of the industrial age is typical for the Holocene, and the current rate of increase in temperatures is unusual but not unprecedented. Past periods of consistently increasing (or decreasing) temperatures have not persisted much longer than the current interval, so temperature trends may well reverse in the near future. There are distinct cyclic patterns in temperatures recorded in the GISP2 ice core, including a pronounced sawtoothed 560-year sequence of relatively abrupt change followed by a gradual reversal. The present trend may be the initial phase of such a pattern. In summary, the present climate does not appear significantly different from the past climate at times prior to industrialization.

INTRODUCTION

Predictions of global climate change are primarily the result of extrapolations made by global climatic models. These models are very large dynamic computer simulations of the interactions between the global atmosphere and the oceans (a typical simulation model is described by Russell et al., 1995). Almost all global climate change models in current use are deterministic, comprised of such familiar modeling elements as diffusion equations, mass-transfer equations, and mass-balance conditions (Hansen et al., 1983). In design, these models consist of an array of two- or three-dimensional cells representing the atmosphere linked to equations representing various inputs and transfers to and from the oceans. The variables in the model include heat, atmospheric gases such as CO_2 and water vapor, airborne particulate material, and other constituents. Interactions between cells are controlled by equations that represent atmospheric circulation induced by pressure gradients that result from the Coriolis force, differences in topography, and differences in air density caused by insolation and other factors. The models are conditioned to the historic record of weather conditions and other proxy variables measured at stations scattered around the world (Mann et al., 1998). Both the number of stations and the lengths of the historic records are very limited.

Global climate change models have many parallels with models for petroleum reservoir production and groundwater flow. These geologic models are also deterministic and attempt to predict the movement of fluids in response to pressure gradients (reviews of reservoir modeling are provided by Mattax, 1990, and Peaceman, 1977; a review of hydrologic flow models is given in Anderson and Woessner, 1991). The geologic models are similar to climatic models in complexity, size, and structure, because they also are composed of a very large number of cell elements linked by fluid-flow and mass-balance equations. They are subject to the same scaling problems, because conditions measured at points—whether they represent weather stations or wells—must be extended somehow over very large cells. The models are tuned by the same history-matching process (Boberg et al., 1990), which consists of comparing the behavior of the model to known conditions recorded at observation points. Finally, the geological models used to forecast the future state of a reservoir or aquifer serve the same purpose as climate models that predict atmospheric changes (Jacks, 1990). Because of these broad similarities, the experience of petroleum geologists and hydrologists may be relevant for addressing the problems of global climate change modeling.

All models, which by their very nature are nonunique, share certain inherent problems and limitations. The failure of a model to match an observed history demonstrates that the model is incorrect, but a successful match indicates only that a specific model is among the class of feasible solutions. Even if a model produces a perfect simulation of the observed history, it is always possible to specify an alternative model that will match equally well.

All models are based on highly simplified conditions and physical assumptions. When compared with reality, whether in the myriad details of a reservoir or the atmosphere, the models are seen to be extremely crude in both resolution and behavior. Models suffer from a "degrees-of-freedom problem"—there are many parameters, but little data from which to estimate the values of the parameters. Finally, in most models, the forecasts are either unconstrained or only partially constrained. Once the model has been run past the end of the history on which its parameters have been conditioned, there are few, if any, limitations on the output. Good behavior during the history-matching phase of modeling is no guarantee that a model will continue to behave in a reasonable manner when extrapolations beyond the limits of the data are attempted.

Petroleum geologists and hydrologists have learned the value of experience in modeling. Creating and carrying out thousands of simulations have demonstrated the value of reservoir and aquifer characterization studies that examine spatial and temporal variations in real reservoirs and aquifers to help specify realistic parameters and set limits on models. Climate change models would benefit from similar studies, but unfortunately there is only one earth and only a short historic record to examine. Rather than relying on meager information gathered by human observers, climate modelers must identify and use more complete sources of data.

THE HOLOCENE RECORD

As a geochronologic unit, the Holocene is defined as the youngest epoch of the Cenozoic. Although the Holocene has no defined initial boundary, by general agreement it represents the interval of modern "postglacial" conditions following the rapid, final collapse of continental ice sheets in Scandinavia and North America (Jackson, 1997). The Holocene began approximately 10,000 years ago. From natural records formed during the Holocene, we can infer climatic conditions that could be used to condition global climate change models. Such natural records are almost two orders of magnitude longer than the historical records of meteorological measurements, so they should provide valuable insight into the changes we see in global climate at the present time if the natural records can be interpreted with sufficient accuracy. Natural records include ice cores such as those from Greenland and the Antarctic, varved lake and seafloor deposits, and growth rings from trees and corals—all of which preserve time records of some aspect of the climate (see Mann and Bradley, 1999).

The most extensive natural records that reflect Holocene climate change are contained in the ice cores drilled through the Greenland ice sheet and in the Antarctic. The Greenland Ice Sheet Project (GRIP) ice core was drilled through the ice cap in central Greenland in 1991 by a consortium of European research institutions. The GISP2 core was drilled somewhat later, in 1993, by the U.S. National Science Foundation. Both drilling sites are located near the summit of the Greenland ice sheet where lateral ice flow was expected to be at a minimum. Annual layers in the ice are discernable as cloudy bands caused by microscopic inclusions, allowing precise age dating of the cores by simply counting layers downward from the surface. (Ice-core ages are expressed in years Before Present (B.P.), where the "Present" is defined as 1950; thus, the year 2000 in the Gregorian calendar is −50 B.P.)

The Greenland ice sheet probably became established about 2.4 million years ago, but the ice-core record does not extend nearly so far into the past. The GISP2 core is almost 3000 m in length, representing about the last 240,000 years of ice deposition. However, the reliably interpretable portion of the ice cores (where annual layers can be counted) includes only the last 110,000 years. The perfect correlation between the GISP2 and GRIP ice cores down to this depth confirms that the ice has been preserved in a relatively undisturbed state in the uppermost part of the cores.

The ice cores have been meticulously sampled and analyzed for many chemical and physical properties, but the most significant of these, from a climatic point of view, is the ratio between the isotopes of oxygen, ^{16}O and ^{18}O. These two isotopes occur in an almost constant ratio in seawater, but are fractionated during evaporation and subsequent precipitation in a manner that is temperature dependent. The ratio of the two oxygen isotopes in snow records the temperature of formation of the snow, a record that is preserved as the snow is buried and turns to glacial ice (Robin, 1983). The measure of the isotopic ratio is expressed in relation to the mean ratio of seawater and is called $\delta^{18}O$. The relationship between $\delta^{18}O$ and temperature in Greenland has been empirically calibrated by comparing the ratios measured in snowfall to air temperatures at the time of snowfall.

In the GISP2 core, annual values of $\delta^{18}O$ have been measured for the last 4000 years. Below this depth, diffusion of water molecules makes it impractical to determine variations in $\delta^{18}O$ at an annual scale. To extend the temperature record beyond this limit, analyses have been determined as averages for successive 20-year intervals. It is this record of 20-year average values of $\delta^{18}O$ from the GISP2 ice core that we have used for our study.

Both long- and short-term trends in temperature patterns are detectable in the Holocene record. If the climate has changed as a result of anthropogenic causes, this should be expressed as different trends before and after the industrial revolution. Great significance has been attached to an apparent increase of global temperatures over the past 30 years, which is interpreted to be the result of a "greenhouse effect" attributed to higher levels of atmospheric CO_2 caused by the burning of fossil fuels.

Have there been comparable intervals of increasing temperature in the past when anthropogenic causes were not operative, and how long did these increases persist? In spite of the noise in the ice-core records, it should be possible to detect recurring patterns (regardless of their form) in the Holocene portion of the ice cores and to estimate the probability of occurrence of specific patterns. The "memory" or persistence of the patterns of temperature change can be estimated, which should help judge the reliability of forecasts made by global climate change models.

The Pleistocene epoch of the Cenozoic is characterized by several well-known climatic fluctuations, the most prominent of which are correlated with orbital phenomena—the Milankovitch cycles—that affected insolation (see Imbrie and Imbrie, 1980; Paul and Berger, 1999). More subtle, short-term fluctuations (Dansgaard-Oeschgar events) can be detected within the glaciated phases of the Pleistocene and may be the result of time lag in oceanic heat storage (Grootes et al., 1993). There are faint indications of similar fluctuations in the Holocene.

The Greenland ice cores have been analyzed not only for $\delta^{18}O$, but also for deuterium, CO_2, ammonia, sulfates, nitrates, various chemical elements, dielectric properties, laser light scattering, volcanic ash, ice-crystal size, layer thickness, and numerous other variables (National Snow and Ice Data Center User Services, 1997). Relationships among these variables may substantiate both the physical and statistical significance of features detected in the record. Ice cores from different localities also may be compared with each other to determine the persistence and consistency of features or patterns. Finally, the ice cores may be compared with other environmentally sensitive records from the Holocene, such as seabottom and lacustrine cores from stratified water bodies where varves were deposited, growth rings from corals and trees, and other records of the recent past (a survey of these records is provided by Lowe and Walker, 1997).

Because of the great length of natural records in comparison with the brief interval of manmade recordings of meteorological fluctuations, sources such as ice cores should provide invaluable information for studies of climate change. Statistical analyses of these data should provide better input for climate change models than may be derived from the historical record. Although numerous statistical analyses have been made of the ice-core data, most of these have either examined entire cores or have concentrated on specific intervals within the Pleistocene. Representative statistical studies are referenced in Yiou et al. (1998). By confining our analyses solely to the Holocene portion of the ice-core record, we have avoided many of the problems of nonstationarity and nonuniform sampling that have forced others to resort to "exploratory" methods. Also, we believe that results based on Holocene data are most pertinent to present climatic conditions.

LONG-TERM TRENDS IN TEMPERATURE

The present relationship between the $\delta^{18}O$ ratio in glacial ice and temperature, as determined by measurements in Greenland (Johnsen et al., 1992), is

$$\delta^{18}O = 0.67T - 13.7$$

or approximately 1°Celsius change in temperature for every 1.5 parts-per-thousand change in the ^{18}O ratio. We can examine the 20-year average data from the GISP2 ice core for the Holocene interval to see if present-day temperatures are unusually high. The record of 20-year averages of $\delta^{18}O$ consists of 502 values ranging in age from –30 B.P. (A.D. 1980) to 9990 B.P. (on the ice-core age scale, (present) is defined as 1950). Because these values are averages over 20-year intervals, the data are considerably smoother than annual $\delta^{18}O$ measurements; nonetheless, the ice-core record is highly erratic.

If we plot the values as a cumulative distribution on a normal probability scale **(Figure 1)**, we can see how typical (or atypical) our modern temperatures are. The data fall on an almost perfectly straight line, indicating that temperatures throughout the Holocene have followed a normal, or Gaussian, distribution—an interpretation confirmed by statistical tests. Individual 20-year average ^{18}O values measured on the GISP2 core since the start of the industrial revolution are indicated. Most of these measurements lie below the Holocene median value of $\delta^{18}O = -34.77$ (equivalent to 31.76°C at Summit, Greenland), indicating that modern temperatures have tended to be in the low part of the temperature range for the entire Holocene. The variation in 20-year average temperatures during the industrial age has been less than 1°C, and the median temperature during this time has not been significantly different from the median temperature throughout the Holocene. These statistical tests assume that temperatures have been stationary during the Holocene, when actually they exhibit a trend of progressively lower temperatures with time. The $\delta^{18}O$ data can be made stationary by converting the 20-year averages to deviations from a linear trend fitted by least squares. This transformation does not alter the test results.)

Some investigators (e.g., Mann and Bradley, 1999) have expressed concern that although temperatures in the twentieth century were not unusually high, there is an ominous trend toward higher temperatures. We can examine this possibility by using the time series formed by 20-year average $\delta^{18}O$ values from the GISP2 ice core. A scatter plot can be constructed of $\delta^{18}O$ 20-year averages against the year B.P. which each average represents, as shown in **Figure 2**. Then, using linear regression, we can calculate lines of best fit from the past to the present over successively shorter intervals. These fitted lines, as shown in **Figure 2,** estimate the slopes of any trends in temperature over time (the time scale has been transformed so that positive slopes indicate increasing temperatures). A comparison of the different fitted lines indicates how temperature trends have changed since the start of the Holocene.

As recorded in the GISP2 ice core, the Holocene began with a change in ratios that corresponded to an abrupt increase of about 7.5°C in average temperature. Starting at about 10,000 B.P., $\delta^{18}O$ ratios have tended downward at a rate of –0.000024 $\delta^{18}O$ per year (line A on **Figure 2**). This trend toward lower temperatures is statistically significant,[1] meaning that the trend is highly unlikely to be a result

---

[1] "Significant" has its usual statistical meaning of $p = 0.05$; the probability is 0.05 or less that an equal or greater test result will occur by chance when the hypothesis that the regression slope is zero is true. Sequential tests of shorter intervals have lower significance because the intervals are not independent.

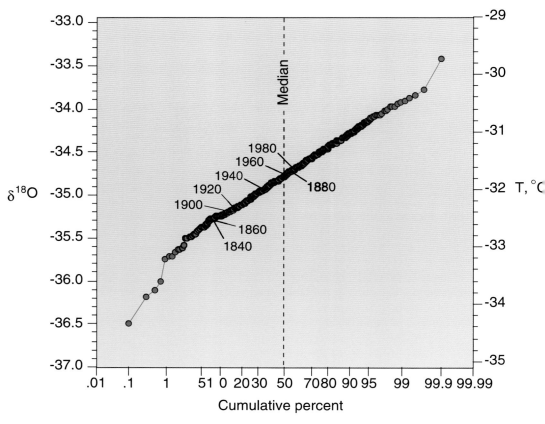

Figure 1 Cumulative distribution of 20-year average $\delta^{18}O$ ratios measured on the GISP2 Greenland ice core, plotted on a normal probability scale. Measurements of $\delta^{18}O$ for 20-year intervals since the start of the industrial revolution are indicated. Equivalent Greenland temperatures are determined by using the equation of Johnsen et al. (1992).

of random fluctuations in the record. Since the start of the Christian era, at about 2000 B.P., there ha[s] been an even steeper trend (–0.000196 $\delta^{18}O$ per year) toward lower temperatures (line B); this tren[d] also is statistically significant. In the period since the start of the "Little Ice Age" at approximately 7C[0] years B.P., temperatures have remained essentially constant (line C). The trend in $\delta^{18}O$ ratios is not sig[-] nificantly different from zero (slope = –0.000067 $\delta^{18}O$).

The industrial revolution began approximately 100 years B.P., and over the past century and half, the 20-year average temperatures have tended toward higher values (as shown in **Figure** [2,] line D, slope = +0.00351 $\delta^{18}O$ per year). However, the trend is not statistically significant becaus[e] there are so few observations in this short interval. In summary, throughout most of the Holocen[e] values and temperatures have declined. In recent years this trend has reversed, but the interval [of] time is too brief to determine whether the change is statistically meaningful.

We can determine whether the modern trend of increasing temperature is unusual by exam[-] ining the Holocene for similar trends over periods of time whose lengths are equal to the duratio[n]

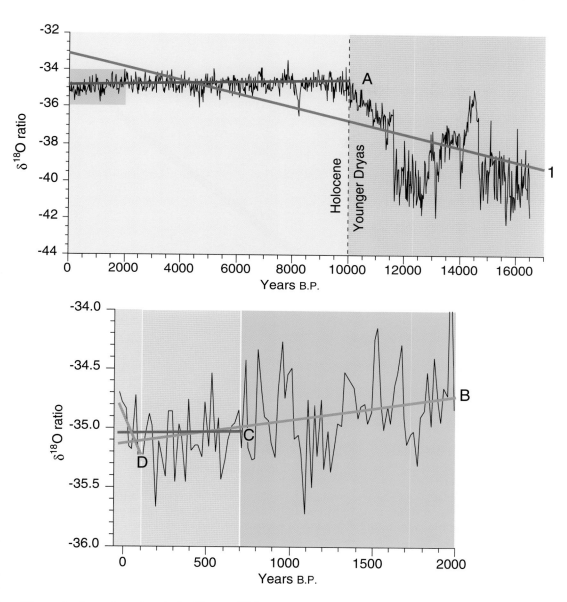

Figure 2 Long-term trends in 20-year $\delta^{18}O$ ratios measured on the GISP2 Greenland ice core. A = Trend since start of Holocene, 10,000 B.P. Slope = –0.000024**. B = Trend since start of Christian era, 2000 B.P. Slope = –0.000196**. C = Trend since start of Little Ice Age, 700 B.P. Slope = –0.000067[NS]. D = Trend since start of industrial revolution, 100 B.P. Slope = +0.003518[NS]. Positive slope indicates rising temperatures. ** = highly significant (p = 0.01), [NS] = not significant.

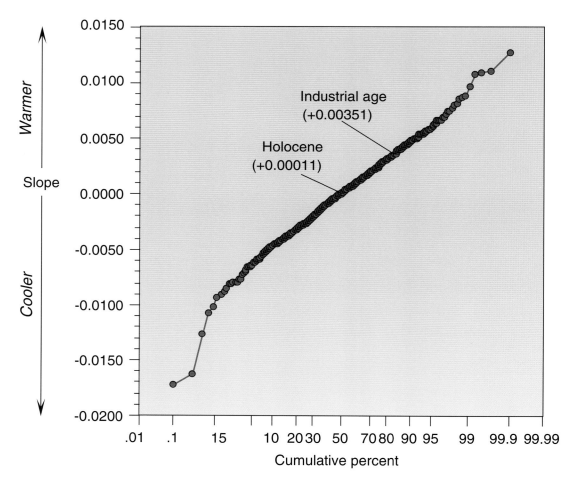

Figure 3 Cumulative distribution of slopes (in $\delta^{18}O$ per year) of trends over 140-year intervals in the record of 20-year $\delta^{18}O$ ratios measured on the GISP2 Greenland ice core. Slope of interval during the industrial age (1840–1980) is indicated, as is median of all slopes over 140-year intervals in Holocene.

of the industrial age (which we have taken as 1980 – 1840 = 140 years, or seven 20-year average intervals). **Figure 3** shows the cumulative distribution of the slopes of lines fitted over all such periods; the slope for the industrial age (+0.00351) is indicated, as is the median of the slopes of all periods of the same length in the Holocene. About 20%, or one out of every five intervals of time in the Holocene that were equal in length to the duration of the industrial age, were characterized by temperatures that increased at least as rapidly as during the present.

An alternative way of examining the uniqueness of recent trends in temperature can be made by looking at the record of "first differences." We start at the beginning of the Holocene at 10,000 years B.P. and compare each 20-year average $\delta^{18}O$ value to the average value during the preceding 20-year interval. If an interval is warmer than the interval that preceded it (that is, the interval has

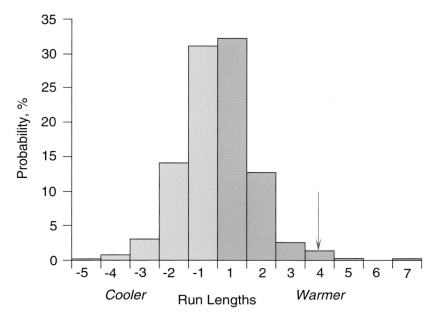

Figure 4 Histogram of lengths of runs of increasing or decreasing $\delta^{18}O$ ratios for the GISP2 20-year average ice-core record. Negative runs indicate increasing temperatures; positive runs indicate decreasing temperatures.

higher average $\delta^{18}O$ value than the preceding interval), it is assigned a plus (+) sign. If an interval is cooler than the preceding interval, the interval is assigned a minus (−) sign. We then count the number of "runs," or successions of the same sign. The histogram in **Figure 4** summarizes the runs of changes in temperature for the Holocene, based on 20-year average $\delta^{18}O$ values. The most recent run consists of four successive 20-year periods of increasing average $\delta^{18}O$, which is a positive run of length four, or 80 years (indicated by the arrow in **Figure 4**).

There are simple nonparametric tests to determine if a sequence of runs is random (see Conover, 1980). In the Holocene record of 20-year average $\delta^{18}O$ values, there are many more observed runs than would be expected in a random sequence, indicating that relatively warm and cool 20-year intervals alternate far more frequently than would be expected by random chance. In addition, periods of persistently increasing or decreasing $\delta^{18}O$ ratios (and hence temperatures) are rare. Longer runs do occur, however. If we assume the Holocene record provides a guide to the present, we can estimate there is about a 5% probability that an interval of rising average $\delta^{18}O$ ratios will last for more than 60 years, and a 2% probability that such an interval will last for more than 80 years. The longest interval of increasing temperatures in the Holocene had a duration of 140 years.

PERSISTENCE OF CONDITIONS

Although a runs analysis provides a simple way of examining the persistence of trends, runs suffer from the shortcoming that even a slight reversal will break what otherwise might be a long-term pattern of increasing or decreasing values. Another approach is to borrow a procedure familiar to

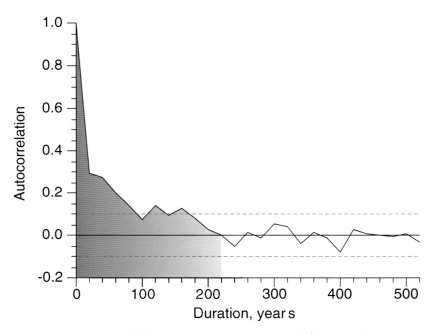

Figure 5 Autocorrelogram of GISP2 20-year average record of $\delta^{18}O$ ratios. Shaded interval indicates duration of statistically significant (95% probability) autocorrelation.

geophysicists and estimate the autocorrelation function. Autocorrelation is a basic measure of temporal continuity in time-series analysis as well as seismology. It quantifies the similarity in a time series between observations and other observations that are separated by equal increments of time. The higher the autocorrelation, the greater the similarity between observations being compared. Because the time series is observed at uniform intervals (averages over 20-year segments in the Holocene ice-core data used here), the autocorrelation can be plotted against lags, or the number of intervals separating the observations being compared. Because the autocorrelation for a specific interval of time is calculated over all possible intervals of the same duration in the record, the characteristics of a specific, individual interval have a limited impact on the computed autocorrelation. The autocorrelation function expresses the average behavior of the $\delta^{18}O$ ratio as a function of time.

Figure 5 shows the autocorrelation of 20-year average $\delta^{18}O$ ratios for the Holocene portion of the GISP2 ice core. To produce an unbiased estimate of the autocorrelation function, the record should be stationary, or have the same average value over its entire length. A nonstationary time series can be leveled by subtracting a linear trend such as that shown by line A in **Figure 2**. The autocorrelation is calculated based on the residuals that remain after the fitted line has been removed. A theoretical autocorrelation function for an infinitely long random sequence will begin at 1.0 for lag 0, then immediately drop to a value of 0.0 for all successive lags. This means there is no relationship between values at a given time and values at the immediately following time, or at any other time. Even though in theory we expect the autocorrelation of a random sequence to be 0.0 for all lags, a real sequence of limited length will yield an empirical autocorrelation function that fluctuates around zero. By considering the sampling error, which is a function of the number of observations in the sequence and their variability, we can estimate confidence intervals for the autocorrelation function. The dashed horizontal lines above and below the zero line in **Figure 5** represent upper and lower 95%

confidence limits. If the process that gave rise to the observed sequence of 20-year average $\delta^{18}O$ values were truly random and created a large number of alternative realizations, we would expect the autocorrelation functions of 95% of these sequences to fall within the confidence limits.

The most obvious characteristic of the autocorrelation function shown in **Figure 5** is that lags 1 through 4 (20 through 80 years) exceed the upper 95% confidence limit. This indicates that there is a statistically significant tendency for similar $\delta^{18}O$ values to persist over an interval of about 80 years. The autocorrelation function does not drop to zero until about lag 10 (200 years), although the autocorrelations at longer lags are not significant. How can these findings be reconciled with results from the runs analysis? First, the autocorrelation reflects persistence without regard to whether $\delta^{18}O$ values are increasing or decreasing, so runs up and runs down both contribute to the magnitude of the autocorrelation. Second, a temporary reversal in the middle of what otherwise would be a sequence of persistently increasing or decreasing $\delta^{18}O$ values will not greatly affect the autocorrelation but will terminate a run. Perhaps a collective interpretation of these two analyses is that the Holocene climate has tended to maintain stable patterns of 80 to 200 years' duration, but superimposed on these are rapid alternations of increasing or decreasing temperature.

RECURRENCE OF PATTERNS

To search ice-core records for patterns of greater complexity than simple long- or short-term trends, we must resort to methods that are mathematically more sophisticated. We have employed two methods in an attempt to identify recurring patterns in the 20-year average $\delta^{18}O$ data from the Holocene portion of the GISP2 ice core. The first procedure is spectral, or Fourier, analysis, another technique familiar to geophysicists and to statisticians as well. The record of $\delta^{18}O$ ratios can be considered a "signal" that we can decompose by a Fourier transformation into a set of sinusoids, each having a specific frequency and phase angle. The ice-core record is characterized by a total amount of variation, which is apportioned among the different sinusoids. The power, or proportion of variation associated with each of the different periodicities, can be estimated and displayed as a spectrum, or plot of power versus periodicity.[2]

Figure 6 is the power spectrum of the 20-year average $\delta^{18}O$ ratios from the Holocene portion of the GISP2 ice core. **Figure 7** is the spectrum for the record of 2-m average $\delta^{18}O$ data from the GISP2 ice core in which measurements of $\delta^{18}O$ are averaged in sets representing 2 m of core. (This second ice-core record contains values averaged over fixed intervals of depth rather than fixed intervals of time. Both records are based on measurements made on the same physical ice core.) The periods corresponding to selected peaks in the spectrum are indicated on the figures. Note that the periods identified on the spectrum of the 20-year average data **(Figure 6)** are not integer multiples of 20 years, as might be expected as a result of the 20-year interval between observations. The use of "oversampling" in the frequency domain permits this to occur.

---

[2]The spectral analyses presented here have been made using Schulz and Statteger's (1997) program, SPECTRUM, which averages several Welch-overlapped segments to produce smooth spectra. This procedure divides the sequence of observations into overlapping subsegments, performs a Fourier transform on each subsegment, and averages the resulting spectra. Each subsegment is individually detrended and windowed prior to computing the Fourier transforms. Windowing consists of multiplying each subsegment by a function (in this instance, a Hanning window) which smoothly tapers the data at the ends of the subsegments in order to reduce frequency leakage that results from segmenting the entire time series. Random variation in the individual spectra tends to cancel out when averaged together, so the resulting average transform is smoother and it is easier to identify significant peaks. This procedure avoids many of the problems caused by nonstationarity and nonuniform sampling in conventional spectral analysis.

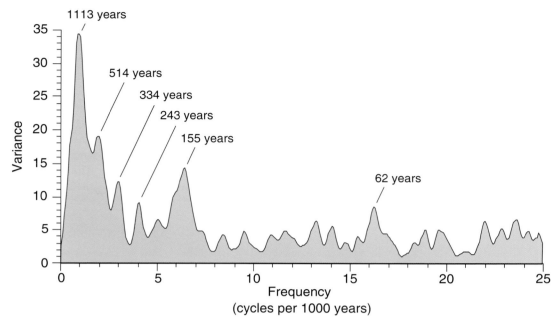

Figure 6 Power spectrum of 20-year average record of $\delta^{18}O$ ratios measured on the Holocene portion of the GISP2 ice core. Wavelengths, in years, of dominant peaks are labeled.

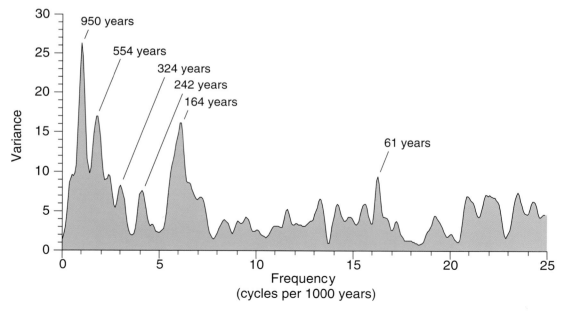

Figure 7 Power spectrum of 2-m average record of $\delta^{18}O$ ratios measured on the Holocene portion of the GISP2 ice core. Wavelengths, in years, of dominant peaks are labeled.

Clearly prominent peaks on both **Figures 6** and **7** represent periods of 1113 or 950 years, 155 or 164 years, and 61 or 62 years, depending on the data series. The peaks for periods of 334 and 243 years on the spectrum calculated from the 20-year average data **(Figure 6)** seem to correspond to less prominent peaks for 324-year and 242-year periods on the spectrum calculated from 2-m average data **(Figure 7)**. The prominent peak at 554 years on the spectrum of the 2-m average data appears to correspond to the less pronounced 514-year peak on the spectrum from the 20-year average data. The resolution of spectral analysis decreases with decreasing frequency, so these two spectra seem to contain essentially the same fundamental periodicities, as we might anticipate, because the data come from the same ice core. It is possible that the 334-year peak (324-year peak on the 2-m spectrum) is a multiple of the 155-year (164-year) peak, and the 242-year (243-year) peak is a multiple of the 61-year (62-year) peak. If we consider both spectra to be estimates of an underlying signal recorded in the ice core, we can summarize the results of the spectral analysis as revealing periodicities in the signal at approximately 1000 years, 160 years, and 60 years, with possible harmonics at about 320 years and 240 years.

Figure 8 shows the spectrum for the 1-year average $\delta^{18}O$ data from the GISP2 Holocene core, computed in a similar manner to spectra shown in **Figures 6** and **7**. Because of increasing diffusion in the ice, it is possible to determine 1-year average $\delta^{18}O$ only to 1132 years B.P., so the earliest year in the annual record is A.D. 818. Periods are identified for several of the prominent peaks on the spectrum. The approximate 160-year peak in the spectrum for the 20-year averaged $\delta^{18}O$ data **(Figures 6** and **7)** appears to be present as a 156-year peak in the spectrum for the one-year average $\delta^{18}O$ data. Peaks at higher frequencies represent periodicities of 71, 40, and 28 years. There is no peak corresponding to the 60-year periodicity apparent in the 20-year or 2-m average records.

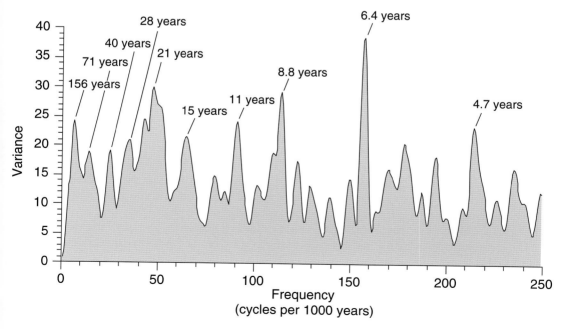

Figure 8 Power spectrum of annual average record of $\delta^{18}O$ ratios measured on the GISP2 ice core for years between A.D. 818 and A.D. 1980. Wavelengths, in years, of dominant peaks are labeled.

Other methods may be used to discern climatic patterns in natural records. In Fourier analysis, any continuous signal is regarded as the sum of a series of sinusoidal waveforms whose contributions are expressed in the power spectrum. If the original signal contains periodic constituents that are not sinusoidal in form (square or sawtoothed waves, for example), an inordinately large number of high-frequency sinusoids may be required to duplicate the abrupt changes in the signal. This problem is avoided in the optimal eigenstate method proposed by Ralston (1998), which can decompose a time series such as the record of 20-year average $\delta^{18}O$ ratios into a set of orthogonal basis functions (that is, functions that are mutually independent of each other) of arbitrary form.

In Fourier analysis, the basis functions are the set of sines and cosines which define the spectrum, while in the optimal eigenstate method the basis functions are found from patterns within the data themselves. Because the eigenstate functions are not constrained to be sinusoids (and, in fact, can be of any shape), fewer functions may be required to reproduce the signal. Calculations for the optimal eigenstate method begin by dividing the record of $\delta^{18}O$ ratios into nonoverlapping segments. Next, each segment is premultiplied by its transpose and the resulting square matrices are summed to form a composite matrix. The eigenvalues and eigenvectors of this matrix are extracted and examined. If there are recurring patterns in the ice-core record, the durations of which correspond to the length of time represented by the segments, the patterns will be expressed by the eigenvectors associated with the largest eigenvalues of the composite density matrix. By specifying a succession of segments of different lengths, the optimal eigenstate method can be used to extract from the record characteristic patterns that correspond to these lengths. In essence, the segment lengths correspond roughly to the periods or wavelengths of spectral analysis.

We used Ralston's optimal eigenstate method to search the 20-year average $\delta^{18}O$ ice-core record for patterns corresponding approximately to the three most pronounced peaks of the spectra shown in **Figures 6** and **7**, which represent periods of about 160, 560, and 960 years. The optimal eigenstate method works best on a stationary time series with a mean of zero, so the data were transformed into residuals from a linear trend of the 20-year average $\delta^{18}O$ data. **Figure 9A** is a plot of the successive eigenvalues extracted using a segment length of 28, corresponding to a period of 560 years. The first eigenvalue is the largest of any extracted from the ice-core data, and it accounts for 27% of the variation in the entire Holocene record of 20-year average $\delta^{18}O$ values. The second and third eigenvalues account for substantially less variation (13% and less than 10%, respectively).

If we plot the loadings on the eigenvectors associated with each eigenvalue, we see the patterns that the optimal eigenstate method has found in the ice-core record. The plots are constructed so that younger dates appear on the left. For example, **Figure 9B** is the first, or dominant, pattern found within segments of 560 years' duration. The pattern is a "sawtooth" with an abrupt increase in the $\delta^{18}O$ ratio, followed by a gradual decrease. This pattern may be either positively or negatively present in any particular segment. About half of the 560-year segments in the Holocene record are positively correlated with this eigenvector, meaning that they tend to show an abrupt increase in $\delta^{18}O$ values followed by a gradual decrease. The remaining segments are negatively correlated, indicating that they tend to show an abrupt decrease in $\delta^{18}O$ followed by a gradual increase.

The patterns formed by the second and third eigenvectors of the 560-year segments **(Figures 9C and 9D)** are much less distinctive. However, the second eigenvector accounts for only about 13% of the total variation in the record, and the third eigenvector accounts for less than 10% of the total variation. Thus, we do not anticipate that their patterns will be conspicuous in the record. **Figure 10** shows the Holocene sequence of $\delta^{18}O$ ice-core residuals in comparison with a reconstructed sequence built from the three eigenvectors of **Figure 9**. The reconstruction accounts for 50% of the total variation in the original record and essentially reflects the sawtooth, "abrupt change–gradual recovery" pattern seen in **Figure 9B**.

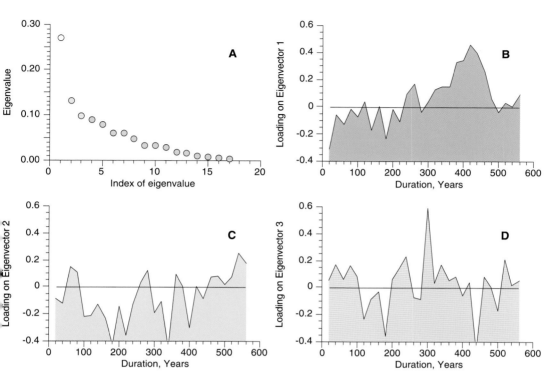

Figure 9 Optimal eigenstate analysis for 560-year periods in GISP2 20-year record of average $\delta^{18}O$ data. (A) Magnitudes of eigenvalues 1 through 17. (B) Loadings on eigenvector 1. (C) Loadings on eigenvector 2. (D) Loadings on eigenvector 3.

SUMMARY

National environmental policies are being formulated on the basis of predictions made by global climate change models that are, for the most part, conditioned only on brief historical records of climate (e.g., see Office of Science and Technology Policy, 1997). These models emphasize anthropogenic causes and sometimes extrapolate dire predictions of extreme climate change in the near future (Wigley, 1989). Global climate modelers are limited to one world—the earth—and have chosen to use only the short, man-made historical record for history-matching. Experience with reservoir and geohydraulic models, which are comparable to global climate change models in both complexity and construction but which have been applied to thousands of reservoirs and aquifers, suggests that global climate forecasts lack sufficient constraints due to the limited amount of data available for model conditioning. Unfortunately, there are no other worlds comparable to earth, no "laboratory" where we can test global climate simulation models. But there are much more extensive climatic histories at our disposal from our own planet. Ice cores and other natural records of climatic variables provide such histories. A cursory examination of one ice-core record for the Holocene shows that there are trends, patterns, and other features that must be accounted for if any climate change model is to be deemed convincing.

The temperature record of the Holocene epoch, based on 20-year average $\delta^{18}O$ values measured on the GISP2 ice core, indicates that present-day temperatures are typical for the Holocene.

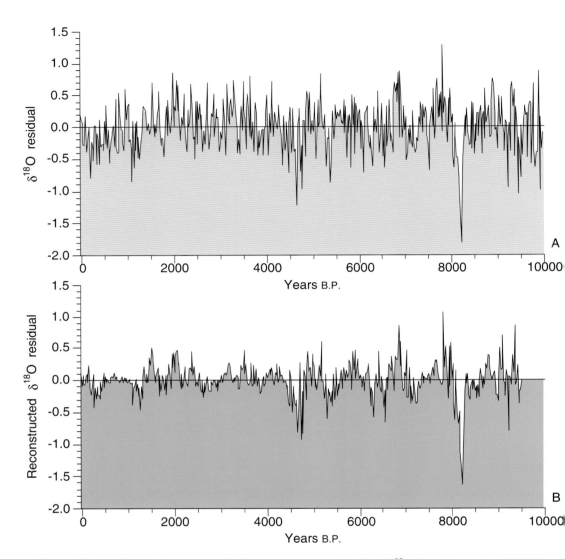

Figure 10 (A) Original sequence of 20-year average residuals of δ¹⁸O from linear trend. (B) Reconstructed sequence of estimated 20-year average residuals of δ¹⁸O is based on an optimal eigenstate analysis for a segment length of 560 years.

Over the past 140 years, Greenland temperatures, as recorded in the ice core, have risen at an unusually steep rate. However, such an increase is not unprecedented, because periods of similar or greater increase in temperature and of similar or greater duration have occurred before in the Holocene. Autocorrelation analysis suggests that the present trend may already be past its midpoint. The modern weather record may be the latest instance of patterns that have repeated throughout geologic history, as demonstrated by pre-industrial age Holocene examples. Blaming the current warming trend of earth's climate on human activity may simply be anthropocentric hubris.

REFERENCES CITED

Anderson, M. P., and Woessner, W. W., 1991, Applied Groundwater Modeling: Simulation of Flow and Advective Transport: Academic Press, New York, 381 p.

Boberg, T. C., Dalton, R. L., and Mattax, C. C., 1990, History matching: Testing the validity of the reservoir model, *in* Mattax, C. C., and Dalton, R. L., Reservoir Simulation: Society of Petroleum Engineers, Monograph, v. 13, p. 87–98.

Connover, W. J., 1980, Practical Nonparametric Statistics, 2nd. ed.: John Wiley & Sons, Inc., New York, 493 p.

Grootes, P. M., M. Stuiver, J. W. C. White, S. Johnsen, and J. Jouzel, 1993, Comparison of oxygen isotope records from the GISP2 and GRIP Greenland ice cores: Nature, v. 366, p. 552–554.

Hansen, J. E., Russell, G., Rind, D., Stone, P., Lacis, A., Lebedeff, S., Ruedy, R., and Travis, L., 1983, Efficient three-dimensional global models for climate studies: Models I and II: Monthly Weather Review, v. 111, p. 609–662.

Imbrie, J., and Imbrie, J. Z., 1980, Modeling the climatic response to orbital variations: Science, v. 202, p. 943–953.

Jacks, H. H., 1990, Forecasting future performance, *in* Mattax, C. C., and Dalton, R. L., Reservoir Simulation: Society of Petroleum Engineers, Monograph, v. 13, p. 99–110.

Jackson, J. A., ed., 1997, Glossary of Geology, 4th ed.: American Geological Institute, Alexandria, Virginia, 769 p.

Johnsen, S. J., Clausen, H. B., Dansgaard, W., Fuhrer, K., Gundestrup, N. S., Hammer, C. U., Iverson, P., Jouzel, J., Stauffer, B., and Steffensen, J. P., 1992, Irregular glacial interstadials recorded in a new Greenland ice core: Nature, v. 359, p. 311–313.

Lowe, J. J., and Walker, M. J. C., 1997, Reconstructing Quaternary Environments, 2nd. ed.: Addison Wesley Longman Ltd., Essex, U.K., 446 p.

Mann, M. E., and Bradley, R. S., 1999, Northern Hemisphere temperatures during the past millennium: Inferences, uncertainties, and limitations: Geophysical Research Letters, v. 26, no. 6, p. 759–762.

Mann, M. E., Bradley, R. S., and Hughes, M. K., 1998, Global-scale temperature patterns and climate forcing over the past six centuries: Nature, v. 392, p. 779–787.

Mattax, C. C., 1990, Modeling concepts, *in* Mattax, C. C., and Dalton, R. L., Reservoir Simulation: Society of Petroleum Engineers, Monograph, v. 13, p. 6–12.

NSIDC User Services, 1997, The Greenland Summit Ice Cores CD-ROM, GISP-2/GRIP: World Data Center A for Glaciology, CIRES, Univ. Colorado, Boulder.

Office of Science and Technology Policy, 1997, Climate Change State of Knowledge: Executive Office of the President, Washington, D.C., 18 p.

Paul, A., and Berger, W. H., 1999, Climate cycles and climate transitions as a response to astronomical and CO_2 forcings, *in* Harff, J., Lemke, W., and Stattegger, K., eds.: Computerized Modeling of Sedimentary Systems, Springer-Verlag, Berlin, p. 223–245.

Peaceman, D. W., 1977, Fundamentals of Numerical Reservoir Simulation: Elsevier Science Pub. B.V., Amsterdam, 176 p.

Ralston, J., 1998, The optimal eigenstate method: Unpublished report, Department of Physics, Univ. of Kansas, Lawrence, 21 p.

Robin, G. deQ., ed., 1983, The Climatic Record in Polar Ice Sheets: Cambridge Univ. Press, 212 p.

Russell, G. L., Miller, J. R., and Rind, D., 1995, A coupled atmosphere-ocean model for transient climate change studies: Atmosphere–Ocean, v. 33, p. 683–730.

Schulz, M., and K. Statteger, 1997, SPECTRUM: Spectral analysis of unevenly spaced paleoclimatic time series, Computers & Geosciences, v. 23, no. 9, pp. 929–945.

Wigley, T. M. L., 1989, Measurement and prediction of global warming, *in* Jones, R. R., and Wigley, T. M. L., eds., Ozone Depletion: Health and Environmental Consequences: John Wiley & Sons Ltd., London, p. 85–97.

Yiou, P., Fuhrer, K., Meeker, L. D., Jouzel, J., Johnsen, S. J., and Mayewski, P. A., 1998, Paleoclimatic variability inferred from the spectral analysis of Greenland and Antarctic ice-core data: Journal of Geophysical Research, v. 102, p. 26441–26454.

Harff, J., A. Frischbutter, R. Lampe, and M. Meyer, Sea-level change in the Baltic Sea: Interrelation of climatic and geological processes, 2001, *in* L. C. Gerhard, W. E. Harrison, and B. M. Hanson, eds., Geological perspectives of global climate change, p. 231–250.

12 | Sea-Level Change in the Baltic Sea: Interrelation of Climatic and Geological Processes

Jan Harff
Baltic Sea Research Institute
Warnemünde, Germany

Alexander Frischbutter
GeoForschungsZentrum
Potsdam, Germany

Reinhard Lampe
Geographical Institute
University of Greifswald
Greifswald, Germany

Michael Meyer
Institute of Geological Sciences
University of Greifswald
Greifswald, Germany

ABSTRACT

The development of the Baltic Sea during the Holocene was mainly controlled by climatically driven eustatic sea-level change and vertical crustal movements. Both factors affect sea-level changes interpreted from sedimentological investigations at coastal locations. Comparison of relative sea-level curves with a eustatic curve reveals the contribution of vertical crustal movements to coastal changes as expressed via a "coast index." A coast index c_i for the Baltic region can be derived, by which a location can be allocated to either a "crustal-uplift/subsidence type" or a "climate-controlled type." Coastal locations investigated along the Fennoscandian Shield belong to the crustal-uplift type and locations along the southern and southwestern coast belong to the climate-controlled type, regardless of whether they are on the East European Platform or the West European Platform. Data on recent vertical crustal movements show a broad, predominantly subsiding zone between the rising blocks of Scandinavia (glacio-isostatic uplift) and the Carpathians (northward drift of the African Plate) to the west of the Tornquist-Teisseyre Zone (TTZ). Movements may additionally be influenced by processes initiated along the North Atlantic Mid-Ocean Ridge. The subsiding belt contiguous to the Fennoscandian Shield is interpreted as a collapsing asthenospheric bulge originally surrounding the Pleistocene ice shield. The analysis of relative sea level changes leads to the assumption that the process of collapse reached a steady state during the Late Litorina Stage. Crustal movement data, together with data from modeling of future sea-level change, can be used for calculating scenarios of relative sea-level development, providing a background for long-term planning of human activities in coastal areas.

INTRODUCTION

Research over the last few decades on the effect of global climate change on sea-level variations has attracted broad interest. Coastline changes have economic and social impact on a human population concentrated along coastlines. The investigation of coastal change processes becomes important for future planning, and the derivation of scenarios must be based on an understanding of the driving processes. Because it is well known that coastal change is a complex result of an interaction of climatically driven eustatic sea-level change and vertical crustal movements, both aspects have to be taken into consideration. Recent questions raised by planning agencies on future relative sea level change require dealing with both the problems of climate change processes *and* crustal dynamics in order to merge the results for comprehensive answers. For the exogenic component numerical climate models provide scenarios of sea-level change for the next 700 to 1000 years. For endogenically caused vertical crustal movement, we are still in need of reliable results from corresponding models. The reason for the lack of such models with a subregional resolution lies in the complex nature of geodynamics, which allows us to predict vertical crustal movements on a regional scale but not at the resolution needed for subregional sea-level scenarios. A way out of this dilemma can be provided by a combination of the analysis of changes in paleo-sea-level curves and of measurements of recent crustal movements and their extrapolation to the future, along with eustatic-change scenarios based on climate models.

Problems of long-term sea-level change in the Baltic Sea are being investigated in the framework of a project launched in 1999 by the German Science Foundation. The semienclosed Baltic Sea is an area on the boundary between the East and West European Platforms, where the problems of interrelations between eustatic change and vertical crustal movements can be studied in a unique manner.

This paper describes the project's concept and scientific results: the analysis of existing data delivered by other international and national research projects. Sea-level data were taken mainly from the International Geological Correlation Program (IGCP) Project No. 274, "Coastal Evolution during the Quaternary," and from the project "Klimaänderungen und Küste" (KLIBO), funded by the Governmental Department of Science and Technology, Federal Republic of Germany. The Project "Baltic Sea System Study" (BASYS), sponsored by the European Commission, provided data for the Holocene history of the Baltic Sea. The data on recent vertical crustal movements were compiled in the framework of IGCP-Project No. 346, "Neogeodynamica Baltica," and used to estimate and predict future sea-level and coastline changes.

THE BALTIC SEA: GEOTECTONIC SETTING AND HYDROGRAPHY

The Baltic Sea is a semienclosed intracontinental sea surrounded by the landmasses of Scandinavia, northern central Europe, and northeastern Europe. **Figure 1** shows the main tectonic units schematically. The Baltic Sea basin tectonically bridges the East European Platform and the West European Platform, separated by the deep fracture of the northwest-southeast-striking Tornquist-Teisseyre Zone (TTZ) and its northwestern prolongation, the Sorgenfrei-Tornquist Zone (STZ). Northeast of this zone, we distinguish between the Fennoscandian Shield with outcrops of Precambrian crystalline rocks and the Russian Plate with flat-lying Phanerozoic sediments on a Precambrian basement. An exception is formed by the Baltic Syneclise, a Paleozoic basin structure occupying the northwestern part of the Russian Plate. West of the TTZ, the Central European

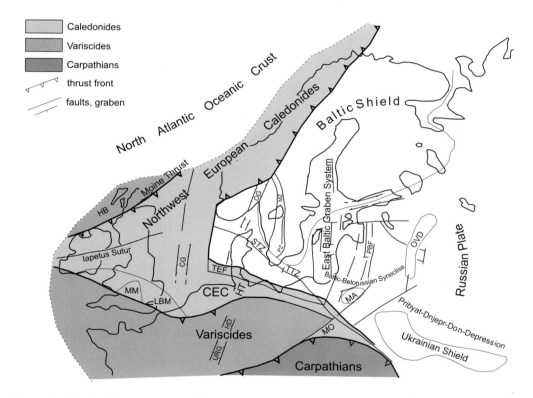

Figure 1 Regional tectonic units of northern Europe. The deep-fracture Tornquist-Teisseyre Zone (TTZ) and its northwestern prolongation, the Sorgenfrei-Tornquist Zone (STZ), separate the main tectonic units of the Baltic area: the East European Platform and the West European Platform. The first one consists of the Baltic (Fennoscandian) Shield with outcropping Precambrian crystalline rocks and the Russian Plate with flat-lying Phanerozoic sediments on a Precambrian basement. The Northwest European Caledonides border the Baltic Shield in the northwest. West of the TTZ the Central European Caledonides and Variscides together form the Central European Depression.

HB: Hebridic Shield; MM: Midland Massif; LBM: London Brabant Massif; CG: Central Graben; MZ: Mybonite Zone; PZ: Protegin Zone; MA: Masurian Anteclise; OVD: Orsha-Valday Depression; PBF: Pribaltic Faults; MO: Moravosilesian; TEF: Trans-European Fault; CEC: Central European Caledonides; URG: Upper Rhein Graben; HD: Hessian Depression; HT: Hamburg Trog; OG: Oslo Graben

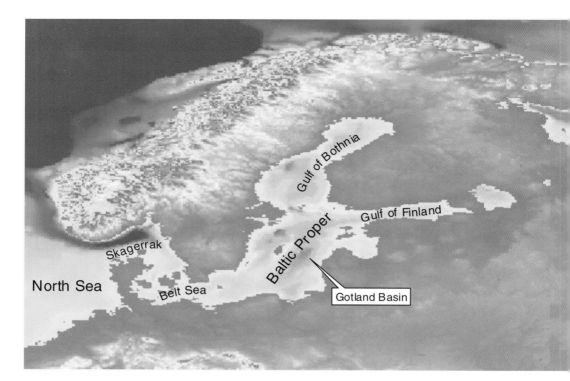

Figure 2 Digital elevation model (DEM) of the Baltic Sea area, data from ETOPO5 (Edwards, 1989). Quaternary glaciations created the morphology of the Baltic region. The relief of the Northwest European Caledonides and the surface of crystalline Precambrian rock of the Baltic Shield were shaped by glacial erosion, whereas the lowlands of the Russian Plate and West European Platform are covered by glacial sediments. The Baltic Sea Basin (a series of subbasins separated by shallow sills) was formed by glacial excavation. The Belts and the Sound connect the Baltic Sea to the world ocean.

Caledonides (CEC) and Variscides together form the Central European Depression, a deep sedimentary basin. In the northwest, the Fennoscandian Shield is bordered by the Northwest European Caledonides. Quaternary glaciations created the morphology of the Baltic region **(Figure 2)**. The relief of the Northwest European Caledonides, with elevations up to 2470 m, and the surface of crystalline Precambrian rock of the Fennoscandian Shield, were shaped by glacial erosion, whereas the lowlands of the Russian Plate and West European Platform are covered by glacial sediments. Glaciers also have excavated the Baltic Sea Basin itself (with an average water depth of 55 m) and formed a series of subbasins (Mecklenburgian Bight, 25 m; Arkona Basin, 50 m; Bornholm Basin, 100 m; Gotland Basin, 230 m; Gulf of Bothnia, 120 m) separated by shallower sills. The Baltic Sea is connected with the North Sea by the Belts and the Sound, which form a bottleneck for exchange of water with the world ocean. The humid climate and a positive water balance produce estuarine circulation and a stratified water body with remarkable vertical and horizontal differences in salinity, density, and temperature (Matthäus and Franck, 1992; Wulff et al., 1990). The halocline prevents vertical water exchange and leads to anoxic conditions below a permanent redoxcline, particularly in the deep central basin. **Figure 3** shows the oxygen content within the Baltic Sea in a

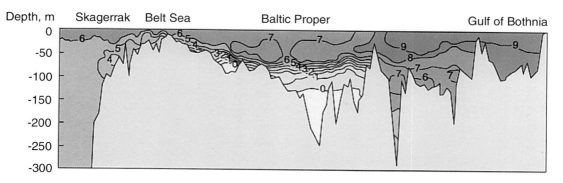

Figure 3 Vertical oxygen concentration in the Baltic Sea from the Skagerrak to the Gulf of Bothnia, summer 1988 (ml/l), after Sjöberg (1992). An estuarine circulation is typical for the Baltic Sea as a semienclosed marginal sea in the humid climate zone. The stratified water body prevents vertical water exchange and leads to anoxic conditions below a permanent redoxcline in the deep central basin. The absence of bioturbation causes laminated sediment sequences recording environmental change with high resolution.

schematic cross section from the Skagerrak, through the Belt and the Baltic proper to the Gulf of Bothnia. In the deep basins, the absence of higher benthic biota saves the sediments from bioturbation and causes laminated sequences that record environmental change with high age resolution. Studies of basin sediments and coastline development reveal the history of the Baltic Sea and its driving forces: climate and geological processes interacting in changing proportions.

THE BASIN SEDIMENTS: A MIRROR OF CLIMATE CHANGE

The deposition of freshwater and brackish-marine sediments within the central basins of the Baltic Sea reflects the change of climate after the Weichselian glaciation in northern Europe. Several attempts have been made to subdivide the development of the Baltic Sea Basin into several stages, starting with the early study of Sauramo (1958). The most recent subdivision of basin stages, based on a combination of lithostratigraphic and paleoecological methods, was published by Andren (1999). Another approach based on numerical analysis of sediment-physical multisensor core logging data (MSCL) was introduced by Harff et al. (1999). Three sediment-physical variables (sound velocity, wet bulk density, magnetic susceptibility) from master core 211660-5 within the eastern Gotland Basin (57° 17.00'N, 20°7.13'E) were used to subdivide the sequence into homogeneous statistical units. The subdivision corresponds to the lithofacies of the core and can be correlated with other contiguous sediment cores within the center of the basin. **Figure 4** shows the correlations of the subdivisions with the Baltic Basin Stages of Andren et al. (1999), based on lithological, geochemical, and paleoecological analysis of diatoms in piston-core 211660-1 taken at the same core station cited above. The lower parts of the core (units A-1 and A-2) represent the Baltic Ice Lake (12,000–11,570 calibrated [cal.] ^{14}C-years before present [B.P.]) with varved late glacial clays. The sediments were deposited in a fresh meltwater reservoir in front of the retreating Weichselian ice shield. The globally rising sea level resulted in marine ingressions through south-central Sweden. These reached the Baltic Basin and deposited the brackish-marine sediments (homogenous clay, indistinctly laminated) of the Yoldia Sea Stage (A-4) beginning and ending with freshwater phases (A-3, A-5) between 11,570 and 10,700 cal. [14]C B.P. Subsequently, uplift in response to isostatic rebound sev-

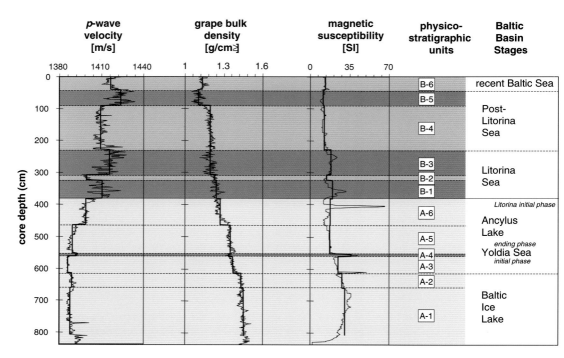

Figure 4 Correlation of physicostratigraphic units of Holocene sediments within the Gotland Basin core 211660-5, after Harff et al. (1999), and Baltic Sea Stages, after Andren (1999) and Andren et al. (1999). The subdivision of sediments into physicostratigraphic units reflects the change of hydrographic conditions during the late Pleistocene and Holocene. Units A-1 and A-2 represent the Baltic Ice Lake (12,000–11,570 cal. [14]C B.P.) with varved late glacial clays. Sea-level rise by melting ice shields has caused a marine ingression mirrored by the brackish marine Yoldia Sea Stage between 11,570 and 10,700 cal. [14]C B.P. (Units A3–A5). Subsequently, uplift in response to isostatic rebound severed the connection to the world ocean and the Baltic became the freshwater Ancylus Lake (10,700 to 8300/8100 cal. [14]C B.P.) (Unit A6). Next, rapidly rising sea level caused the Litorina transgression through the Danish Straits, which ever since have served as the continuous connection to the world ocean. After an initial phase (8300/8100 to 7400/7000 cal. [14]C B.P.) the brackish-marine Litorina Sea is replaced time-transgressively by laminated gyttja clay reflecting a thermohaline stratified water body. Changes from laminated sequences (B-1, B-3, B-5) to less distinctly laminated to nonlaminated sediments (B-2, B-4) reflect variations in the oxygen supply to the bottom water. The Recent Baltic Sea (B-6) is represented by gyttja clays of unit B-6.

ered the connection to the world ocean and the Baltic became the freshwater Ancylus Lake (10,700 to 8,300/8,100 cal. [14]C B.P.). Next, rapidly rising sea level caused the Litorina transgression through the Danish Straits, which have subsequently served as the continuous connection to the world ocean. After an initial phase (8300/8100 to 7400/7000 cal. [14]C B.P.) the brackish marine Litorina Sea is replaced time-transgressively by laminated gyttja clay reflecting a thermohaline stratified water body. Anoxic bottom water prohibited any higher benthic life and prevented bioturbation. Changes from laminated sequences (B-1, B-3, B-5) to less distinctly laminated to nonlaminated sediments (B-2, B-4) reflect variations in the oxygen supply to the bottom water. This alternation of anoxic and oxic bottom-water conditions was triggered by climatic fluctuations (Andren and Andren, 1999). Cold periods with a southward-moving polar front brought storms and marine water inflow to the Baltic Sea, producing a stratified water body and laminated sediments. Investigating the ecological proxies, Andren (1999) separated the brackish-marine Litorina Sea Stage (7400/7000 to 3700/3500 cal.^{14}C B.P.) from the more brackish Post-Litorina Sea Stage (3700/3500 to 700/850 cal. ^{14}C B.P.). The recent Baltic Sea (B-6) is represented by gyttja clays of unit B-6.

COASTLINE EVOLUTION: COMPETITION OF CLIMATE CHANGE AND VERTICAL CRUSTAL MOVEMENTS

Regional Aspects

The coastline change during the Holocene has been investigated at many locations around the Baltic Sea, and relative sea-level (rsl) change curves have been published, mostly during the 1970s and 1980s. Many of those are reproduced by Pirazzoli (1991) as major products from the IGCP-Project No. 274, "Coastal Evolution during the Quaternary." The curves giving the altitude of a location relative to the sea level are derived from mapping ancient coastlines and ^{14}C-dating of carbon-rich terrestrial sediments in outcrops and sediment cores. In order to describe the different behavior of the regional tectonic units influencing the development of the Baltic Basin, the following curves were selected (corresponding locations are marked on the map in **Figure 5**):

Fennoscandian Shield:
    location 1 (Åse and Bergström, 1982)
    location 2 (Mörner, 1979)
    location 3 (Donner, 1983)
    location 4 (Eronen and Haila, 1982)

Transition Fennoscandian Shield to East European Platform:
    location 5 (Kessel and Raukas, 1979)

East European Platform:
    location 6 (Mojski, 1982)

West European Platform:
    location 7 (Kliewe and Janke, 1982)
    location 8 (Duphorn, 1979)

Transition Fennoscandian Shield to West European Platform:
    location 9 (Björck, 1979).

For every curve, the time is given here in conventional ^{14}C-years B.P.

Every rsl curve shows the joint effect of eustatic change and vertical crustal movement, whereby the proportions of both causes differ remarkably. To reveal the proportions of eustasy and isostasy, each rsl curve was compared with the eustatic curve (esl) for the Holocene published by Mörner (1976). This curve was selected because it was derived for the Kattegat area at the mouth of the permanent connection of the Baltic Sea with the world ocean since the Litorina transgression.

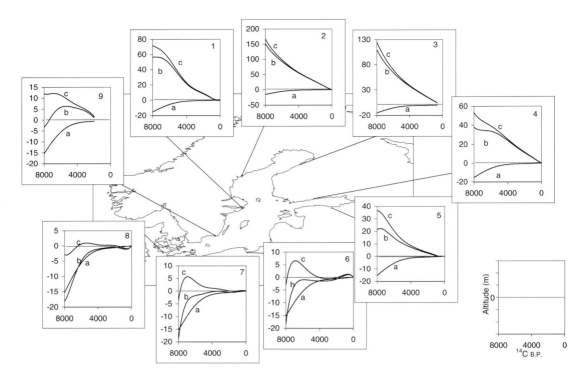

Figure 5 Relative sea-level change curves (rsl$_i$, marked by "b") compared to a eustatic curve (esl, marked by "a," based on Mörner's 1976 curve) and the isostatic component (d_i, marked by "c") at nine locations in the Baltic Sea area. The form of curves 1 to 4 qualifies them as clearly controlled by isostatic uplift, whereas curves 6 to 8 show the form for coasts controlled by climatically driven eustatic sea-level rise. For further explanation, see the text.

Since the initial phase of the Litorina Stage, this connection has been continuously open (Björck, 1995; Lemke et al., 1997, 1998) so that sea-level changes of the North Atlantic reached the Baltic Sea directly. Despite Mörner's (1980) and Shennan's (1989) subsequent criticism of regional eustatic models, Mörner's (1976) Kattegat curve still provides the most faithful basis for regional comparisons and the detection of trends within the Baltic area. Local transgressions and regressions cannot be examined at the regional scale investigated here. For comparison of the curves, only the time span between 8000 ^{14}C B.P. and today (0^{14}C B.P. = A.D. 1950) was taken into account. Fluctuations were eliminated from the eustatic curve and the rsl curves cited above by fitting with polynomial functions of the sixth degree. Mörner's (1976) dates given in sidereal years were converted to conventional ^{14}C years B.P. using a transformation function published by Stuiver and Becker (1993). **Figure 6** shows the transformed and fitted eustatic data, after Mörner (1976). A difference ($d_i(t)$) curve describing the noneustatic component of the rsl curves was calculated and is regarded as having been caused principally by crustal movement. Isostasy plays the most important role, but it is modified by tectonic, halokinetic, and compaction processes.

$$d_i(t) = rsl_i(t) - esl(t), \; i \in \{1, ..., 9\}, \; t \in \{-8000, 0\},$$

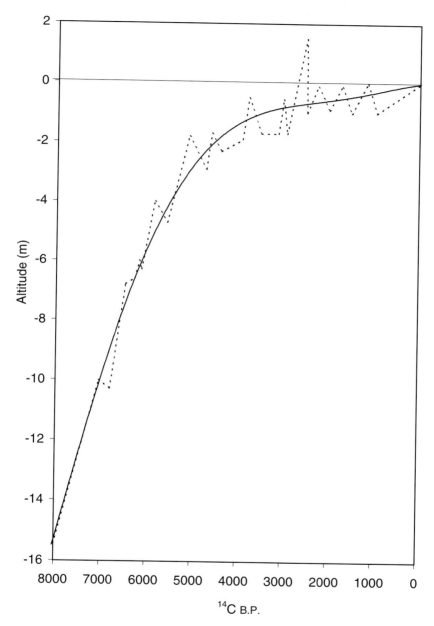

Figure 6 Eustatic curve for the Kattegat area by Mörner (1976) (dotted line) and sixth-degree polynomial fit (full line). In order to reveal the proportions of eustasy and isostasy, each rsl curve in **Figure 5** was compared to the eustatic curve (esl) for the Holocene published by Mörner (1976) for the Kattegat area. The polynomial trend neglects local fluctuation.

where $rsl_i(t)$ describes the relative sea-level curves at the nine locations cited above and esl stands for the eustatic curve. **Figure 5** shows the esl curve, rsl_i curves, and d_i curves at each location. In general the curves extend from $t_2 = 8000$ ^{14}C B.P. to $t_1 = 0$, except that curve 9, which ends at $t_1 = 1960$ ^{14}C B.P., and curve 3 which ends at $t_1 = 630$ ^{14}C B.P., form exceptions. The difference in sea-level change at the coasts on the Fennoscandian Shield and the East and West European Platforms is clearly visible. Taking into account a eustatic sea-level rise of 15 m during the last 8000 years, a relative sea-level fall of 150 m at the west coast of the Gulf of Bothnia on the one hand and a sea-level rise of 20 m at the German Coast of the State of Mecklenburg-Vorpommern can be explained only by differences in vertical crustal movements. For a quantitative comparison, we introduce the metric

$$\rho(i) = \frac{1}{t_2 - t_1} \int_{t_1}^{t_2} |d_i(t)| dt$$

This metric expresses the average deviation in altitude from sea level, caused by vertical crustal movements at a location i, since 8000 ^{14}C years B.P.

Defining a eustatic index

$$e = \frac{1}{t_2 - t_1} \int_{t_1}^{t_2} |esl(t)| dt$$

one can derive a "coast index" $c(i) = \rho(i)/e$.

Provided the esl curve reflects the general effect of climate change (melting ice shields, thermal expansion of the world ocean) and $\rho(i)$ expresses the vertical crustal movement at a location, the index $c(i)$ gives an estimate of the controlling factor at the location i. Generally, coasts can be classified as "crustal-uplift/subsiding coasts" if $c(i) > 1$, and "climate-controlled coasts" if c(i) < 1.

For the data used here the eustatic index was determined ($e = 3.57$), and $\rho(i)$ and $c(i)$ were calculated. Results are given in **Table 1**, which also classifies the locations based on the coastal index $c(i)$.

Locations 1 (Åse and Bergström, 1982), 2 (Mörner, 1979), 3 (Donner, 1983), and 4 (Eronen and Haila, 1982) are dominated by isostatic uplift. The climatically controlled eustatic rise influences the coastal development to a minor extent, as one can see from the continuously decreasing $d_i(t)$ curves and the value of function $\rho(i)$. Within this coastal type, we find archipelagoes, fjords, and very flat coasts of emerging sea bottom. Coast locations 5 (Kessel and Raukas, 1979) and 9 (Björck, 1979) represent a transition type mostly controlled by vertical crustal uplift, but more moderate. Cliffs and archipelagoes form the coast here. Quite different coastal types (straightened coasts, cliffs, barrier coasts) are represented by locations 6 (Mojski, 1982), 7 (Kliewe and Janke, 1982), and 8 (Duphorn, 1979) along the southern coast. Despite the fact that between locations 6 and 7 the TTZ separates the East European Platform and the West European Platform, each $rsl_i(t)$ and $d_i(t)$ curve shows the same behaviour and is quite different from the crustal uplift type. In every case, the esl change compensates or increases the value of the vertical crustal movements and causes continuously retreating coastlines. The $d_i(t)$ curves show a relatively rapidly subsiding crust between 8000 and 6000 ^{14}C years B.P., whereas during the last 5000 years this process seems to have ended, showing only moderate gradients in recent time. This result agrees with assumptions about the collapse of an asthenospheric bulge in front of the retreating Weichselian ice shield (Fjeldskaar, 1994).

Due to the rapidly rising sea level at the beginning of the Litorina Stage, the glacigenically shaped land relief was drowned without any notable longshore transport processes. Only after the

Table 1 Classification of nine coastal locations at the Baltic Sea based on the index $c(i)$. The coast index $c(i)$ quantifies the results visible in **Figure 5**. High values of the coast index for curves from central parts of the Baltic Shield confirm the assumption for isostatically controlled coasts. Curves from the southern Baltic are allocated by the index to climatically controlled coasts. The southern Gulf of Finland and southern Sweden (see locations of curves 5 and 9) occupy a transitional position.

Authors	$p(i)$	$c(i)$	Coastal Type
Åse and Bergström (1982)	30.1	8.34	crustal uplift
Mörner (1979)	64.1	17.9	
Donner (1983)	45.9	12.8	
Eronen and Haila (1982)	25.4	7.11	
Kessel and Raukas (1979)	12.5	3.5	transition
Mojski (1982)	2.0	0.56	climate controlled
Kliewe and Janke (1982)	1.7	0.47	
Duphorn (1979)	0.7	0.19	
Bjorck (1979)	7.3	2.0	

sea-level rise slowed at the end of the Litorina Stage and during the Post-Litorina Stage coastal features did form by transport, producing the spit and barrier coast seen today. This process is dominantly controlled by climatic factors such as the wind-driven hydrographic regime.

A Local Study

To understand the formation of the climatically controlled coastline along the southern Baltic Sea, vertical crustal movements must be considered as neotectonic processes at a local scale. Here the vertical movement of tectonic blocks can lead to total uplift and subsidence, as has been reported recently from northeastern Germany. A detailed analysis of the depositional history of the barriers separating lagoons and estuaries from the Baltic Sea, by Janke and Lampe (1999), confirmed the rapid sea-level rise during the period between 8000 and 5800 ^{14}C B.P., as described earlier by Kliewe and Janke (1982). At the onset of the Litorina transgression, the water level rose at a rate of 25 mm/yr (eustatic rise added to crustal subsidence, as described above), which later slowed to about 3 mm/yr. Whether sea level before 5800 B.P. rose intermittently, i.e., interrupted by brief regressions or retardations, as described by Mörner (1976), Kolp (1979), and Tooley (1982), could not be detemined. Only at the beginning of a major decrease in the rate of rise could some oscillations be detected and attributed to climate variations.

The sea-level rise was connected with the development of basal peats in coastal areas due to a rising groundwater table. A first maximum of sea level was reached at about 5800 ^{14}C B.P. The altitude reached is indicated by peats that today underlie many beach ridges and marsh areas at depths of –0.5 m to 0.0 m below sea level. These beach ridges, marsh muds, and peats are younger than 1000 ^{14}C B.P., which leads to the assumption of a hiatus of about 4800 years caused by a low sea-level stand between 5800 and 1000 ^{14}C B.P. Especially between 5800 and 4500 ^{14}C B.P. the dates are scattered through a wide depth interval. The lowest locations, between –2.5 and –10.5 m below

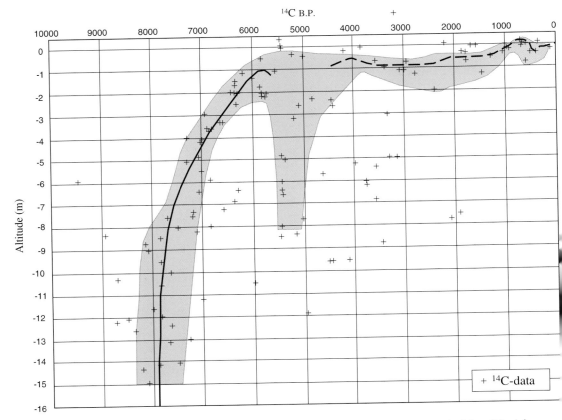

Figure 7 Preliminary curve of relative sea-level change for western Pomerania (simplified from Janke and Lampe, 1999), shaded area: interval of "data confidence," solid line: local trend for western Pomerania; dashed line: estimated trend. The curve shows fast subsidence until 6000 [14]C B.P. and long-lasting subsidence after 4500 [14]C B.P. In the time between, data lead to the assumption of tectonically controlled local uplift.

sea level, are occupied by peats and fossil soils in lagoonal sediment columns. The sea-level low stand was connected with a lower groundwater table, which can also be detected in terrestrial se quences farther inland where smaller lakes dried out or bogs stopped growing. Because the exten of the sea-level fall cannot be explained by the Late Atlantic regression, one has to assume that nec tectonic uplift was responsible.

These results coincide with assumptions of Schumacher and Bayerl (1997), who described local sea-level curve from the Schaabe barrier, Rügen Island, which shows a sudden tectonic uplit in the Late Atlantic, followed by long-lasting subsidence. The curve given in **Figure 7** is mor comprehensive and was constructed using all reliable ^{14}C-data available from spit, barrier, lagoon al, lake, river, and wetland sediment sequences from the entire western Pomeranian coast. Dat scatter is caused by differential sediment compaction, differences in the vertical movemen of distinct coastal areas, and uncertainties of measurements. A confidence interval is marked o the graph by a shaded zone. Within this zone the rsl curve represents the local trend for wester

Pomerania. For the most likely time of uplift, 4500 ^{14}C B.P., the curve is blanked out, but was continued for the time after 5400 ^{14}C B.P. The tendency of crustal uplift after the period of subsidence is recorded in each $d_i(t)$ curve for the southern Baltic Sea **(Figure 5)**. This tendency seems to be intensified by neotectonic uplift events on the western Pomeranian coast, resulting in the observed relative sea-level fall. Detailed studies in the near future shall address the questions of causes, intensity, and spatial differentiation of the Late Atlantic uplift.

FROM RELATIVE TO ABSOLUTE DATA

The timescales of the processes studied require different methods of data interpretation. In the previous section, data on eustatic variation and vertical crustal movements were derived from proxy data, which can only give comparative insights into the process. It was shown that the controlling factors of relative sea-level change differ remarkably around the Baltic Sea. Uplift processes caused emergent seafloor along the coast of the Fennoscandian Shield, and climatically triggered eustatic rise, together with meteorologically forced hydrography, produced the retreating coast of the southern Baltic Sea. Particularly for subregional and local studies, knowledge of the detailed pattern of vertical crustal movements is needed for an understanding of coastal processes. Modern data provide only a snapshot of the process studied, but they can be used to adjust the trends derived from paleodata. Recent crustal movement data give insight into the subregional to local coastline's behaviour needed for coastal protection planning.

Data on recent vertical crustal movements were compiled for IGCP-Project No. 346, "Neogeodynamica Baltica," spanning the area between 4° and 36° E longitude and 47° and 65° N latitude. Although the data are incomplete and not uniform with regard to benchmarks used for individual areas, the map shown in **Figure 8** was compiled as a summary of the present knowledge. The sources used and the procedure of map compilation are given in the Appendix. Relationships between uplift and regional structure can be seen by comparing **Figure 8** with **Figure 1** and Grigelis (1978), Krauss (1994), Erlström et al. (1997), Berthelsen (1994), Karabanov et al. (1994), and Hofmann and Franke (1997).

Depending on the scale of local map sources, the degree of the compiled map's precision increases from Fennoscandia through eastern Europe and Denmark to western Europe. In its northern part, the map is dominated by the extensive northeast-southwest-stretched uplift of Fennoscandia with maximum values of >8 mm/yr (northern Gulf of Bothnia). Differential movement rates are to be expected from the distribution pattern of seismic events, especially following the north-south-trending structures such as the Oslo Graben, the Mylonite Zone, the Protegin Zone, and the eastern flank of the Stockholm Range. The Fennoscandian uplift is surrounded by a subsidence zone, most distinctly developed in an area from the Baltic States to northern Poland. This zone is divided by north-south–trending structures (for instance, the Pribaltic Faults) and blends into the more subtle anomaly pattern of central Europe. The north-south trends determine the anomaly pattern from the Gulf of Finland to the Masurian Block and possibly farther, to the eastern border of the Carpathian Arc. The intersection of the Pribaltic Faults with east-west-trending faults coincides with a northeast-southwest-stretched block of inversion within the Belorussian Syneclise. A spatial sequence of single movement maxima may confirm the assumption of a structural connection between the north-south-trending East Baltic Graben System southward through the Mazovian Depression and up to the Carpathian Foreland. This part of the old platform is aseismic. The few recent stress field measurements for southeastern Poland and the Balticum indicate a north-south orientation.

Maximum subsidence of up to 6 mm/yr is observed within a belt surrounding the Fennoscandian uplift, especially where it is cut by the north-south-trending Orscha and Waldai Depressions.

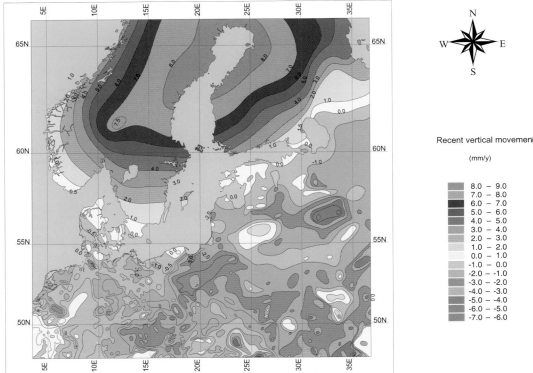

Figure 8 Recent vertical movement of the earth's crust. In the northern part, the map is dominated by the extensive, northeast-southwest-stretched uplift of Fennoscandia with maximum values of >8 mm/yr (northern Botnian Bay). The subsidence of a belt surrounding the Baltic Shield is less differentiated over the old East European Plate and mirrors the pattern of an old block structure. The West European (young) Plate is distinguished by a more differentiated pattern of vertical recent crustal movement, possibly additionally influenced by the opening processes of the Atlantic. The southern margin of the Baltic Shield is enclosed by a belt of subsidence, which could be associated with a collapsed structure of a circum-Fennoscandian ring bulge of the upper mantle after melting of the Weichselian ice shield. The Tornquist-Teisseyre Zone is obviously inactive. For further explanation, see the text and the Appendix.

This circum-Fennoscandian depression can be followed to the west, crossing the Polish-Lithuania Depression with moderate rates of subsidence (2.5–3 mm/yr) up to the Tornquist-Teisseyre Zor This depression has been discussed as a collapse structure of an upper mantle ring bulge, form as a reaction to the glacio-isostatic uplift of the Fennoscandian Block. Fjeldskaar (1994) estimate the conditions of recent uplift and subsidence on the basis of the lithosphere's rheological prope ties. He was able to show that the last glaciation of Scandinavia (15,000 B.P.) generated a ring bul of +60 m height related to the equilibrium state. This bulge collapsed gradually, and without a appreciable lateral migration, after the ice shield vanished.

The Tornquist-Teisseyre Zone is not reflected in the recent vertical movement pattern. An indication of its existence can be seen only in more or less strong gradients of different polarity between the Carpathians and the Baltic Sea. The TTZ area is aseismic and does not influence the recent stress field (Grünthal and Stromeyer, 1995).

The isoline pattern of Denmark seems to fit between the TTZ and the frequently discussed Trans-European Fault (TEF). In general, the region follows the movement of the Fennoscandian Shield, but it seems to be additionally influenced by the denser pattern of anomalies typical for central and western Europe. Moreover, the Danish pattern seems to be rotated from Northwest-southest to north-northwest-south-southeast and north-south, compared to the Fennoscandian pattern of isolines. This could be an indication of a relatively independent behaviour of the crustal block between TTZ and TEF, but it might also be caused by superposition of the north-south–trending structures of the Mediterranean-Mjösa Zone.

The importance of north-south directions for recent vertical crustal movements west of the TTZ has been emphasized by Ellenberg (1992). The north-south direction is most distinct within the area showing differentiated patterns of isolines in a zone of weak subsidence, starting from the Upper Rhine Graben and extending through the Hessian Depression up to Schleswig-Holstein (Hamburg Basin). Interpretations of remote sensing data (Meteor 25) confirm north-south-trending lineations of supraregional character (Bankwitz et al., 1982).

CONCLUSION: CAN WE PREDICT THE FUTURE?

Because the process of eustatic rise and trends of crustal movement during the late Holocene are known, the question of predicting future increases in relative sea level needs to be addressed.

Secular eustatic sea-level variations are related to the problem of future climate change. Hay et al. (1997) have discussed in detail the question of using the "past as the key for the future" and have discussed the fundamental assumptions to be taken into account for numerical modeling of climate and consequently sea-level change. Of the numerous attempts to simulate future increases in sea level, only that of Gregory and Oerlemans (1998), who concentrate on the effect of melting glaciers, is cited here. Another approach is represented by Cubasch et al. (1994, 1995) and Storch et al. (1997), who use the ECHAM/LSG global atmosphere-ocean circulation model for simulating global warming and the effect on thermal expansion of the ocean water and sea-level rise. Voß et al. (1997) have used this model to simulate global sea-level rise for the next 700 years as experiments based on Intergovernmental Panel on Climate Change (IPCC) scenario A for the rise of atmospheric CO_2 concentration due to anthropogenic impacts (Houghton et al., 1990). Results of the experiments are available as gridded data with a spatial resolution of 5.6°. Data for the western North Atlantic and the North Sea can be applied for estimating the change of eustatic sea level in the Baltic Sea using appropriate transfer functions, which have yet to be developed. The regionalized sea-level change has to be added to the vertical crustal movement referred to as a recent digital elevation model (DEM). For the derivation of scenarios for elevation change during the coming centuries covered by climate modeling (Voß et al., 1997), the map of crustal movement gradients given in this publication can be used as a guide. The steady-state character of the process of vertical crustal movement remains a question.

As the studies in the previous section have shown, there is an extensive region of recent subsidence between the Fennoscandian Block, characterized by glacial-isostatic uplift, and the uplift area of the Carpathians (thought to be due to the northward drift of the African Plate). It may be less distinct over the old East European Plate, which responds as old structural blocks. The younger West European Plate is distinguished by a more differentiated pattern of recent vertical crustal movement, possibly additionally influenced by the opening processes of the Atlantic. The southern margin of the Fennoscandian Block is enclosed by a belt of subsidence, which could be associated with collapse of a circum-Fennoscandian ring bulge of upper mantle material. The Tornquist-Teisseyre Zone is obviously inactive. Consequently, for the Baltic Sea coast area one has to concentrate on the glacio-isostatic uplift of the Fennoscandian Shield and the southerly, relatively narrow belt of subsidence caused by the collapse of the former ring bulge surrounding the ice shield. For both areas, the investigation of relative sea-level curves and separation of the component of vertical crustal movement indicate that the isostatic process reached a steady state during the Late Litorina Stage. In the framework of the research project introduced here, linear forward extrapolations of a recent digital elevation model DEM(t), $t = 0$, for the Baltic area, based on the map of vertical crustal movement **(Figure 8)**, are planned as numerical experiments.

Future scenarios, DEM$_i$(t), can be derived by overlaying the corresponding earth-surface elevation data with regionalized sea-level data from climate modeling (for instance, Voß et al., 1997). These scenarios can be produced for different distinct future time points $t \in T$, and different assumptions $i \in I$ for future CO_2 emissions. For subregional and local studies of the climatically controlled coasts, additional investigations are needed. These must include the architecture of the tectonic elements and their local vertical movements as well as processes of formation of coastal morphology. Scenarios are not to be regarded as predictions of the future; they only visualize in a cause-and-effect relationship the possible reactions of the system—here the coastal zone of the Baltic Sea—to natural and man-made changes of the forcing conditions. However, such scenarios may aid long-term planning of human activities in the Baltic Sea coastal zone.

ACKNOWLEDGMENTS

The authors thank Dr. H. Dietrich (University of Greifswald), Prof. Dr. G. Katzung (University of Greifswald), Dr. W. Lemke (Baltic Sea Research Institute Warnemünde, [IOW]), Dr. H. Montag (Brandenburg), Dr. G. Schwab, and Dr. P. Vyskocil (Prag) for helpful discussions.

Thanks must be expressed to Dr. J. Pilz (IOW), who digitized and prepared the map of recent vertical crustal movements for printing, and to Heidemarie Brendel and Marion Sussujew (IOW) for the computerized design of graphics.

The German Science Foundation (DFG) deserves thanks for supporting the study.

APPENDIX

Data sources for the compilation of the map of recent vertical crustal movements in **Figure 8**.

1. Vyskocil (1990), benchmark: mean sea level (calculated from Baltic Sea, Black Sea, Asow Sea).
2. Kaschina (1989), benchmark: mean sea level.
3. Joo (1992), benchmark: mean sea level (calculated from Baltic Sea, Black Sea, Adria).
4. Höggerl (1986), benchmark: Freistadt (N. Austria).
5. Mälzer (1990), benchmark: Schernfeld (Jurassic), Saldenburg (basement).
6. Gubler et al. (1992), benchmark: Wallenhorst (near Osnabrück), Wahlenau (Rheinland-Pfalz), Freudenstadt (Baden-Würtemberg), Schernfeld (Bayern).
7. Pissart and Lambot (1989), benchmark: Uccle (Brüssel).
8. Lorenz (1994), benchmark: Normaal Amsterdam Peil (model 01).
9. Leonhard (1988), benchmark: Wallenhorst (near Osnabrück).

10. Andersen (1992), benchmark: mean sea level (Arhus, Kattegat).

11. Gubler et al. (1992), benchmark: mean sea level.

12. Ihde et al. (1987), for completion of (1) by original intermediate data, benchmark: mean sea level (calculated from Baltic Sea, Black Sea, Asow Sea).

13. Wyrzykowski (1985), for completion of (1) by original intermediate data, benchmark: mean sea level (Baltic Sea).

14. Forniguet, J. (1987), benchmark: related to the calculated difference of benchmark altitudes (mean sea level Marseille) used during two adjacent levelings.

More than three-quarters of the map are covered by data from only two single maps, each related to the benchmark "mean sea level": eastern and middle Europe (1) and Scandinavia (11). The database for the western and southern parts of the map is extremely inhomogeneous. For the comparatively small part of western Europe, the map must be constructed from seven regional maps; the connection among them is problematical because of the different benchmarks used and because of different leveling fundamental networks. The isolines for Austria were interpolated from velocity data of vertical crustal movements within first- and second-order junctions (4) (Höggerl, 1986). The course of isolines for the territory of Belgium was calculated from data of altitude changes within the junctions of the levelings from 1946–1948 and 1976–1980 (7) (Pissart and Lambot, 1989). For the area of Lower Saxony (Germany), no data are available up to now. An adjustment of isolines was calculated between the regions based on the original maps cited. The mean difference of the isoline values along the borders of the maps was used as a quantity for correction. The procedure started from map (1) (Vyskocil et al.)—which already included the compilation from (2) (Kaschina) (12) (Ihde et al.), and (13) (Wyrzykowski)—was calculated with (3) (Joo). The interpolative combination of (1) and (3) was straigthforward; both maps were related to mean sea level and the quantity for correction became zero. The fit procedure was continued west and northward. The adjustment for the southwestern, western, and northern parts of the map (Austria, western Germany, Belgium to Denmark) is problematic, not only regarding the adjustment of isolines, but also regarding the homogeneity of the resulting pattern of anomalies. Especially for these regions, the map cannot comply with geodetic requirements for accuracy and has a more hypothetical character.

REFERENCES CITED

Andersen, N., 1992, The hydrostatic levellings in Denmark: Geophysical Monograph 69, IUGG, v. 11, p. 107–111.

Andren, E., 1999, Holocene environmental changes recorded by diatom stratigraphy in the southern Baltic Sea: Doctoral thesis, Meddelanden Stockholms Universitets Insitution Geologi och geokemi, v. 302.

Andren, E., and T. Andren, 1999, Largescale climatic influence on the Holocene history of the Baltic Proper: Doctoral thesis, Meddelanden Stockholms Universitets Insitution Geologi och geokemi, v. 302.

Andren, E., T. Andren, and H. Kunzendorf, 1999, Holocene history of the southwestern Baltic Sea as a background for assessing records of human impact in the sediments of the Gotland Basin: Doctoral thesis, Meddelanden Stockholms Universitets Insitution Geologi och geokemi, v. 302, p. 24.

Åse, L. E., and E. Bergström, 1982, The ancient shorelines of the Uppsala esker around Uppsala and the shore displacement: Geographical Annales, v. 64 A, p. 229–244.

Bankwitz, P., E. Bankwitz, and A. Frischbutter, 1982, Lineamentij na territorii Germanskoi Demokraticheskoj Respubliki.: Iccledovanija. Semli is Kosmosa, v. 2, p. 25–26.

Berthelsen, A., 1994, Europrobe's 2nd Trans-European Suture Zone: Europrobe News, Issue No. 5, April, v. 1–3.

Björck, S., 1979, Late Weichselian Stratigraphiy of Blekinge, SE Sweden, and Water Level Changes in the Baltic Ice Lake: Department Quaternary Geology University Lund, Thesis 7, p. 248.

Björck, S., 1995, A review of the history of the Baltic Sea: Quaterny International, v. 27, p. 1940.

Cubasch, U., G. C. Hegerl, A. Hellbach, H. Höck, U. Mikolajewicz, B. D. Santer, and R. Voss, 1995, A climate change simulation starting from 1935: Climate Dynamics, v. 11, p. 7184.

Cubasch, U., B. D. Santer, A. Hellbach, G. Hegerl, H. Höck, E. MaierReimer, U. Mikolajewicz, A. Stössel, and R. Voß, 1994, Monte Carlo climate change forecasts with a global coupled ocean atmosphere model: Climate Dynamics, v. 10, p. 119.

Donner, J. J., 1983, The identification of Eemian interglacial and Weichselian interstadial deposits in Finland: Annales Academiae Scientiarum Fennicae, v. A (136), p. 138.

Duphorn, K., 1979, The Quaternary history of the Baltic, The Federal Republic of Germany, in V. Gudelis and L. K. Königsson, eds., The Quaternary History of the Baltic, Univ. Uppsala, p. 195–206.

Edwards, M. O., 1989, Global Gridded Elevation and Bathymetry (ETOPO5). Digital raster data on a 5-minute Geography (lat/lon) 2160 x 4320 (centoid-registered) grid. 9-track tape, Boulder: National Oceanographic and Atmospheric Administration (NOAA) National Geophysical Data Center, 18.6MB.

Ellenberg, J., 1992, Recent fault tectonics and their relations to the seismicity of East Germany: Tectonophysics, v. 202, p. 117–121.

Erlström, M., S. A. Thomas, N. Deeks, and U. Sivhed, 1997, Structure and tectonic evolution of the Tornquist Zone and adjacent sedimentary basins in Scandia and the southern Baltic Sea area: Tectonophysics, v. 271, p. 191–215.

Eronen, M., and H. Haila, 1982, Shoreline displacement near Helsinki, southern Finland, during the Ancylus Lake stage: Annales Academiae Scientiarum Fennicae, v. AIII (134), p. 111–129.

Fjeldskaar, W., 1994, The amplitude and decay of the glacial forebulge in Fennoscandia, Norsk Geologisk Tidsskrift, p. 2–8.

Forniguet, J., 1987, Géodynamique actuelle dans le Nord et el NordEst de la France: Mémoires Bulletin du Bureau de Recherches Geologiques et Minneres, v. 127, p. 173.

Frischbutter, A., and G. Schwab, 1995, Karte der rezenten vertikalen Krustenbewegungen in der Umrahmung der OstseeDepression. Ein Beitrag zum IGCPProjekt Nr. 346: Neogeodynamica Baltica: Brandenburgische Geowissenschaftliche Beiträge, v. 2, p. 59–67.

Gregory, J. M., and J. Oerlemans, 1998, Simulated future sea-level rise due to glacier melt based on regionally and seasonally resolved temperature changes: Nature, v. 391, p. 474–476.

Grigelis, A., ed., 1978, Tectonical Map of the Soviet Baltic Republics: Ministry of Geology of the USSR, 1:500 000.

Grünthal, G., and D. Stromeyer, 1995, Rezentes Spannungsfeld und Seismizität des baltischen Raumes und angrenzender Gebiete ein Ausdruck aktueller geodynamischer Prozesse: Brandenburgische Geowissenschaftliche Beiträge, v. 2, p. 69–78.

Gubler, E., S. Arca, J. Kakkuri, and K. Zippelt, 1992, Recent vertical crustal movement, in S. Freeman and S. Müller, eds., A continent revealed: The European Geotraverse: Atlas of compiled data, Cambridge University, p. 2024.

Harff, J., G. C. Bohling, R. Endler, J. C. Davis, and R. A. Olea, 1999, Gliederung holozäner Ostseesedimente nach physikalischen Eigenschaften: Petermanns Geographische Mitteilungen, v. 143, p. 50–55.

Hay, W. W., R. M. DeConto, and C. N. Wold, 1997, Climate: Is the past the key to the future?: Geologische Rundschau, v. 86, p. 471–491.

Hofmann, N., and D. Franke, 1997, The Avalonia-Baltica Suture in NE Germany—New Constraints and Alternative Interpretations, Zeitschrift Geologische Wissenschaften, v. 25, p. 3–14.

Höggerl, N., 1986, Report on Austrian efforts in the field of high precision levelling and recent crustal movements between 1983 and 1986 and future activities, in H. Pelzer, and W. Niemeier, eds., Determination of Heights and Heights Changes: p. 729–735.

Houghton, J. T., G. J. Jenkins, and J. J. Ephraums, 1990, Climate change: The IPCC scientific assessment, Cambridge, Cambridge Univ. Press.

Ihde, J., J. Steinberg, J. Ellenberg, and E. Bankwitz, 1987, On recent vertical crustal movements derived from relevelings within the territory of the GDR: Gerlands Beiträge zur Geophysik, v. 96, p. 206–217.

Janke, W., and R. Lampe, 1999, The Sea-Level Rise on the South Baltic Coast over the past 8000 Years: New Results and new Questions: Terra Nostra, v. 4, p. 126–128.

Joo, I., 1992, Recent vertical surface movements in the Carpathian Basin: Tectonophysics, v. 202, p. 129134.

Karabanov, A.K., R.G. Garetzki, E.A. Levkov and R.E. Aizberg (1994), Zur neotektonischen Entwicklung des südöstlichen Ostseebeckens (Spätaligozän-Quartär), Zeitschrift Geologische Wissenschaften, v. 22, p. 271-274.

Kaschina, L. A., 1989, Karta sovremennych vertikalnych dvizenij Zemnoj kory po geodeziceskim dannym po territoriju SSSR, 1:5 000 000, Moskva.

Kessel, H., and A. Raukas, 1979, The Quaternary history of the Baltic, Estonia, in V. Gudelis and L. K. Königsson, eds., The Quaternary History of the Baltic, Univ. Uppsala, p. 127–146.

Kliewe, H., and W. Janke, 1982, Der holozäne Wasserspiegelanstieg der Ostsee im nordöstlichen Küstengebiet der DDR: Petermanns Geographische Mitteilungen, v. 126, p. 65–74.

Kolp, O., 1979, Eustatische und isostatische Veränderungen des südlichen Ostseeraumes: Petermans Geographische Mitteilungen, v. 123, p. 177–187.

Kooi, H., P. Johnston, K. Lambeck, C. Smither, and R. Molendijk, 1998, Geological causes of recent (~100 yr) vertical land movement in the Netherlands: Tectonophysics, v. 299, p. 297–316.

Krauss, M. (1994), The tectonic structure below the southern Baltic Sea and its evolution: Zeitschrift Geologische Wissenschaften, v. 22, p. 19–32.

Lemke, W., J. B. Jensen, O. Bennike, A. Witkowski, and A. Kuijpers, 1997, Ancylus Lake overflow: The Dana River speculations and facts, in A. Grigelis, ed., The fifth marine geological conference, "The Baltic": Lithuanian Institute of Geology, p. 59.

Lemke, W., R. Endler, F. Tauber, J. B. Jentzen, and O. Bennike, 1998, Late and Postglacial Sedimentation in the Tromper Wieck northeast of Rügen (western Baltic): Meyniana, v. 50, p. 155–173.

Leonhard, T., 1988, Zur Berechnung von Höhenänderungen in Norddeutschland Modelldiskussion, Lösbarkeitsanalyse und numerische Ergebnisse: Wissenschaftliche Arbeiten Fachrichtung Vermessungswesen Universität Hannover; 152, 158 p.

Lorenz, G. K., 1994, International Federation of Surveyors, XX Congress, Commission 5, Survey Instruments and Methods: The first primary leveling in the Netherlands (1875–1885), Melbourne, Australia.

Mälzer, H., 1990, DGKArbeitskreis für rezente Krustenbewegungen Berechnungen von Höhenänderungen im Bayrischen Haupthöhennetz unter Verwendung unterschiedlicher Modelle, München.

Matthäus, W., and H. Franck, 1992, Characteristics of major Baltic inflow a statistical analysis: Continental Shelf Research, v. 12, p. 137–140.

Mojski, J. E., 1982, Geological section across the Holocene sediments in the northern and eastern parts of the Vistula deltaic plain: Geological Studies, Polish Academy of Sciences, Institute of Geography and Spatial Organization, v. 1, p. 149–169.

Mörner, N. A., 1976, Eustatic changes during the last 8000 years in view of radiocarbon calibration and new information from the Kattegat region and other Northwestern European coastal areas: Palaegeography, Palaeoclimatology, Palaeoecology, v. 19, p. 63–85.

Mörner, N. A., 1979, The Fennoscandian uplift and late Cenozoic geodynamics: Geological evidence: Geo-Journal, v. 3 (3), p. 287–318.

Mörner, N. A., 1980, The Fennoscandian Uplift: Geological Implication, in Mörner, N. A., ed., Earth Reology and Eustasy, Wiley & Sons, p. 251–283.

Pirazzoli, P. A., 1991, World Atlas of Holocene Sea-Level Changes: Elsevier Oceanography Series, v. 58: Amsterdam, Elsevier, 280 p.

Pissart, A., and P. Lambot, 1989, Les mouvements actuels du sol en Belgique; comparaison de deux nivellements IGN (1946–1948 et 1976–1980): Annales de la Société géologique de Belgique, v. 112, fasc. 2, p.495–504.

Sauramo, M., 1958, Die Geschichte der Ostsee: Annales Academiae Scientiarium Fennicae AIII, v. 5, 522 p.

Schumacher, W., and K. A. Bayerl, 1997, Die Sedimentationsgeschichte der Schaabe und der holozäne Transgressionsverlauf auf Rügen (Südliche Ostsee): Meyniana, v. 49, p. 151–168.

Shennan, I., 1989, Holocene sea-level changes and crustal movements in the North Sea region: An experiment with regional eustasy, in D. B. Scott, P. A. Pirazzoli, and C. A. Honig, eds., Late Quaternary Sea-Level Correlation and Applications (NATO ASI Series C), Kluwer, p. 125.

Sjöberg, B., 1992, Sea and Coast: Stockholm, Almquist & Wiksell, 128 p.

Storch, J. S. v., V. V. Kharin, U. Cubasch, G. C. Hegerl, D. Schriever, H. v. Storch, and E. Zorita, 1997, A description of a 1260-year control integration with the coupled ECHAM1/LSG general circulation model: American Meteorological Society, v. 10, p. 1525–1543.

Stuiver, M., and B. Becker, 1993, High-precision decadal calibration of the Radiocarbon Time Scale, A.D. 1950–6000 B.C.: Radiocarbon, v. 35, p. 35–65.

Tooley, M. J., 1982, Sea-level changes in northern England: Proceedings of the Geologists Association, v. 93, p. 43–51.

Voß, M., U. Mikolajewicz, and U. Cubasch, 1997, Langfristige Klimaänderungen durch den Anstieg der CO_2-Konzentration in einen gekoppelten Atmosphäre-Ozean-Modell: Annalen der Meteorologie, v. 34, p. 3–4.

Vyskocil, P., 1990, Map of the horizontal gradients of recent vertical crustal movements of the territories of Bulgaria, Czechoslovakia, GDR, Hungary, Poland, Rumania, USSR (European part of the country), 1:2 500 000 Moscow.

Wulff, F., A. Stigebrandt, and L. Rahm, 1990, Nutrient Dynamics of the Baltic Sea: Ambio, v. 19(3), p. 126–133.

Wyrzykowski, T., 1985, Map of the recent vertical movements of the surface of the earth's crust on the territory of Poland: Inst. Geodezij i Kartografii.

Shinn, E.A., Coral reefs and shoreline dipsticks, 2001, *in* L. C. Gerhard, W. E. Harrison, and B. M. Hanson, eds., Geological perspectives of global climate change, p. 251–264.

13 | Coral Reefs and Shoreline Dipsticks

E. A. Shinn

U. S. Geological Survey
Saint Petersburg, Florida, U. S. A.

ABSTRACT

Deep-sea and ice-core data show that sea level, and thus global temperature, fluctuated often during the past 1.8 million years. Fossil coral reefs, tidal flats, and beaches are precise indicators of former sea level. Although large fluctuations are indicated by the geologic record, this paper focuses on the younger, well-documented fluctuations recorded by coral reefs and shoreline deposits that accumulated during the past 140,000 years. During this relatively short period, fossil coral species and depositional processes have remained unchanged, and diagenetic alteration of fossils and sediments was minimal. Coral reefs and shoreline accumulations were selected as sea-level indicators for two reasons: 1) They serve as bathtub rings and/or dipsticks for determining former sea levels, and 2) human activity had no influence on the sea in which they grew or accumulated. Emergent coral reefs worldwide suggest that global sea level was at least 6 m above present during isotope stage 5e approximately 120,000 years (120 ka). Drowned coral reefs and oolitic beaches indicate sea level was about 100 m below present during Stage 2 as little as 12 ka. As many as eight sea-level fluctuations occurred between Stages 2 and 5e, each of which was greater than those projected to result from burning fossil fuels. Ice-core data suggest even more fluctuations between stages 5a and 5e. Because the record is written in stone, geologists, especially sedimentologists, should be well qualified to make future predictions. Geologists have, for the most part, been excluded from official decision making.

INTRODUCTION

The purpose of this chapter is to review, evaluate, and discuss some climate/sea-level changes that occurred during the past 140,000 years and to speculate on the relative importance of anthropogenic versus natural influences. The major focus here is on geologic proxies of past glacially induced sea-level change, namely coral reefs and tropical shorelines (beaches, carbonate tidal flats, and coastal dunes). Many other equally valid approaches, such as study of ice cores (Dansgaard et al., 1993; Davis and Bohling, this publication), submerged speleothems (Dill et al., 1999), tree rings, and sequence stratigraphy also provide useful data for determining late Pleistocene temperatures and the relative positions of sea level over time, but these approaches are beyond the scope of this paper.

Coral reefs and/or tidal-flat and beach accumulations, by their very nature, respond to and provide an accurate record of both climate and sea level. Although the geologic evidence for sea-level fluctuation before the Pleistocene (i.e., before 1.8 million years ago) is abundant and shows equally impressive fluctuations, data from the older rocks are more difficult to interpret and may not serve as a direct model for the present. In addition, associated dissolution, recrystallization, and vastly different suites of fossils make interpretation and comparison of older rock-forming environments with modern ones less certain. For these reasons, the discussions here focus primarily on younger sedimentary accumulations deposited when conditions were more similar to present and fossil species were for the most part extant.

Corals that formed Pleistocene reefs are generally of the same species as those that are constructing reefs today. Growth rates, temperature, and salinity, as well as depth requirements of these species, are well known. Likewise, the sedimentary processes that govern deposition of beach/dunes and tidal-flat accumulations were certainly the same as they are today. Sedimentologists with knowledge of these modern environments and knowledge of how to recognize them in older rocks are therefore uniquely qualified to interpret the preserved climate and sea-level records they contain. It is hoped that interpretation of former sea levels will lead to more accurate predictions of sea-level elevations, duration of cycles, and climate changes in the future.

ISOTOPES AND DEEP-SEA WATER TEMPERATURE: HISTORICAL PERSPECTIVE

The most precise evidence for global temperature changes during the past 1.8 million years has come from study of carbonate-secreting organisms in deep-sea sediment cores. The deep-sea proxy climate record is derived mainly from the isotopic-temperature records preserved in the shells of organisms such as foraminifera. The techniques, universally employed in isotopic-temperature studies of deep-sea accumulations, were established by Emiliani (1955). Even earlier, Arrhenius (1952) had noted chemical changes in deep-sea sediment cores and postulated that these changes were a result of temperature fluctuations associated with glacial and interglacial periods. He assigned stage names to the different sediment zones. The most recent stage he called Stage 1, and the oldest Stage 7. Following the stage name nomenclature of Arrhenius (1952), Emiliani (1955) measured the ratio of the stable isotope of oxygen (^{18}O) to normal oxygen (^{16}O) contained in the skeletons of foraminifera. The method was based on isotope/temperature calculations developed by McCrea (1950) and on carbon-14 (^{14}C) analysis of the shells using the new ^{14}C-dating method pioneered by Libby (1952). The early 1950s was truly a period of discovery with many techniques coming together to form a larger picture. Libby's ^{14}C provided absolute ages; thus, by combining ^{14}C-dating and the isotope-ratio method, Emiliani (1966) later extended the record farther back in time by examining two long cores, and established 17 isotopic stages for the Pleistocene. The following year, Shackleton (1967) showed that not only temperature but also salinity systematically affects the $^{18}O/^{16}O$ ratio. Thus, his work showed that oxygen-isotope ratios in deep-sea cores also reflect variations in ice-sheet volume. Broecker (1966) dated the isotopic stages and proposed that they reflected glacial stages somehow governed by the astronomical theory refined by Milankovitch (1938). Milankovitch first proposed his theory in 1920. Since this early work, these well-established isotopic stages have been discovered in other places and refined many times during the past 50 years. Most recently, Lear et al. (2000) used variations in the magnesium/calcium (Mg/Ca) ratio in benthic foraminiferal skeletons to confirm the ^{18}O-isotope record and extend ocean-temperature data back to the "greenhouse" Cretaceous 70 million years ago. Their Mg/Ca data, along with the isotopic record, indicate that the first major accumulation of antarctic ice began rapidly about 34 million years ago. This new method will doubtless have many applications in the future.

Today, thanks to these pioneers, calculation of seawater temperature from foraminiferal tests is relatively routine, and additional substages have been added to the original seven. Recent ice-core research revealing "sawtooth" temperature changes suggests that still more stages will be discovered. The Milankovitch theory that glaciation is stimulated by variations in insolation in the Northern Hemisphere brought on by changes in the earth's 41,000-year axial tilt and 22,000-year equinoxial precession remains a ruling paradigm for explaining past climate change.

BASICS

The work of the geological giants described above, and many others, clearly shows that during glacial periods, water is locked in polar ice caps, continental ice sheets, permafrost, and giant inland lakes, causing sea temperature and sea level to plummet and climate in general to become more arid. Conversely, during interglacials, polar ice caps melt, seawater temperature rises, sea level rises, the sea becomes less saline, and the climate becomes more humid. In spite of widespread acceptance of the Milankovitch theory, it nevertheless remains somewhat controversial. The prevailing hypothesis is the "ocean conveyor-belt" model. The model states that gradual changes, controlled by Milankovitch forcing, proceed at a relatively slow pace until a critical threshold is reached, causing a geologically sudden change in ocean-current flow. Under normal interglacial conditions, warm, saline, northward-moving currents warm Northern Hemisphere climates. Then, because of high salinity relative to that of fresh meltwater, this denser water sinks to form a cold southward-moving return current. The process is generally referred to as thermohaline circulation. According to this theory, an influx of cold, less dense fresh water could prevent sinking in the North Atlantic, thereby shutting off the return flow of saline water which in turn shuts off northward-flowing warm currents. The Gulf Stream would no longer reach far into the Atlantic and no longer warm Europe. In a relatively short time, the Northern Hemisphere could become colder. Broecker et al. (1998) and Broecker (1997) are major advocates of this model, and recently emphasized the potential drastic outcome if anthropogenic activity disturbs this delicate balance of ocean-current circulation.

Regardless of the mechanisms actually driving thermohaline-induced circulation, sea-level, and temperature changes, the deep-sea record contained in sediments does not provide an accurate measure of the degree of sea-level change. Simply put, there is no "dipstick" or "bathtub ring" preserved in deep-sea sediments. However, there are other sea-level markers that can be correlated with the deep-sea isotopic records. These markers are coral reefs and shoreline deposits that, like bathtub rings, encircle basins and large oceans. Correlating the deep-sea temperature/time record with coral reefs and other shoreline indicators such as erosional surfaces provides a powerful tool for determining the timing of both temperature and past sea levels.

CORRELATING CIRCUMBASIN MARKERS: THE BATHTUB RING

In the late 1960s, Mesolella et al. (1969) showed that uplifted late Pleistocene coral reefs on the island of Barbados could be correlated with the deep-sea isotopic-ratio warm periods that indicate highstands of sea level. However, there were no exposed records for lowstands at Barbados.

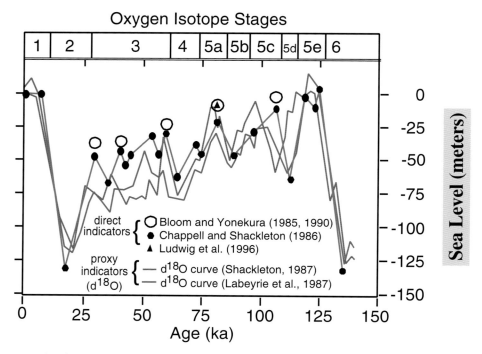

Figure 1 Sea-level fluctuations based on positions determined by Bloom and Yonekura (1985, 1990); Chappell and Shackleton (1986); and Ludwig et al. (1996). The positions are plotted on the deep-sea oxygen-isotopic proxy data of Shackleton (1987) and Labeyrie et al. (1987). Illustration courtesy of David Mallinson.

Bender et al. (1973) and Matthews (1973) extended the Barbados record farther back to the midde Pleistocene by using helium-uranium (He-U) dating methods on corals. On the other side of the plane Bloom and Yonekura (1985, 1990) correlated deep-sea isotopic stages with a series of uplifted coral ree in New Guinea. Chappell and Shackleton (1986) also correlated uplifted New Guinea reefs with th deep-sea record and made calculations of ice-sheet volume to account for sea-level changes **(Figure 1.**

The bathtub ring provided by these uplifted reefs could then be used to determine sea leve more accurately. Certain assumptions had to be made, however, regarding the rate of uplift i order to calculate former sea-level positions. Although fine-tuning of numerous minor fluctuatior is still in progress, the timing and extent of the major fluctuations is universally accepted. But wha about the sea-level lowstands? Numerous authors have tackled this problem: Fairbridge (1961 Shepard (1963), Curray (1965), and Shepard and Curray (1967). Weber (1977) documented a low stand of –175 m 17,000 to 13,000 years ago south of the Great Barrier Reef, based on ^{14}C dating c coral and beachrock collected from a submersible.

A major problem with using reefs/shoreline deposits for determining the precise position c lowstands is accessibility. Lowstand accumulations, unless tectonically uplifted, may be many me ters below the sea surface, and study and sampling require expensive drilling ships and/or re search submersibles. Therefore, only recently have reliable data from drowned reefs and drowned shore deposits been obtained. Even with floating drill ships and research submersibles, a majo limitation remains. Coral-reef and coastal accumulations older than the most recent glacial perio

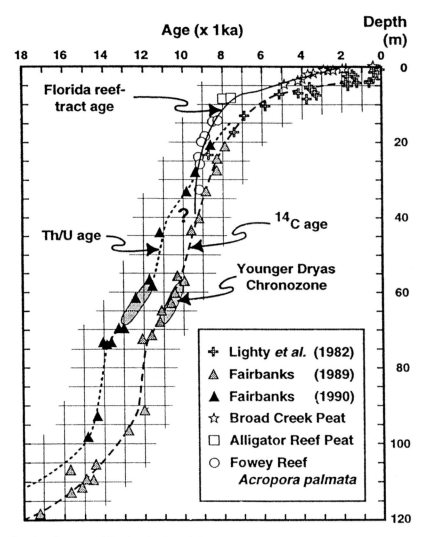

Figure 2 Sea-level curve of Fairbanks (1989) from Bermuda. The upper part of the curve is compared with sea-level data from the Florida reef tract prepared by various authors.

(25,000 to 15,000 years ago) are not only deeply submerged but are usually also covered by younger sequences. New research shows additional data that can be obtained from drowned speleothems. "Tech diving," a newly emerging deep-diving technology, permitted sampling of a stalagmite from –41 m. Uranium-series dating revealed layers correlating with Stages 2, 4, and 6 (Dill et al., 1998; Dill et al., 1999).

Fairbanks (1989) used a drill ship to recover cores from drowned reefs off the west side of Barbados Island. Because the corals sampled were acroporid species that once lived near or at sea level, it was possible to perform isotopic dating and produce a relatively continuous sea-level curve (from below 100 m to present sea level) for the past 18,000 years **(Figure 2)**.

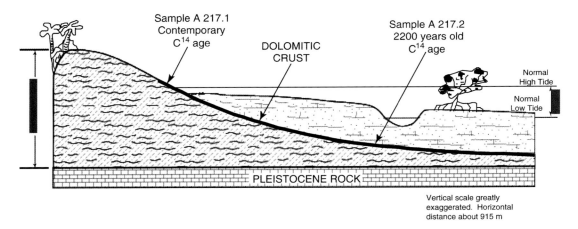

Figure 3 A continuous rise in sea level is traced out by supratidal dolomitic crust on the Andros Island, Bahamas, tidal flat near Williams Island (from Shinn et al., 1965).

The Fairbanks (1989) curve was accurate enough to constrain a period of intense glacial melting and rapid sea-level rise (the Younger Dryas) that occurred about 12,000 years ago. The Fairbanks (1989) sea-level curve, supplemented by other data from shallower and younger drowned reefs in Florida, Lighty et al. (1982), and Robbin (1984), adds refinement to the rate of rise for the upper part of the sea-level curve during isotope Stage 1 **(Figure 2)**. At present, data for other lowstand reef accumulations are limited, but there are new data from drowned carbonate beaches.

Using research submersibles off Florida, Locker et al. (1996) observed and collected specimens from drowned oolitic beaches/dunes. ^{14}C dating showed that these accumulations (at depths ranging from 65 to 80 m) were deposited between 13,800 and 14,500 years ago and thus fit closely on the Fairbanks curve. Beach accumulations are even more "fine-tuned" to sea level than coral reefs; however, the materials that compose a beach usually give ^{14}C ages older than the actual time of accumulation because sediment particles form before deposition and/or are reworked from older accumulations.

Additional coral-reef-based data for sea-level rise in Florida during the Holocene, i.e., the past 10,000 years, are provided by Lighty et al. (1982). Robbin (1984) provide data for the past 7000 years. During this period, the sea initially rapidly flooded shallow shelves worldwide and a little later, between 4000 and 3000 years ago, the rate of sea-level rise slowed drastically. Most modern living coral reefs and active river deltas worldwide began to form between 7000 and 6000 years ago. Unfortunately, siliciclastic beaches provide less precise sea-level information because, unlike carbonate beaches, they are seldom cemented and are often reworked during transgressions.

Accurate sea-level data can also be gleaned by dating plant material such as peat. Mangrove peat is relatively easy to date with the ^{14}C method and has traditionally been considered a precise indicator of sea level. Most true mangrove peat accumulates in the intertidal zone. Accurate sea-level curves for the past 5000 years have therefore been obtained from peats in tidal flats/wetlands, peat islands, and various other shallow nearshore marine environments (Scholl et al., 1969; Enos and Perkins, 1979; Robbin and Stipp, 1984). A unique case for uninterrupted sea-level rise during the past 2200 years (Shinn et al., 1965) is shown in **Figure 3**. In this example, a continuous layer of dolomitic crust that is still forming in the upper intertidal or supratidal zone can be traced downward to a point 1 m below present sea level, where it is covered and no longer forming.

Figure 4 Key Largo Limestone at Adams Cut on Key Largo, Florida. Top of reef is 4 m at this location. Counting of annual coral growth bands in coral to right of ladder showed that the section exposed here accumulated in 360 years during Stage 5e time.

Even more recent information has been gleaned from historical tide-gauge data from Key West, Florida. These data indicate a sea-level rise between 10 and 20 cm during the past 100 years (Maul and Martin, 1993).

DISCUSSION

The highest recorded sea level (see **Figure 1**) occurred during the latter part of the Pleistocene (isotope Stage 5e). Stage 5e reefs, dated by various isotopic methods, range in age from 144,000 to about 117,000 years, according to recent publications by Muhs et al. (1992) and Fruijtier et al. (2000). These older dates are in general agreement with earlier work by Broecker and Thurber (1965) and Osmond et al. (1965) (**Figure 1**, this paper).

The generally accepted age for the Key Largo Limestone is 125,000 years, but more recent work suggests that the true age is 135,000 years (Henderson and Slowey, 2000). Stage 5e reefs occur in many parts of the world. One of the most famous is the Key Largo Limestone, which forms most of the Florida Keys **(Figure 4)**. The highest elevation of the Key Largo Limestone relative to present sea level is 6 m. Because the area is considered relatively tectonically stable, it is assumed that the reef formed when sea level was a minimum of 6 m above present. In the Bahamas, however, Stage 5e reefs **(Figure 5)** occur at lower elevations not exceeding 2 m. Pleistocene 5e coral reefs at remote San Salvador in the eastern Bahamas are also at a 2-m elevation (Carew and Mylroie, 1995; Curran and White, 1995).

Figure 5 Stage 5e coral reef in the Bahamas at Normans Pond Key. Top of reef is level with the back of the investigator in center of photo. Known Stage 5e reefs in the Bahamas do not exceed 2 m above sea level. Beach accumulations at Clifton Pier on New Providence Island, however, indicate a position about 6 m above present.

Various notches in Pleistocene eolian limestone occur throughout the Bahama Islands and are thought to represent sea-level highstands. Unfortunately, these notches, which are 5 to 6 m above sea level, cannot be accurately dated because they lack fossils such as boring clams. However, through correlation, the age of these notches has been inferred to be Stage 5e (Neumann and Hearty, 1996). There is at least one well-documented 5e beach about 6 m above sea level on the west end of New Providence Island (Aurell et al., 1995). Although correlation between Florida and the Bahamas is problematic, the data cited support the widely accepted view that sea level during Stage 5e never exceeded 6 m higher than the present level.

Figure 1 shows several minor so-called fourth- and fifth-order fluctuations since 5e time. These minor fluctuations have been correlated with the deep-sea isotopic-temperature record, but the degree of sea-level fluctuation is still being determined. A minor sea-level highstand that was lower than present, called Stage 5a, allowed growth of a series of linear Pleistocene coral reefs off the Florida Keys about 80,000 years ago (Ludwig et al., 1996). These 80,000-year-old Stage 5a coral reefs, called "outlier" reefs (Lidz et al., 1991 and 1997), indicate that sea-level position at that time was about 6 to 9 m below the present position **(Figure 6)**.

It should be clear from the data and discussion presented here that global sea level fluctuated numerous times during the geologically short period we call the Pleistocene and during the even shorter period we call the Holocene. Historic tide-gauge data indicate that sea level currently is rising at a relatively rapid rate, rivaling that during the Younger Dryas. Clearly, a rapidly rising sea level will profoundly affect human culture and land-use practices. Lowering of sea level, on the other hand, would have equally disastrous effects on both marine organisms and humans.

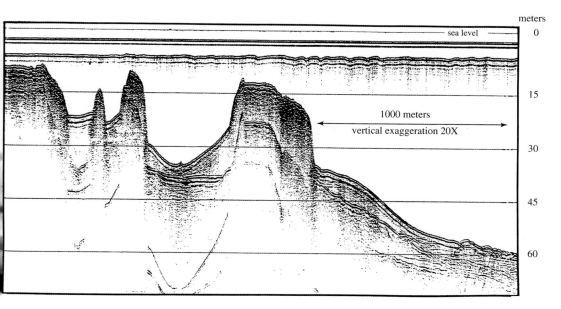

meters

sea level — 0

15

1000 meters

vertical exaggeration 20X

30

45

60

Figure 6 High-resolution north-south seismic line across the platform margin off Key West, Florida. Main reef (Sand Key Reef) forms the shelf margin at left. Features seaward of margin are outlier reefs. The top of the largest outlier reef (in 6 to 7 m of water) has been cored and dated. The top of the outlier is an 80,000-year-old Stage 5a coral reef (from Lidz et al., 1991).

FALLING SEA LEVEL

With a lowering of sea level, coral reefs would die, as they did at the end of Stage 5e and during numerous other fluctuations since 5e time (see **Figure 1**). Wetlands would form seaward of present wetlands, which would be converted to upland forests, prairies, or deserts. If sea level were to drop drastically, as it did during isotope Stage 2 about 18,000 years ago **(Figure 1)**, the areal extent of estuarine nurseries would be sharply reduced and organisms requiring estuaries for their reproductive cycles could be extinguished. Little is known about environmental perturbation when sea level was >100 m below the present, just 10,000 years ago **(Figure 1)**. Examination of world bathymetric charts shows that because of steep slopes along continental margins, the low topography necessary for estuaries and shallow bays was virtually nonexistent. At such low sea levels, estuaries would have been restricted to deep-sea canyons.

RISING SEA LEVEL

Alternatively, a rise in sea level would obviously drown beaches, shorelines, and existing coastal wetlands. Critical environments such as coastal marshlands and forests would shift inland, and new estuaries suitable for marine nursery grounds would be continually evolving and shifting inland. A rapid rise might retard development of marshes and estuaries; thus, nursery ground would not evolve until stabilization allowed establishment of extensive wetlands. Nutrients would be released to the oceans as the sea rose over the land, causing a multitude of ecosystem changes (Hallock and Schlager, 1987). Newly created bays/estuaries would produce turbid water that would affect nearby

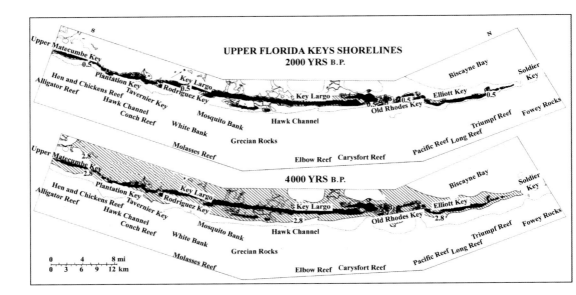

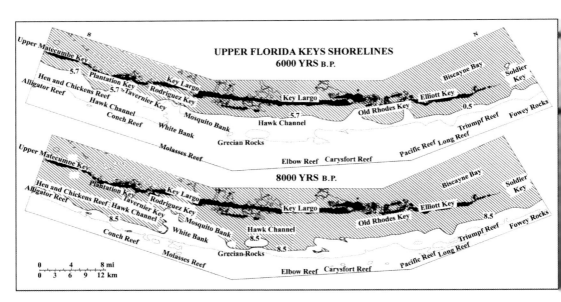

Figure 7 Maps showing approximate paleoshorelines for a portion of the Florida reef tract at 2,000, 4,000, 6,000, and 8,000 years before the present. The Florida Keys (land) are shown in solid black. Crosshatch shows land when sea level was lower. Sea level 2000 years ago was essentially the same as today (from Lidz and Shinn, 1991).

coral reefs, as has happened in the Florida Keys during the past 6,000 years (Ginsburg and Shinn 1964; Shinn et al., 1989). As an example, by constructing a series of relatively accurate sea-level shore line maps for the past 10,000 years, Lidz and Shinn (1991) showed the locations of former shorelines as they marched across the 8- to 10-km-wide shelf seaward of the Florida Keys **(Figure 7)**.

The shelf that now supports the Florida Keys and modern coral reefs was high and dry during isotope Stage 2. The map series reveals many previously unappreciated aspects of Holocene reef evolution, such as offshore topographic ridges that later became sites of reef growth, and topographic lows that with rising sea-level became tidal passes. These tidal passes served to transfer inimical waters from the bay to reef tract, thus suppressing reef development opposite the passes. The maps also suggest scenarios for the future. A 3-ft (1-m) rise would drown about 70% of the existing Florida Keys (Lidz and Shinn, 1991). An 18-ft (6-m) rise, equal to sea level during Stage 5e time, would drown even the highest elevations, providing new areas for coral-reef development on hard-rock surfaces. Such a rise would have profound societal effects worldwide.

THE CHALLENGE

Clearly, the critical question is, because humans obviously played no role in the well-documented past global-warming and sea-level-rise events, and if the past is a true indicator of future cycles, then how much of the most recent rise in temperature and sea level is due to humans? Are we simply victims of Milankovitch-type cycles? We as geologists might reasonably argue that scientists who presently shape public policy (namely, climatologists, sociologists, and economists) rely on computer models that function poorly in the short term in a chaotic atmosphere. The atmosphere/models being used often lack "dipsticks" or "benchmarks" to serve as reference points. Social scientists, although trained to evaluate the enormous social and cascading economic effects of global-climate change, lack the tools to predict when and how fast sea level and climate will change, or whether anthropogenic activities have an overriding influence. Computer models do not lie, but they function correctly only when fed enormous amounts of factual data. The amount of factual data needed to produce reliable predictions may take many years to accumulate. The data needed will probably come from paradigm shifts forced by field observations and measurements, an activity being practiced less and less as we move closer and closer to those "beige boxes" (Shinn, 1986).

The author believes that geologists, especially the ones who specialize in sedimentary rocks (although rarely united in their interpretations), are nevertheless in the best position to discern natural from anthropogenically induced climate and sea-level change. Clearly, no one can accurately foresee the future. We can, however, draw on the solid facts of our geologic past. Just as human history guides us in predicting our cultural and political future, so is "rock-solid" geologic history our most reliable guide to the uncertain geologic future. It is hoped that geologists and sedimentologists will become catalyzed to pursue investigations of previous and future sea level to aid in future decision making. Our only alternative to geologic and oceanographic data is the supercomputer that, no matter how technologically advanced, will function only as it is programmed. We geologists should be guiding the programs that predict our future by feeding them rock-solid data.

CONCLUSIONS

During the past 1.8 million years of geologic history, especially the past 140,000 years, sea level and global climate have fluctuated radically many, many times. Human activities clearly had no influence on those changes. Even older systems, hundreds of millions of years old, record numerous and more drastic sea-level fluctuations than those known from the past 1.8 million years. As yet unproved, nonhuman natural forces, most likely Milankovitch cycles and fluctuations in solar radiation, played the most significant role in changing our planet and in providing the environments we enjoy today.

This chapter only scratches the surface of available data for the Pleistocene/Holocene, but it should be abundantly clear that earlier sea-level changes have been large and numerous. It should also be clear that fluctuations before the industrial revolution began, about 100 years ago, were unaffected by human activity. As an example, **Figure 2** shows that sea level rose many times faster (1–2 m/100 years) between 17,000 and 10,000 years ago than during the past 100 years (maximum 20 cm/100 years) when the industrial revolution and burning of fossil fuels reached record levels.

ENDNOTE

Before *global greenhouse warming* became a universal household phrase, existing evidence pointed toward a return of the Ice Age. Many scientific publications prior to about 1970 predicted a return of the glaciers. Those predictions were based on atmospheric carbon dioxide (CO_2) preserved in polar ice cores that indicates the earth is poised near the CO_2 levels reached at the end of Stage 5e just before glaciers spread toward the equator and sea level fell drastically. Many more ice cores have been obtained since the 1970s, and CO_2 has continued to rise, but curiously, the interpretation has shifted. With all the additional data, it is important that earth scientists have more influence in the decision-making process in the future.

ACKNOWLEDGMENTS

I thank Christopher D. Reich and T. Donald Hickey for preparation of illustrations and especially Barbara Lidz for the many editorial corrections and changes that went into preparing this chapter. I thank Robert F. Dill and Lee Gerhard for suggestions during early reviews and Robert B. Halley and N. Terence (Terry) Edgar for support and critical literature.

REFERENCES CITED

Arrhenius, G., 1952, Sediment cores from the East Pacific, Swedish Deep-Sea Expedition (1947–1948) Reports, 5: Elander, Goteborg, p. 1–207.

Aurell, M., McNeill, D. F., Guyomard, T., and Kindler, P., 1995, Pleistocene shallowing-upward sequences in New Providence, Bahamas: Signature of high-frequency sea-level fluctuations in shallow carbonate platforms: Journal of Sedimentary Research, v. B65, no. 1, p. 170–182.

Bender M. L., Taylor, F. T., and Matthews, R. K., 1973, Helium-uranium dating of corals from middle Pleistocene Barbados reef tracts: Quaternary Research, v. 3, p. 142–146.

Bloom, A. L., and Yonekura, N., 1985, Coastal terraces generated by sea-level change and tectonic uplift: *in* Wodenberg, M. J., ed., Models in Geomorphology: Allen and Unwin, Winchester, Maine, p. 139–154.

Bloom, A. L., and Yonekura, N., 1990, Graphic analysis of dislocated Quaternary shorelines: *in* Revelle, R. M., panel chair, Sea Level Change: Studies in Geophysics, National Research Council, National Academy Press, Washington, D.C., p. 104–115.

Broecker, W. S., 1966, Absolute dating and the astronomical theory of glaciation: Science, v. 151, 21, p. 299–304.

Broecker, W. S., and Thurber, D. L., 1965, Uranium-series dating of corals and oolites from Bahaman and Florida Keys limestones: Science, v. 149, p. 58–60.

Broecker, W. S., 1997, Thermohaline circulation, the Achilles Heel of our climate system: Will man-made CO_2 upset the current balance? Science, v. 278, p. 1582–1588.

Broecker, W. S., Bond, G., Klas, M., Bonani, G., and Wolff, W., et al., 1990, A salt oscillator in the glacial Atlantic? 1. The Concept: Paleoceanography, v. 5, p. 469–477.

Carew, J. L., and Mylroie, J. E., 1995, Depositional model and stratigraphy for the Quaternary geology of the Bahamian Islands: *in* Curran, H. A., and White, B., eds., Terrestrial and Shallow Marine Geology of the Bahamas and Bermuda: Geological Society of America, Special Paper 300, p. 5–32.

Chappell, J., and Shackleton, N. J., 1986, Oxygen isotopes and sea level: Nature, v. 324, p. 137–140.

Curran, H. A., and White, B., 1995, Bahamas geology: *in* Curran, H. A., and White, B., eds., Terrestrial and Shallow Marine Geology of the Bahamas and Bermuda: Geological Society of America, Special Paper 300, p. 1–3.

Curray, J. R., 1965, Late Quaternary history, continental shelves of the United States: in The Quaternary of the United States, Wright, H. E., and Frey, D. C., eds., Princeton University Press, Princeton, N.J., p. 723.

Dansgaard, W., Johnsen, S. J., Clausen, H. B., Dahl-Jensen, D., Gunderstrup, N. S., Hammer, C. U., Hvidbergk, C. S., Steffensen, J. P., Sveinbjornsdottir, A. E., Jouzel, J., and Bond G., 1993, Evidence of general instability of past climate from a 250-kyr ice-core record. Nature, v. 364, p. 218–219.

Dill, R. F., Land, L. S., Mack, L. E., and Schwarcz, H. P., 1998, A submerged stalactite from Belize: Petrography, geochemistry and geochronology of massive marine cementation. Carbonites and Evaporites, v. 13, no. 2, p. 189–197.

Dill, R. F., Jones, A., and Yename, A., 1999, Speleothems in submerged cave systems determine timing of geological origin, global changes in oceanographic processes, and Pleistocene sea level variations: American Association of Petroleum Geologists, Abstracts with Program, p. A32.

Emiliani, C., 1955, Pleistocene temperatures: Journal of Geology, v. 63, no. 6, p. 538–578.

Emiliani, C., 1966, Paleotemperature analysis of Caribbean cores P6304-8 and P6304-9 and a generalized temperature curve for the past 425,000 years: Journal of Geology, v. 74, p. 109–126.

Enos, Paul, and Perkins, R. D., 1979, Evolution of Florida Bay from island stratigraphy: Geological Society of America Bulletin, Part 1, v. 90, p. 59–83.

Fairbanks, R. G., 1989, A 17,000-year glacio-eustatic sea level record: Influence of glacial melting rates on Younger Dryas event and deep ocean circulation: Nature, v. 342, p. 637–642.

Fairbanks, R. G., 1990, The age and origin of the "younger Dryas climate event" in Greenland ice cores, Paleoceanography, v. 5, no. 6, p. 937–948.

Fairbridge, R. W., 1961, Eustatic changes in sea level: *in* Physics and Chemistry of the Earth, vol. 4, Ahrens, H. F., Press, F., Rankama, K., and Runcorn, S. K., eds., Pergamon, NY, p. 99–185.

Fruijtier, C., Elliott, T., and Schlager, W., 2000, Mass-spectrometric ^{234}U-^{230}Th ages from the Key Largo Formation, Florida Keys, United States: Constraints on diagenetic age disturbance: Geological Society of America Bulletin, v. 112, no. 2, p. 267–277.

Ginsburg, R. N., and Shinn, E. A., 1964, Distribution of the reef-building community in Florida and the Bahamas, American Association of Petroleum Geologists Bulletin, Abstracts with Program, v. 48, p. 527.

Hallock, P., and Schlager, W., 1987, Nutrient excess and the demise of coral reefs and carbonate platforms: PALAIOS, v. 1, p. 389–398.

Henderson, G. M., and Slowey, N. C., 2000, Evidence from U-Th dating against Northern Hemisphere forcing of the penultimate deglaciation: Nature, v. 404, p. 61–66.

Labeyrie, L., Duplessy, J. C., and Blanc, P. L., 1987, Variation in the mode of formation and temperature of ocean deep waters over the past 125 thousand years. Nature, vol. 327, p. 477–482.

Lear, C. H., Elderfield, P. A., and Wilson, P. A., 2000, Cenozoic deep-sea temperatures and global ice volumes from Mg/Ca in benthic foraminiferal calcite: Science, v. 287, no. 5451, p. 269–272.

Libby, W. F., 1952, Radiocarbon Dating: University of Chicago Press, p. 124.

Lidz, B. H., Hine, A. C., Shinn, E. A., and Kindinger, J. L., 1991, Multiple outer-reef tracts along the south Florida bank margin: Outlier reefs, a new windward-margin model: Geology, v. 19, p. 115–118.

Lidz, B. H., and Shinn, E. A., 1991, Paleoshorelines, reefs, and a rising sea: South Florida, U.S.A.: Journal of Coastal Research, v. 7, no. 1, p. 203–229.

Lidz, B. H., Shinn, E. A., Hine, A. C., and Locker, S. D., 1997, Contrasts within an outlier-reef system: Evidence for differential Quaternary evolution, South Florida windward margin, U.S.A.: Journal of Coastal Research, v. 13, no. 3, p. 711–731.

Lighty, R. G., Macintyre, I. G., and Stuckenrath, R., 1982, *Acropora palmata* reef framework: A reliable indica tion of sea level in the Western Atlantic for the past 10,000 years: Coral Reefs, v. 1, p. 125–130.

Locker, S. D., Hine, A. C., Tedesco, L. P., and Shinn, E. A., 1996, Magnitude and timing of episodic sea-level rise during the last deglaciation: Geology, v. 24, no. 9, p. 827–830.

Ludwig, K. R., Muhs, D. R., Simmons, K. R., Halley, R. B., and Shinn, E. A., 1996, Sea-level records at 80 ka from tectonically stable platforms: Florida and Bermuda: Geology, v. 24, no. 3, p. 211–214.

Maul, G. A., and Martin, D. M., 1993, Sea level rise at Key West, Florida, 1846–1992: America's longest instru ment record? Geophysical Research Letters, v. 20, p. 1955–1958.

McCrea, J. M., 1950, On the isotope chemistry of carbonates and a paleotemperature scale: Journal of Chem istry and Physics, v. 18, p. 849–857.

Mesolella, K. J., Matthews, R. K., Broecker, W. S., and Thurber, D. L., 1969, The astronomical theory of climatic change: Barbados data: Journal of Geology, v. 77, p. 250–274.

Milankovitch, M., 1938, Astronomische Mittel zur Erforschung der erdgeschichtlichen Klimate: in Gutten berg, B., ed., Handbuch der Geophysik, v. 9, Berlin, p. 593–698.

Muhs, D. R., Szaabo, B. J., McCartan, L., Maat, P. B., Bush, C. A., and Halley, R. B., 1992, Uranium-series age estimates of corals from Quaternary marine sediments of southern Florida: Florida Geological Survey Special Publication 36, p. 41–49.

Neumann, A. C., and Hearty, P. J., 1996, Rapid sea-level changes at the close of the last interglacial (substage 5e) recorded in Bahamian island geology: Geology, v. 24, no. 9, p. 775–778.

Osmond, J. K., Carpenter, J. R., and Windom, H. L., 1965, Th^{230}/U^{234} age of the Pleistocene corals and oolites of Florida: Journal of Geophysical Research, v. 70, p. 1843–1847.

Robbin, D. M., 1984, A new Holocene sea level curve for the upper Florida Keys and Florida reef tract: in Gleason, P. J., ed., Environments of South Florida, Present and Past, II: Miami Geological Society, Miami Fla., p. 437–458.

Scholl, D. W., Craighead, F. C. Sr., and Stuiver, M., 1969, Florida submergence curve revised: Its relation to sed imentation rates: Science, v. 163, p. 562–564.

Shackleton, N. J., 1967, Oxygen isotope analyses and Pleistocene temperature re-assessed: Nature, v. 215, p. 15–17.

Shackleton, N. J., 1987, Oxygen isotopes, ice volume and sea level. Quaternary Science Review, vol. 6, p 183–190.

Shepard F. P., 1963, Thirty-five thousand years of sea level, in Clements, T., ed., Essays in Marine Geology in Honor of K. O. Emery: University of Southern California Press, p. 1–10.

Shepard, F. P., and Curray, J. R., 1967, Carbon-14 determination of sea level changes in stable areas: *in* Sears M., ed., Progress in Oceanography: London, Pergamon Press, p. 283–291.

Shinn, E. A., 1986, Paradigm disease: PALAIOS, v. 1, no. 3, p. 205.

Shinn, E. A., 1988, The geology of the Florida Keys: Oceanus, v. 31, no. 1, p. 47–53.

Shinn, E. A., Ginsburg, R. N., and Lloyd, R. M., 1965, Recent supratidal dolomite from Andros Island, Bahamas: *in* Pray, L. C., and Murray, R. C., eds., Dolomitization and Limestone Diagenesis, a Sympo sium: Society of Economic Paleontologists and Mineralogists, Special Publication 13, p. 112–123.

Shinn, E. A., Lidz, B. H., Kindinger, J. L., Hudson, J. H., and Halley, R. B., 1989, Reefs of Florida and the Dry Tortugas: IGC Field Guide T176: American Geophysical Union, Washington, D.C., p. 1–53.

Weber, J. N., 1977, Use of corals in determining glacial-interglacial changes in temperature and isotopic com position of sea water. American Association of Petroleum Geologists, Studies in Geology No. 4, p. 289–295.

Part IV | Policy Drivers

The way people feel about certain issues often becomes the basis for governmental policies. Human-designed policies developed to address the dynamic systems of the earth would benefit by commentary from those who best understand such systems. Those policies will likely be ineffective unless they are tempered by scientific investigations directed toward a better understanding of earth's cause-and-effect dynamics.

Yates, K. K., and L. L. Robbins, Microbial lime-mud production and its relation to climate change, 2001, *in*
L. C. Gerhard, W. E. Harrison, and B. M. Hanson,
eds., Geological perspectives of global climate
change, p. 267–283.

14 | Microbial Lime-Mud Production and Its Relation to Climate Change

K. K. Yates
L. L. Robbins
U.S. Geological Survey
Center for Coastal and Marine Geology
Saint Petersburg, Florida, U.S.A.

ABSTRACT

Microbial calcification has been identified as a significant source of carbonate sediment production in modern marine and lacustrine environments around the globe. This process has been linked to the production of modern whitings and large, micritic carbonate deposits throughout the geologic record. Furthermore, carbonate deposits believed to be the result of cyanobacterial and microalgal calcification suggest that the potential exists for long-term preservation of microbial precipitates and storage of carbon dioxide (CO_2). Recent research has advanced our understanding of the microbial-calcification mechanism as a photosynthetically driven process. However, little is known of the effects of this process on inorganic carbon cycling or of the effects of changing climate on microbial-calcification mechanisms.

Laboratory experiments on microbial cellular physiology demonstrate that cyanobacteria and green algae can utilize different carbon species for metabolism and calcification. Cyanobacterial calcification relies on bicarbonate (HCO_3^-) utilization while green algae use primarily CO_2. Therefore, depending on which carbonate species (HCO_3^- or CO_2) dominates in the ocean or lacustrine environments (a condition ultimately linked to atmospheric partial pressure PCO_2), the origin of lime-mud production by cyanobacteria and/or algae may fluctuate through geologic time. Trends of cyanobacteria versus algal dominance in the rock record corroborate this conclusion. These results suggest that relative species abundances of calcareous cyanobacteria and algae in the Phanerozoic may serve as potential proxies for assessing paleoclimatic conditions, including fluctuations in atmospheric PCO_2.

INTRODUCTION

Large deposits of nonfossiliferous lime mud comprise significant portions of the rock record throughout geological time at various localities around the globe (Cloud, 1961; Meeder, 1979; Horodyski and Mankiewicz, 1990; Knoll and Swett, 1990; Kazmierczak et al., 1994; Davis et al., 1995). Deposits such as these represent a significant portion of the carbonate sediment budget, and their formation has inevitably influenced atmospheric PCO_2. The origin for both modern and ancient nonfossiliferous lime-mud deposits (Drew, 1914; Lowenstam and Epstein, 1957; Horodyski and Mankiewicz, 1990; Knoll and Swett, 1990; Robbins and Blackwelder, 1992; Boss and Neuman, 1993; Morse and He, 1993; Kazmierczak et al., 1994) and the effect of climate change on their formation (Berner, 1990; Opdyke and Walker, 1992; Sundquist, 1993; Sarmiento and Bender, 1994) have been highly debated. Much evidence, however, suggests that photosynthetic microbes played a role in production of such deposits (Horodyski and Mankiewicz, 1990; Awramik, 1991; Barattolo, 1991; Flugel, 1991; Mamet, 1991; Riding, 1991; Riding and Guo, 1991; Roux, 1991; Kazmierczak et al., 1994). Several modern species of both freshwater and marine unicellular green algae and cyanobacteria capable of precipitating calcium carbonate have been identified (Stabel, 1986; Thompson and Ferris, 1990; Merz, 1992; Hartley et al., 1995; Yates and Robbins, 1995, 1998). Results discussed in Yates (1996) indicate that a single species of unicellular green algae may be capable of producing several thousand kilograms of calcium-carbonate sediment per day.

Resolving controversies on the effects of climate change on lime-mud deposition requires an understanding of the links between resulting changes in seawater chemistry and their effect on calcification processes. Biocalcification, whether highly controlled (as the formation of intricate carbonate skeletons by mollusks and corals) or induced as a by-product of biological activity, is necessarily linked to cellular metabolism. Many primary metabolic cell functions serve to regulate intracellular chemistry, and change in response to environmental conditions. It is well known that microbial metabolism responds quickly to environmental perturbations resulting from daily and seasonal climate change. For example, rates of microbial photosynthesis increase or decrease with temperature and variation in light intensity. Microbes respond to changes in salinity, pH, and other changes in water chemistry resulting from rainfall or evaporation by altering the rates at which ions such as (Na^+) and (H^+) are transported into and out of cells (Kalinkina and Strongonov, 1986; Broun et al., 1992). Changes in inorganic carbon sources and availability result in changes in rates of microbial photosynthesis (Badger et al., 1993). Research on microbial-calcification mechanisms indicates that calcium-carbonate precipitation is linked to photosynthetic and ion transport physiology (Morita, 1980; Merz, 1992). Thus, it is likely that microbial calcification is strongly influenced by environmental perturbations. However, little is known of the effects of cell physiology or of short- and long-term climate change on microbial lime-mud production. The objective of this review is to critically examine key arguments on the influence of photosynthetic microbial cell physiology on calcium-carbonate precipitation, discuss the potential effects of climate change on microbial lime-mud production, and identify potential environmental triggering mechanisms for inducing microbial calcification.

MICROBIAL PRECIPITATION OF LIME MUD— MODERN AND ANCIENT EVIDENCE

The ability of microbes to induce carbonate mineralization has long been recognized, and a number of microbial species capable of biocalcification have been identified. For example, McCallum and Guhathakurta (1970) isolated *Bacillus*, *Pseudomonas*, and *Vibrio* from sediment cores taken from

Bimini, Brown Cay, and Andros Island in the Bahamas. Bacterial cultures precipitated rosettes of aragonite needles in five of six different media tested. Uninoculated control flasks showed no precipitation between pH 6.0 and 8.0, and a few dumbbell-shaped crystals between pH 8.0 and 9.0. Precipitation of aragonite needles on cell surfaces of marine heterotrophic bacteria was also observed under aerobic and anaerobic, agitated and non-agitated conditions at pH 6.9 to 8.7 from seawater on the addition of 0.01–0.1% organic matter (Krumbein, 1974). In separate experiments with marine bacteria collected from the Great Barrier Reef and Bermuda, bacterial precipitation of calcite or aragonite occurred on the addition of trimethylamino-N-oxide (TMAO) or TMAO and organic matter to seawater cultures (Morita, 1980; Novitsky, 1981). No precipitation occurred in controls without bacteria. Through meticulous examination of carbonate grains by scanning electron micrography (SEM), Chafetz (1986) and later Folk (1993) provided evidence that a variety of carbonate sediments consisted of products of bacterial precipitation. In addition, various actively photosynthesizing green algae, including *Chlorococcum* and *Chlorella* sp., were observed to initiate heterogeneous nucleation of calcite (Stabel, 1986; Hartley et al., 1995). Isolated cultures of cyanobacteria *Synechococcus* sp. from the freshwater Fayetteville Green Lake, New York, showed epicellular biomineralization of calcite, gypsum, and magnesite after 24 hours of incubation under fluorescent light (Thompson and Ferris, 1990). Yates and Robbins (1995) also confirmed that isolated cultures of the cyanobacteria *Synechocystis* from Florida Bay and Lake Reeve, Australia (a euryhaline lagoon), were capable of epicellular precipitation of calcite after only 3 hours of incubation in growth media, while no precipitation was observed in uninoculated or dead cell controls. Similar experiments by Yates (1996) using the unicellular green alga *Nannochloris atomus* (also isolated from Lake Reeve, Australia) demonstrated its ability to precipitate calcite within a few hours of incubation **(Figure 1)**.

(a)   (b)

Figure 1 (a) Scanning electron micrograph of a cyanobacterial cell with a calcium-carbonate seed crystal forming on the cell membrane. Laboratory observations show that microbial mineralization begins with the formation of seed crystals followed by encrustation of the cell with larger crystals. (b) Transmission electron micrograph of a cyanobacterial cell (approximately 2 microns in length) encrusted with calcium-carbonate crystals. Cells collected from whitings near Andros Island, Bahamas.

Despite the widespread occurrence and distribution of calcifying microbial species, few attempts have been made to quantify the potential contribution of modern microbial lime-mud production to carbonate sedimentation. Results of calcification potential experiments on cultures of calcifying *Nannochloris atomus* cells (Yates and Robbins, 1998) indicate that blooms of a single microbial species are capable of generating several thousand kilograms of carbonate sediment per day. They calculate that *N. atomus*, at a cell population density of 1.0×10^5 cells/mL, precipitates 1.55mg/L of calcium carbonate in 12 hrs or 4960 kg of calcium carbonate per day in a single bloom of 0.64 km^2 and 5 m depth. Recent research on marine (Robbins and Blackwelder, 1992; Robbins et al., 1996) and freshwater (Thompson et al., 1990; Hodell et al., 1998) whitings, large patches (1–2+ kilometers in length and width) of water column containing suspended calcium-carbonate sediment, indicates that their production is linked to microbial calcification **(Figure 2)**. Field observations of marine whitings on the Great Bahama Bank indicate that whitings are associated with blooms of cyanobacterial species such as *Synechococcus* and *Synechocystis*. Comparable observations have been made on freshwater whitings in Fayetteville Green Lake (Thompson et al., 1990; Hodell et al., 1998). Robbins and Tao (1996) estimate that whitings in the waters of the Great Bahama Bank cover approximately 70 km^2 per day and may account for 40% of lime-mud volume on the bank top. Sediment production by a bloom of *N. atomus* is comparable to the amount of sediment contained in a Bahamian whiting of similar size. Because species of *Nannochloris* and other

Figure 2 (a) Satellite image of whitings west of Andros Island, Bahamas (photograph courtesy of NASA). Arrows indicate whitings.

Figure 2 *(continued)* (b) Aerial photograph of a whiting approximately 2 km x 0.5 km wide. (c) Whiting near Andros Island with well-defined boundaries, as shown by color contrast near top of image.

calcium-carbonate-precipitating unicellular green algae and cyanobacteria are globally abundant in freshwater and marine environments (Stockner, 1988) and blooms of these species are likely to cover thousands of square kilometers, the contribution of microbial calcification to the global carbonate sediment budget has probably been greatly underestimated.

The geologic record, from the Proterozoic through the Holocene, is characterized by extensive deposits of nonfossiliferous micrite at various localities around the globe, the origins of which have been attributed to microbes. For example, the Akademikerbreen Group, Spitsbergen, dated at 800–700 million years ago(Ma) (late Proterozoic age), comprises 2000 m of carbonates of which 60% is micrite (Knoll and Swett, 1990). Knoll and Swett (1990) suggest that one possible origin for this micrite is precipitation of carbonates in microbial mats and subsequent disintegration to mud, or precipitation by phytoplankton blooms in the water column. The Pahrump Group of southeastern California is characterized by a middle unit of relatively nonfossiliferous dolomite of middle to late Proterozoic age. Horodyski and Mankiewicz (1990) have identified a calcifying microalga in this deposit that may have contributed to sediment deposition. Kazmierczak et al. (1994) describe pure micritic and peloidal limestones (10–20 m thick) from the Upper Jurassic of the Holy Cross Mountains near central Poland. They suggest that these sediments were derived entirely from *in vivo* calcified benthic coccoid cyanobacterial mats. Riding (1996) suggests that whitings may have been a major source of micrite, particularly in the Proterozoic. Lake Reeve, Victoria, Australia, and Coorong Lagoon, Australia, are characterized by modern deposits of calcite mud and calcite and aragonite mud, respectively, consisting of calcium-carbonate crystals associated with cyanobacterial cells (Davis et al., 1995), some of which may result from microbial precipitation in the water column and as microbial mats. The Bahama Banks and Florida Bay, both locations where whitings occur, are also characterized by thick sequences of Holocene limestone deposits (Cloud, 1961). Results from Hawkins (1995) indicate that monoclonal antibodies specific to modern *Synechocystis* reacted downcore with intracrystalline proteins from aragonite needles in cores collected from Florida Bay (recent to 3800 yrs). This indicates that similar bacteria were present during formation of this sediment and may have contributed to its production. Furthermore, fossils of calcareous cyanobacteria and algae have been documented in rocks from the Proterozoic to Cenozoic, and notable trends in the appearance of cyanobacterial and algal species and variation in dominant calcifiers through geologic time have been observed (Horodyski and Mankiewicz, 1990; Awramik, 1991; Barattolo, 1991; Flugel, 1991; Mamet, 1991; Riding, 1991; Riding and Guo, 1991; Roux, 1991; and Kazmierczak et al., 1994). It is evident from these examples that microbial calcification has contributed significantly to the sediment budget throughout the geologic record. However, although many micrite deposits show fossil evidence of microbial activity, there are currently no accurate methods for distinguishing modern or ancient whitings deposits from nonmicrobial carbonates.

THE EFFECTS OF CELL PHYSIOLOGY ON MICROBIAL CALCIFICATION

Studies by Morita (1980) and Yates (1996) are among the very few carried out to specifically investigate microbial-calcification mechanisms. However, several theoretical models for calcification by cyanobacteria and green algae have been proposed (Merz, 1992; McConnaughey, 1991; Thompson and Ferris, 1990). Although many of the mechanistic details of these models remain controversial, all the mechanisms are based on the common theme of photosynthetically driven calcification, and they link carbonate mineral precipitation reactions to cellular physiological processes.

Cellular physiology in general is the study of the fundamental activities of plant cells, animals, and microorganisms, including nutrition, response to the environment, growth, and reproduction (Giese, 1962). The primary cellular physiological aspects of photosynthetic microbes most likely to influence calcification fall under the categories of nutrition (uptake and preparation of food, release of energy in cells, and elimination of waste products (Giese, 1962) and response to the environment (both chemical and physical aspects). Upon close examination of these aspects of microbial physiology, it is evident that they are driven by a system of molecular and ionic pumps and by chemical gradients. For example, the driving force for production of adenosine triphosphate (ATP), the primary energy source for cellular metabolism of all living organisms, is production of a proton motive force, or H^+ electrochemical gradient, across cellular membranes. The primary food source for photosynthetic microbial cells is inorganic carbon, primarily in the form of CO_2 and HCO_3^-, which must be transported into the cell from the external environment for photosynthetic fixation. In addition, transport of ions such as Ca^{2+} and Na^+ across cell membranes plays a vital role in many physiological aspects of cells, including production of energy-generating electrochemical potentials, transport of carbon across cell membranes, and maintenance of highly regulated intracellular chemical environments. As calcium carbonate chemistry is defined by reactions involving these chemical species (CO_2, HCO_3^-, CO_3^{2-}, H_2CO_3, Ca^{2+}, and H^+ [pH]), it is clear that microbial physiological processes have the potential to affect carbonate chemistry because these cellular organisms are capable of manipulating vital chemical components. Additionally, generation of electrochemical gradients that drive cell physiology requires that cells remain out of equilibrium with their ambient environment. Thus, carbonate mineralization reactions associated with microbial physiology may not occur in equilibrium with ambient water chemistry.

For example, experimental results of Yates and Robbins (unpublished, 1995) indicate that increased medium Ca^{2+} concentrations induce calcification in cultures of cyanobacteria and microalgae (but not in uninoculated media), and result in rapid death of cells upon calcium-carbonate ($CaCO_3$) precipitation. All cells must maintain low internal calcium concentrations (approximately 10^{-7} mols $liter^{-1}$ (M)) to survive (Kretsinger, 1983). Elevated medium Ca^{2+} may require that cells expel intracellular calcium more rapidly. Cells can expel Ca^{2+} through the electroneutral exchange of Ca^{2+} out of the cell and simultaneous transport of H^+ ions into the cell (Niggli et al., 1982). This Ca^{2+}-out/2H^+-in exchange may result in elevation of pH and calcium concentrations, and thus the saturation state with respect to carbonate minerals, in very small, localized areas (or microenvironments) near the external cell membrane. Evidence that cellular export of calcium plays a role in microbial and algal calcification has been demonstrated by Yates and Robbins (1999) and McConnaughey and Falk (1991). The observation that cells expire rapidly on calcification induced by increased Ca^{2+} concentrations indicates that $CaCO_3$ precipitation may be a lethal side effect from an attempt to expel toxic levels of Ca^{2+}. In a "last-gasp" effort, cells may be unable to export Ca^{2+} fast enough to avoid accumulating toxic levels, or may become encased in $CaCO_3$ to the extent that the cell is isolated from external media and unable to take up vital nutrients. Calcium concentrations ranging from 0.245 millimols $liter^{-1}$ (mM) (Yates and Robbins, 1995) to 14.2 mM, approximately equal to or less than seawater calcium concentrations of 10^{-2}M, are high enough to induce the "last-gasp" effect. However, all microbes in the oceans clearly do not calcify to the point of extinction. This may be because of a combination of environmental factors, not elevated calcium concentrations alone, is required to induce calcification. For example, microbial calcification in laboratory cultures occurs to a greater extent when a combination of elevated calcium concentrations, temperature, and light intensity is used. Similar combinations may be required to induce calcification in natural systems as well. This is supported by the fact that extensive microbial calcification is apparently localized to specific natural environments (Thompson and Ferris, 1990; Robbins and Blackwelder, 1992; Davis et al., 1995).

Similarly, membrane potential across microbial cellular membranes is generated as Na^+ and is expelled from cells by an electrogenic 3Na^+-out/ATP (adenosine triphosphate) pump. This in turn drives H^+ out of cells through an H^+-out/2Na^+-in exchange mechanism (Ritchie, 1992). Expulsion of H^+ out of the cell results in production of an acidic microenvironment near this region of the cell membrane. In the Yates (1996) model, calcification occurs as CO_2 generated in an acidic microenvironment is taken up as the primary source of inorganic carbon for photosynthesis **(Figure 3)**. Some

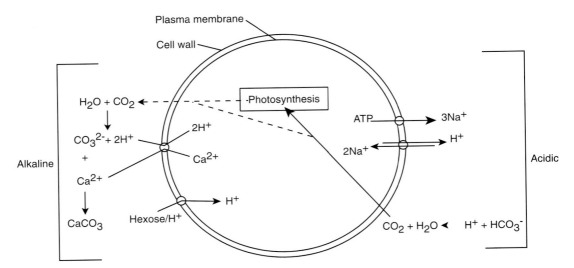

Figure 3 Theoretical model for microbial calcification, modified from Yates (1996). Calcification occurs in an extracellular alkaline microenvironment generated as Ca^{2+} is exported from the cell in exchange for uptake of $2H^+$. Na^+ and H^+ exchange across the cell membrane may generate a microacidic environment. Microbial calcification relies on uptake of either CO_2 (microalgae) or HCO_3^- (cyanobacteria). Excess intracellular inorganic carbon is absorbed from the cell into the microalkaline environment, providing an additional source of carbon for calcification.

of this carbon is subsequently absorbed by an extracellular alkaline microenvironment, converted to CO_3^{2-}, and precipitated as $CaCO_3$. This is in contrast to the mechanism of cyanobacterial calcification proposed by Yates and Robbins (1995). They suggest that HCO_3^- is taken up as the primary source of inorganic carbon for photosynthesis and transported to the carboxysome, where it is converted to CO_2 and accumulated as a dissolved inorganic carbon (DIC) pool. Excess CO_2, available as ribulose bisphospate carboxylase (Rubisco, an enzyme that catalyzes the addition of CO_2 to the 5-carbon sugar ribulose bisphosphate, the first step of photosynthesis), becomes saturated, diffuses from the cell to an alkaline extracellular microenvironment where it is converted to CO_3^{2-}, and precipitates as calcium carbonate. Previous research on photosynthetic uptake of inorganic carbon by both microalgae and cyanobacteria (Badger et al., 1993) supports the results of Yates and Robbins (1995) and Yates (1996), indicating that calcification by microalgae depends primarily on CO_2 uptake while cyanobacterial calcification relies on HCO_3^- uptake. Merz-Preiss and Riding (1999) observed two morphologically different styles of calcification in filamentous, sheathed cyanobacteria from freshwater streams. They suggest that the morphology and location of precipitates on cells depends on whether cells are utilizing HCO_3^- or CO_2 for photosynthesis. Cellular uptake of HCO_3^- leads to mineralization within cyanobacterial sheaths, while uptake of CO_2 leads to passive encrustation of filaments and negligible photosynthetic influence on calcification. In addition, results from Yates (1996) and Yates and Robbins (1995) support evidence that both microalgae and cyanobacteria are capable of utilizing either CO_2 or HCO_3^- for photosynthesis and that the species of inorganic carbon used by cells depends on external CO_2 and HCO_3^- concentrations.

The detrimental effect that calcification has on microbes suggests that calcification is a by-product of cellular metabolism. The ion transport and inorganic carbon uptake processes described above must occur, regardless of whether or not cells are actively calcifying, to regulate intracellular chemical environments. Thus, carbonate mineral precipitation associated with cellular ion transport processes occurs in a nonequilibrium chemical environment. For example, as inorganic calcification proceeds via $Ca^{2+} + 2HCO_3^- \rightarrow CaCO_3 + CO_2 + H_2O$, pH decreases due to generation of H^+ ions from production of carbonic acid, calcification eventually ceases, and dissolution may begin as the system attempts to reach equilibrium. However, as calcification occurs in an alkaline microenvironment near a cell membrane (generated as Ca^{2+} is pumped out of the cell and H^+ is pumped in), pH is maintained as H^+ ions generated by mineralization are sequestered by the cell, Ca^{2+} is replenished through cellular export, and CO_2 generated by calcification is rapidly hydrated or hydroxylated to HCO_3^- or CO_3^{2-} and recycled back into the calcification reaction. This mechanism suggests that microbial calcification may not generate the predicted molar ratio of 0.6 moles of CO_2 for each mole of calcium carbonate, as has been suggested for seawater calcification processes.

The ultimate fate of CO_2 sequestered by both photosynthesis and calcium-carbonate production depends on the preservation potential of mineral and organic material. As organic material is easily oxidized on the death of the organism, releasing CO_2, and calcium-carbonate minerals (while still subject to dissolution) are more likely to be preserved than organics, effective sequestration and storage of CO_2 requires that the mass of carbon in calcium carbonate exceeds the mass of carbon in organic material. The cyanobacterial precipitation model predicts that cells accumulate an order of magnitude more CO_2 than can be sequestered by Rubisco (Ribulose-1,5 -bis phosphate carboxylase, an enzyme used in converting carbon dioxide into organic carbon via photosynthesis). This excess CO_2 has the potential to leak from cells and become incorporated into carbonate minerals. Research by Badger et al. (1993) supports leakage of inorganic carbon from cyanobacterial cells and indicates that leakage of inorganic carbon from microalgal cells is greater than from cyanobacteria. Incorporation of excessive amounts of CO_2 into calcium carbonate should result in a carbonate-carbon to organic-carbon ratio of much greater than one. It is interesting to note that SEM and transmission electron micrograph (TEM) analysis of whitings sediment, a potential product of microbial calcification (Robbins and Blackwelder, 1992), reveals cyanobacterial cells encased in massive deposits of calcium-carbonate crystals. Although ratios of calcium-carbonate inorganic carbon to organic carbon in whitings have yet to be determined, it is possible that this ratio exceeds one. Furthermore, paleontological evidence for preservation of Jurassic (Kazmierczak et al., 1994) and Proterozoic (Horodyski and Mankiewicz, 1990) carbonate deposits believed to be the result of cyanobacterial and microalgal calcification suggests that the potential exists for long-term preservation of microbial precipitates and storage of CO_2.

Theoretical calcification models for cyanobacteria and microalgae indicate that microbial calcification is dependent on environmental conditions, and a number of potential triggering mechanisms for this process exist. For example, fluctuations in salinity may induce microbial calcification. Sodium concentrations of 25–30 mM enhance electrophoretic uptake of H^+ by cyanobacteria (Broun et al., 1992), resulting in enhancement of microalkaline regions near the cell membrane. In addition, increased Na^+ concentrations induce HCO_3^- uptake by cyanobacteria, resulting in accumulation of a large intracellular DIC pool (Espie et al., 1988) and a source of excess CO_2 for leakage to alkaline sites of calcification. In microalgae, increased salinity increases photorespiration (Kalinkina and Strogonov, 1986), which may also enhance diffusion of CO_2 to alkaline regions. Increased calcium concentrations can also induce microbial calcification, as described in the previous section by the "last-gasp" hypothesis. In microalgae, addition of calcium to salinized cultures in which photorespiration was increased and photosynthetic oxygen evolution was inhibited improved photosynthetic oxygen evolution (Ahmed et al., 1989). Thus, a combination of high Na^+ and Ca^{2+} concentrations may induce calcification in microalgae. As high Na^+ enhances photorespiration, more photorespiratory CO_2 is generated and is available for diffusion to sites of calcification. It is possible that under Na^+ stress, cells take up Ca^{2+} in an attempt to maintain photosynthesis rates. This would require increased expulsion of calcium to avoid toxic intracellular levels. Thus, again, alkaline regions become enhanced, calcium concentrations increase in these regions, and an excess source of CO_2 is available for leakage to this site. In microalgal cells, the addition of calcium to cultures increases uptake of hexose via hexose/H^+ symport (Komor, 1973), also resulting in enhancement of alkaline regions.

Of particular interest is the dependency of cyanobacterial and microalgal calcification mechanisms on external CO_2 and HCO_3^- concentrations. Although both cell types are capable of taking up either HCO_3^- or CO_2 for photosynthesis, cyanobacterial calcification relies on HCO_3^- uptake while microalgal calcification relies on CO_2 uptake. Low extracellular CO_2 concentrations result in induction of the HCO_3^- uptake mechanism in cyanobacteria and, according to the calcification model (Yates, 1996), induction or enhancement of calcium-carbonate precipitation. In contrast, low extracellular CO_2 concentrations may inhibit or limit microalgal calcification as these cells become dependent on HCO_3^- and calcification relies on CO_2 uptake (Yates, 1996). Alternatively, high CO_2 may limit cyanobacterial calcification and promote microalgal calcification. It has been suggested that the general difference between cyanobacteria and green algae is that cyanobacteria primarily take up HCO_3^- into cells and green algae mainly take up CO_2 (Tsuzuki and Miyachi, 1991). Thus variations in aqueous CO_2 concentrations may be the driving force for transitions in dominance of calcifying cyanobacteria over green algae and vice versa in local environments, through seasonal variations, and over longer timescale CO_2 fluctuations. Further investigations into calcifying microbial species distributions in various environments, and shifts in relative species abundance with CO_2 fluctuations, are required to confirm this hypothesis.

ATMOSPHERIC CO_2 FLUCTUATIONS VERSUS LIME-MUD DEPOSITION—TRENDS THROUGH GEOLOGIC TIME

Fluctuations in environmental parameters such as seawater chemistry and CO_2 may have been the driving force for evolution of calcifying microbes and large-scale precipitation of nonfossiliferous lime mud through geologic time. Stromatolites, believed to have formed as the result of microbial activity, were present as early as 3500 Ma (Awramik, 1991), approximately the same time as the appearance of microbial fossils in rocks from Western Australia (Lowe, 1980; Walter et al., 1980; Awramik, 1991) and South Africa (Byerly et al., 1986; Awramik, 1991). Calcareous cyanobacteria appeared suddenly near the base of the Cambrian (Riding, 1982) and dominated the calcareous flora of that period. Their sudden appearance is attributed to a significant change in seawater chemistry (Riding, 1991). Kempe et al. (1989) discuss the ramifications of the evolution of a soda ocean on the evolution of biomineralizing organisms. They suggest that the early ocean was highly alkaline, with a high pH, low calcium and magnesium concentrations, and high dissolved $NaCO_3$ concentrations. Gradually, as silicates weathered, calcium concentrations increased from possibly as low as 10^{-7} M in the Archean to 10^{-5} or 10^{-4} M during the Archean to Proterozoic transition, and to present day levels of 10^{-2} M by the late Precambrian. The drastic increase in seawater Ca^2 during the Proterozoic required evolution of calcium regulation and extrusion systems in cells and by the end of the Precambrian, resulted in the development of biocalcification as a strategy to deal with otherwise toxic calcium levels. Thus, the mechanism used in the "last-gasp" hypothesis for induction of calcification in microbes may have existed as early as the Proterozoic to Precambrian.

Because the earliest calcifying cells may have evolved under conditions of high alkalinity (Kempe and Degens, 1985) in the Proterozoic, (570–2500 million years before the present) with the primary source of inorganic carbon species in the form of HCO_3^- and CO_3^{2-}, cyanobacterial cells may have already had the ability to take up HCO_3^- as a primary inorganic carbon source for photosynthesis by the Cambrian. That calcareous cyanobacteria flourished during the Cambrian with a photosynthetic physiology dependent on bicarbonate uptake may seem contradictory because atmospheric PCO_2 was 80–600 times greater than today's concentrations (Sarmiento and Bender, 1994). However, one must consider the effects of other environmental perturbations resulting from high atmospheric CO_2. It is well known that increased global temperatures are associated with high CO_2 due to the greenhouse effect, and that this is reflected in warmer seawater temperatures. The saturation partial pressure of CO_2 decreases by 4.3% for every degree centigrade of seawater warming (Sarmiento and Bender, 1994). Thus, warming of seawater decreases the solubility of CO_2, resulting in outgassing and partitioning of this inorganic carbon species to the atmosphere (Sarmiento and Bender, 1994). In addition, the rate of continental weathering rises as CO_2 increases, resulting in consumption of CO_2 and production of Ca^{2+}, HCO_3^-, and (SiO_2), which dissolve in rivers and flow back into the ocean (Sarmiento and Bender 1994). Therefore, periods of high atmospheric PCO_2, characterized by outgassing of CO_2 from warmer seawater and increased HCO_3^- and CO_3^{2-} concentrations, may promote calcification by cyanobacteria, which relies on cellular uptake of HCO_3^-. Applying similar reasoning, periods of low atmospheric PCO_2, accompanied by lower seawater temperatures and weathering rates, may promote algal calcification as seawater CO_2 increases and algal calcification relies on uptake of CO_2.

Comparison of dominance trends for cyanobacterial and algal calcifiers from the Cambrian to the Cenozoic to atmospheric CO_2 fluctuations reveals some interesting correlations which support the hypothesis that CO_2 fluctuations influenced the evolution of calcifying microbes **(Figure 4)**. Berner (1990) presents a model for calculating atmospheric CO_2 throughout the Phanerozoic, which, in general, predicts high CO_2 levels during the early Paleozoic and Mesozoic and low levels during the Permian-Carboniferous and late Cenozoic. The Cambrian was dominated by calcareous cyanobacteria (Riding, 1991, 1996). However, the Ordovician through Devonian was dominated by chlorophytes and saw the appearance and diversification of a variety of algal groups (Roux, 1991). Episodes of abundant calcified cyanobacteria also occurred during the Middle-Late Devonian and Middle-Late Triassic (Riding, 1992, 1996), corresponding to periods of elevated PCO_2. By the Carboniferous, when CO_2 concentrations were at their lowest for the Paleozoic, cyanobacteria were represented by only seven genera, while chlorophyte algae comprised at least 48 genera (Mamet, 1991). Calcified cyanobacteria were of minor importance during the Permian with only five representative genera (Riding and Guo, 1991). The Early Permian remained dominated by chlorophytes, with 47 genera followed by a decline in the Late Permian. The general decrease in calcareous cyanobacteria and transition in dominance to algal calcifiers from the Ordovician to the Late Permian corresponds to a general decrease in atmospheric CO_2 concentrations during the same period of time. The correlation between smaller CO_2 fluctuations relative to the Cambrian to Permian transition through the Mesozoic is also evident. The Late Permian shows the decline of primary groups of chlorophyte calcifiers (Riding and Guo, 1991) which continues into the Early Triassic (Barattolo, 1991), corresponding to an increase in PCO_2. Increases in primary groups of calcareous chlorophytes corresponding to decreasing atmospheric CO_2 levels are seen in the Middle to Late Triassic, Late Jurassic to Early Cretaceous, and beginning of the Cenozoic (Barattolo, 1991). Decreases in primary chlorophyte groups corresponding to increasing atmospheric CO_2 levels occur during the Early Triassic and Early Middle Jurassic (Barattolo, 1991). While the Late Cretaceous shows a collapse of chlorophytes (Barattolo, 1991) which corresponds to decreasing PCO_2, this may be attributed to events resulting in global mass extinctions at the K-T boundary as opposed to CO_2 fluctuations. Calcareous cyanobacteria are well represented throughout the Mesozoic (possibly

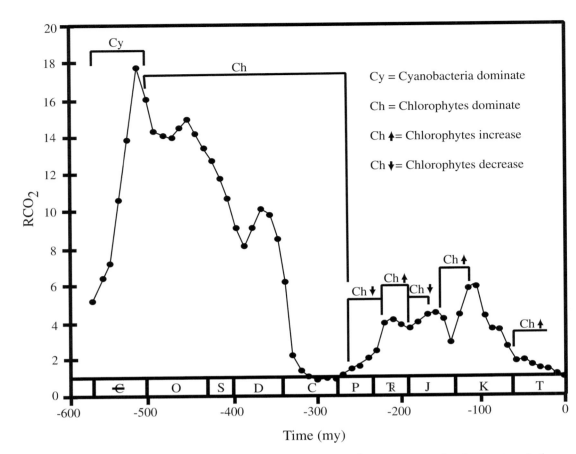

Figure 4 Comparison of Phanerozoic atmospheric PCO$_2$ fluctuations to dominance trends for calcareous cyanobacteria and algae. Note that cyanobacteria dominate during periods of high CO$_2$ while algae dominate periods of low CO$_2$. RCO$_2$ = ratio of the mass of CO$_2$ in the atmosphere at the indicated time to that in the present atmosphere. (Figure modified from Berner, 1990, American Association for the Advancement of Science, publisher).

due to generally high CO$_2$ levels throughout this time period), but they undergo a significant numerical reduction in the Cenozoic (Barattolo, 1991) as PCO$_2$ decreases. Although additional research is required to confirm the relation between CO$_2$ fluctuations and dominance trends in calcareous cyanobacteria and algae, this comparison supports such a relationship and suggests that cyanobacteria and algae of the past may have employed mechanisms for calcification similar to those of modern organisms.

The occurrence of extensive microbial lime-mud deposits during periods of high atmospheric PCO$_2$ (Horodyski and Mankiewicz, 1990; Knoll and Swett, 1990; and Kazmierczak et al., 1994) may result from the environmental advantage that cyanobacterial calcification has over microalgal with regard to environmental triggering mechanisms. Conditions necessary for production of a high HCO$_3^-$ environment capable of promoting cyanobacterial calcification also result in an increase in the saturation state of calcium carbonate, enabling inorganic precipitation as well. Outgassing of

CO_2 due to an increase in seawater temperature results in an increase in $CO_3{}^{2-}$, and increased weathering rates enhance seawater Ca^{2+}, both of which increase the saturation state with respect to $CaCO_3$. It is likely that once microbes precipitate seed crystals, inorganic precipitation is induced in $CaCO_3$-saturated solutions. Thus, the contribution of cyanobacterial calcification to lime-mud deposition during periods of high atmospheric PCO_2 may be enhanced by concurrent deposition of inorganic $CaCO_3$. This seeding effect for inorganic precipitation is less significant during microalgal calcification because these cells require higher CO_2 concentrations for $CaCO_3$ precipitation, and environmental conditions necessary for higher aqueous CO_2 decrease the $CaCO_3$ saturation state.

Stanley and Hardie (1998) present evidence that variations in mineralogy of nonskeletal carbonates and dominant reef-building organisms during the Phanerozoic correlate well with variations in seawater calcium concentrations and Mg/Ca ratios resulting from fluctuations in spreading rates at midocean ridges. The effect of calcium concentrations and Mg/Ca ratios on the mineralogy of modern microbial carbonates remains to be quantified. However, comparison of cyanobacterial and chlorophyte dominance trends throughout the Phanerozoic to calcium concentrations and Mg/Ca ratios calculated by Hardie (1996) **(Figure 5)** shows no clear correlation.

Examination of the effects of environmental perturbations on calcification enables placement of constraints on the external environmental conditions required to initiate calcification, and may provide insight into ancient microbial environments. For example, the threshold levels of CO_2 and $HCO_3{}^-$ required to initiate calcification by microalgae and cyanobacteria, respectively, are unknown. However, determination of minimum extracellular inorganic carbon requirements for

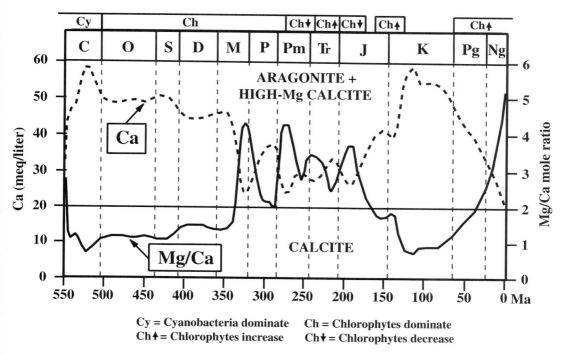

Cy = Cyanobacteria dominate   Ch = Chlorophytes dominate
Ch↑ = Chlorophytes increase   Ch↓ = Chlorophytes decrease

Figure 5 Secular variation in the Mg/Ca ratio and Ca concentration in seawater at 25°C (modified from Hardie, 1996, the Geological Society of America, Inc., publisher).

initiating calcification mechanisms of these organisms may provide insight into aqueous inorganic carbon concentrations throughout the past. Because inorganic carbon species distributions are pH dependent, one may begin to estimate pH and alkalinity required to achieve specific inorganic carbon concentrations. Likewise, cyanobacteria and algae may require minimum Na^+ and Ca^{2+} concentrations for calcification (Yates, 1996). Thus, examination of species distributions may provide information on fluctuations of these seawater constituents.

CONCLUSIONS

Close examination of the mechanisms of microbial calcification indicates that these processes occur in response to fluctuation in environmental parameters, including DIC concentrations and species distributions, pH, alkalinity, salinity, and calcium and sodium ion concentrations. These parameters are strongly influenced by local and global short-term and long-term climate change. Alteration of water chemistry due to daily and seasonal environmental perturbations (rainfall, evaporation, temperature, etc.) may provide triggering mechanisms for inducing microbial calcification or enhancing rates of lime-mud production. The ability to control microbial calcification in laboratory environments will enable future investigations into maximum and minimum levels of these parameters required for induction of calcification. Once constraints can be placed on modern microbes in laboratory settings, predictions on natural environments of microbial calcification can be made, followed by investigations to verify their accuracy. As we are able to accurately predict the onset of microbial calcification in modern natural environments, it is likely that we will be able to use this information, combined with calcareous fossil species distribution information, to provide insight into ancient environments of precipitation.

Cyanobacterial and algal calcification depend on the concentration and species of inorganic carbon available for photosynthesis and calcification. Seawater and freshwater DIC is strongly linked to atmospheric PCO_2. Examination of fluctuations in dominance trends for calcifying cyanobacteria and algae correlates with variations in atmospheric CO_2 from the Cambrian to the Holocene. This provides evidence that climate change may have influenced the evolution of cyanobacterial and algal calcification mechanisms and the origin of micritic lime mud deposits. Further, the relationship of algal/cyanobacterial abundance and PCO_2 provides an extremely valuable tool in the assessment of paleoclimatic conditions in the rock record.

Because we are just beginning to realize the extent of microbial calcification, it is likely that its contribution to CO_2 cycling and sedimentation has been greatly underestimated. Many species of microbes capable of precipitating $CaCO_3$ are dominant components of the microbial flora in a wide range of both freshwater and marine environments, and are capable of generating copious amounts of carbonate sediments (Yates and Robbins, 1998). It is likely, as investigations continue into the future, that numerous additional calcifying microbes will be identified as well. Understanding calcification mechanisms and their driving forces will provide the basis to predict environments of microbial calcification and to assess paleoclimatic conditions.

REFERENCES CITED

Ahmed, A. M., A. R. Radi, M. D. Heikal, and R. Abdel-Basset. 1989. Effect of Na-Ca combinations on photosynthesis and some related processes of Chlorella vulgaris. Journal of Plant Physiology 135: 175–178.

Awramik, S. M. 1991. Archaean and Proterozoic stromatolites. *In* Calcareous algae and stromatolites, ed. R. Riding, 289–304. New York: Springer-Verlag.

Badger, M. R., K. Palmqvist, and J. Yu. 1993. Measurement of CO_2 and HCO_3^- fluxes in cyanobacteria and microalgae during steady-state photosynthesis. Physiological Plantarum.

Barattolo, F., 1991. Mesozoic and Cenozoic marine benthic calcareous algae with particular regard to Mesozoic Dasycladaleans. In Calcareous algae and stromatolites, ed. R. Riding, 504–540. New York: Springer-Verlag.

Berner, R. A. 1990. Atmospheric carbon dioxide levels over Phanerozoic time. Science 249: 1382–1386.

Boss, S. K. and A. C. Neuman. 1993. Physical versus chemical processes of "whiting" formation in the Bahamas. Carbonates and Evaporites 8:135–148.

Broun, I. I., G. P. Borbik, and O. Y. Mirochnik. 1992. Light-induced Na^+-dependent H^+ uptake by the cyanobacterium *Synechocystis* PCC 6803—detection of a mutant strain lacking Na^+ dependent resistance to protonophores. Biochemistry—Russia 57(10): 1100–1103.

Byerly, G. R., D. R. Lowe, and M. M. Walsh. 1986. Stromatolites from the 3,300–3,500–Myr Swaziland Supergroup, Barberton Mountain Land, South Africa. Nature 319: 489–491.

Chafetz, H. S. 1986. Marine peloids: a product of bacterially induced precipitation of calcite. Journal of Sedimentary Petrology 56:812–817.

Cloud, P. E. Jr. 1961. Environment of calcium carbonate deposition west of Andros Island, Bahamas. U.S. Geological Survey Professional Paper 350: 1–138.

Davis, Richard A., Cathleen Reas, and L. L. Robbins. 1995. Calcite mud in a Holocene back-barrier lagoon: Lake Reeve, Victoria, Australian Journal of Sedimentary Research A65(1): 178–184.

Drew, G. H. 1914. On the precipitation of calcium carbonate in the sea by marine bacteria, and on the action of denitrifying bacteria in tropical and temperate seas. Papers Tortugas Lab, Carnegie Institute of Washington Publications, 182: 7–45.

Espie, G. S., A. G. Miller, and D. T. Canvin. 1988. Characterization of the Na^+ requirement in cyanobacterial photosynthesis. Plant Physiology 88: 757–763.

Flugel, E. 1991. Triassic and Jurassic marine calcareous algae: a critical review. In Calcareous algae and stromatolites, ed. R. Riding, 481–503. New York: Springer-Verlag.

Folk, Robert. 1993. SEM imaging of bacteria and nannobacteria in carbonate sediments and rocks. Journal of Sedimentary Petrology 63(5) 990–999.

Giese, A. C. 1962. Cell physiology. Philadelphia: W. B. Saunders Company, p. 592.

Hardie, L. A. 1996. Secular variation in seawater chemistry: an explanation for the coupled secular variation in the mineralogies of marine limestones and potash evaporites over the past 600 m.y. Geology 24:279–283.

Hartley, A. M., W. A. House, M. E. Callow, and B. S. C. Leadbeater. 1995. The role of green alga in the precipitation of calcite and the coprecipitation of phosphate in freshwater. Internationale Revue der gesamten Hydrobiologie 80(3): 385–401.

Hawkins, Sharon. 1995. Intracrystalline proteins from aragonite-needle mud from Florida Bay: An immunological and biogeochemical study. Master's thesis, University of South Florida.

Hodell, D. A., C. L. Schelske, G. L. Fahnensteil, and L. L. Robbins. 1998. Biologically induced calcite and its isotopic composition in Lake Ontario. Limnology and Oceanography 43 (2), 187–199.

Holligan, P.M., E. Fernandez, J. Aiken, W. M. Balch, P. Boyd, P. H. Burkill, M. Finch, S. B. Groom, G. Malin, K. Muller, D. A. Purdie, C. Robinson, C. C. Trees, S. M. Turner, and P. van der Wal. 1993. A biogeochemical study of the coccolithophore, *Emiliani huxleyi*, in the North Atlantic. Global Biogeochemical Cycles 7:879–900.

Horodyski, R. J., and C. Mankiewicz. 1990. Possible Late Proterozoic skeletal algae from the Pahrump Group, Kingston Range, Southeastern California. American Journal of Science 290: 149–169.

Kalinkina, L. G., and B. P. Strogonov. 1986. Growth and biomass accumulation in marine and freshwater forms of Chlorella during inhibition of the glycolate pathway under conditions of salination. Soviet Plant Physiology 33(3): 445–452.

Kazmierczak, J., M. Gruszczynski, M. L. Coleman, and S. Kempe. 1994. Coccoid cyanobacterial origin of common micritic and peloidal limestones: Jurassic and modern examples. 14th International Sedimentological Congress Abstracts. p. B-6.

Kempe, S., and E. T. Degens. 1985. An early soda ocean?. Chemical Geology 53: 95–108.

Kempe, S., J. Kazmierczak, and E. T. Degens. 1989. The soda ocean concept and its bearing on biotic evolution. *In* Origin, evolution, and modern aspects of biomineralization in plants and animals, ed. R. E. Crick, 29–43. New York: Plenum Press.

Kretsinger, R. H. 1983. A comparison of the roles of calcium in biomineralization and in cytosolic signalling. *In* Biomineralization and biological metal accumulation, eds. P. Westbroek and E.W. de Jong, 123–131. Dordrecht: D. Reidel Publ. Co.

Knoll, A. H., and K. Swett. 1990. Carbonate deposition during the Late Proterozoic Era: An example from Spitsbergen. American Journal of Science 290: 104–133.

Komor, E. 1973. Proton-coupled hexose transport in *Chlorella vulgaris*. FEBS letters 38: 16–18.

Krumbein, W. E. 1974. On the precipitation of aragonite on the surface of marine bacteria. Naturwissenschaften 61:167.

Lowe, D. R. 1980. Stromatolites, 3,400-Myr old from the Archean of Western Australia. Nature 284: 441–443.

Lowenstam, H., and S. Epstein. 1957. On the origin of sedimentary aragonite needles of the Great Bahama Bank. Journal of Geology 65:364–375.

Mamet, B. 1991. Carboniferous calcareous algae. In Calcareous algae and stromatolites, ed. R. Riding, 370–451. New York: Springer-Verlag.

McCallum, M. F., and K. Guhathakurta. 1970. The precipitation of calcium carbonate from seawater by bacteria isolated from the Bahama Bank sediments. Journal of Applied Bacteriology 33:649–655.

McConnaughey, T. A. 1991. Calcification in *Chara corallina*: CO_2 hydroxylation generates protons for bicarbonate assimilation. Limnology and Oceanography 36(4):619–628.

McConnaughey, T. A., and R. H. Falk. 1991. Calcium-proton exchange during algal calcification. Biological Bulletin 180: 185–195.

Meeder, J. F. 1979. The Pliocene fossil reef of southwest Florida. Miami Geological Society Field Guide.

Merz, M. U. E. 1992. The biology of carbonate precipitation by cyanobacteria. Facies 26: 81–102.

Merz-Preiss, M. and R. Riding. 1999. Cyanobacterial tufa calcification in two freshwater streams: Ambient environment, chemical thresholds and biological processes. Sedimentary Geology, 126:103–124.

Morita, R.Y. 1980. Calcite precipitation by marine bacteria. Geomicrobiology Journal 2:63–82.

Morse, J. W., and S. He. 1993. Influences of T, S, and PCO_2 on the pseudo-homogenous precipitation of $CaCO_3$ from seawater: implications for whiting formation. Marine Chemistry 41:291–297.

Niggli, V., E. Sigel, and E. Carafoli. 1982. The purified Ca^{2+} pump of human erythrocyte membranes catalyzes an electroneutral Ca^{2+}-H^+ exchange in reconstituted liposomal systems. The Journal of Biological Chemistry 257:2350–2356.

Novitsky, J. A. 1981. Calcium carbonate precipitation by marine bacteria. Geomicrobiology Journal 2:375–388.

Opdyke, B. N. and J. C. G. Walker. 1992. Return of the coral reef hypothesis: basin to shelf partitioning of $CaCO_3$ and its effect on atmospheric CO_2. Geology 20:733–736.

Riding, R. 1982. Cyanophyte calcification and changes in ocean chemistry. Nature 299:814–815.

Riding, R. 1991. Cambrian calcareous cyanobacteria and algae. *In* Calcareous algae and stromatolites, ed. R. Riding, 305–334. New York: Springer-Verlag.

Riding, R. 1992. Palaeoclimatology; the algal breath of life. Nature 359 (6390), 13–14.

Riding, R. 1996. Long-term changes in marine $CaCO_3$ precipitation. Memoires de la Societe Geologique de France 169:157–166.

Riding, R., and L. Guo. 1991. Permian marine calcareous algae. In Calcareous algae and stromatolites, ed. R. Riding, 452–480. New York: Springer-Verlag.

Ritchie, R. J. 1992. Sodium transport and the origin of the membrane potential in the cyanobacterium *Synechococcus* R-2 (*Anacystis nidulans*) PCC-7942. Journal of Plant Physiology 139(3):320–330.

Robbins, L. L. and P. J. Blackwelder. 1992. Biochemical and ultra-structural evidence for the origin of whitings: a biologically induced calcium carbonate precipitation mechanism. Geology 20: 664–468.

Robbins, L. L. and Y. Tao. 1996. Temporal and spatial distribution of whitings on the Great Bahama Bank and a new lime mud budget. Geology 25(10):947–950.

Robbins, L. L., K. K. Yates, E. A. Shinn, and P. Blackwelder. 1996. Whitings on the Great Bahama Bank: a microscopic solution to a macroscopic mystery. Bahamas Journal of Science 4:2–7.

Roux, A., 1991. Ordovician to Devonian marine calcareous algae. *In* Calcareous algae and stromatolites, ed. R. Riding, 349–369. New York: Springer-Verlag.

Stabel, H. H. 1986. Calcite precipitation in Lake Constance: Chemical equilibrium, sedimentation, and nucleation by algae. Limnology and Oceanography 31(5):1081–1093.

Stanley, S. M., and L. A. Hardie. 1998. Secular oscillations in the carbonate mineralogy of reef-building and sediment-producing organisms driven by tectonically forced shifts in seawater chemistry. Palaeogeography, Palaeoclimatology, Palaeoecology 144:3–19.

Stockner, John G. 1988. Phototrophic picoplankton: an overview from marine and freshwater ecosystems. Limnology and Oceanography 33(4, part 2): 765–775.

Sundquist, E. T. 1993. The global carbon dioxide budget. Science 259:934–941.

Thompson, J. B., and F. G. Ferris. 1990. Cyanobacterial precipitation of gypsum, calcite, and magnesite from natural alkaline lake water. Geology 18: 995–998.

Thompson, J. B., F. G. Ferris, and D. A. Smith. 1990. Geomicrobiology and sedimentology of the mixolimnion and chemocline in Fayetteville Green Lake, New York. Palaios 5: 52–75.

Tsuzuki, M., and S. Miyachi. 1991. CO_2 syndrome in *Chlorella*. Canadian Journal of Botany 69: 1003–1007.

Walter, M. R., R. Buick, and J. S. R. Dunlop. 1980. Stromatolites 3,400–3,500 Myr old from the North Pole area, Western Australia. Nature 284: 443–445.

Yates, K. K. 1996. Microbial precipitation of calcium carbonate: a potential mechanism for lime-mud production. University of South Florida, Ph.D. dissertation. 193 pp.

Yates, K. K., and L. L. Robbins. 1995. Experimental evidence for a $CaCO_3$ precipitation mechanism for marine *Synechocystis*. Bulletin de l'Institut oceanographique, Monaco Special Edition 14(2): 51–59.

Yates, K. K., and L. L. Robbins. 1998. Production of carbonate sediments by a unicellular green alga. American Mineralogist 83:1503–1509.

Yates, K. K., and L. L. Robbins. 1999. Radioisotope tracer studies of inorganic carbon and Ca in microbially derived $CaCO_3$. Geochimica et Cosmochimica Acta 63:129–136.

Bachu, S., Geological sequestration of anthropogenic carbon dioxide: Applicability and current issues, 2001,
in L. C. Gerhard, W. E. Harrison, and B. M. Hanson,
eds., Geological perspectives of global climate
change, p. 285–303.

15 | Geological Sequestration of Anthropogenic Carbon Dioxide: Applicability and Current Issues

Stefan Bachu

Alberta Geological Survey
Alberta Energy and Utilities Board
Edmonton, Alberta, Canada

ABSTRACT

Using the technology and experience already gained by the oil and gas industry and in ground-water resource management, sequestration of carbon dioxide (CO_2) in geological media is an immediately applicable option for the near- to medium-term mitigation of climate-change effects resulting from the release of anthropogenic CO_2 into the atmosphere. Based on its properties and in-situ conditions, CO_2 can be sequestered as a gas, a liquid, or in supercritical state in depleted oil and gas reservoirs, uneconomic coal beds, deep saline aquifers, and salt caverns. Using CO_2 for miscible flooding of oil reservoirs or for methane production from coal beds has an added economic benefit. The main trapping mechanisms responsible for CO_2 sequestration in geological media are geological, solubility, hydrodynamic, mineral, adsorption, and cavern trapping. Basin-scale criteria for CO_2 sequestration, such as tectonic setting, hydrodynamic and geothermal regimes, hydrocarbon potential and basin maturity, and surface infrastructure, should be used in determining sedimentary basins in the world that are suitable for CO_2 sequestration in geological media. Site-specific criteria, such as particular geological media, in-situ conditions, storage capacity, injectivity and flow dynamics, and sequestration efficiency, need to be applied to identify sites, methods, and capacity for CO_2 sequestration. Continental sedimentary basins in North America, foremost in Texas and Alberta, are the prime candidates for CO_2 sequestration in geological media, followed by circum-Atlantic shelf basins. However, a series of major issues still needs to be addressed before proceeding with full-scale implementation, such as identification of specific sites and their capacities; proper characterization of the sequestration medium and of in-situ conditions; predicting and monitoring the fate of the injected CO_2; surface CO_2 capture, transport, and injection; performance assessment; and, finally, general public acceptance. Nevertheless, CO_2 sequestration in geological media is a very promising option for carbon management and is in the stage of development prior to application.

INTRODUCTION

As a result of anthropogenic carbon-dioxide (CO_2) emissions, atmospheric concentrations of CO_2, a major greenhouse gas, have risen from preindustrial levels of 280 parts per million (ppm) to 360 ppm (Bryant, 1997), primarily as a consequence of fossil-fuel combustion for energy production. In a business-as-usual scenario, the Intergovernmental Panel on Climate Change (IPCC) predicts that global emissions of CO_2 to the atmosphere will have increased from 7.4 billion (giga-) tonnes of atmospheric carbon (GtC) per year in 1997 to approximately 26 GtC/yr by 2100 (IPCC, 1996). Increasing concentrations of CO_2 affect the earth-atmosphere energy balance, enhancing the natural greenhouse effect and thereby exerting a warming influence at the earth's surface. However, the detailed response of the climate system is uncertain because of its inherent complexity and natural variability (AGU, 1999). Although the increase in global mean surface temperatures since the beginning of the industrial revolution appears to be unusual in the context of the last few centuries, it is clearly within the range of climate variability of the last few thousand years. The geological record provides evidence of larger climate variations associated with changes in atmospheric CO_2, and the close coupling between climate and the carbon cycle suggests that a change in one would in all likelihood be accompanied by a change in the other (AGU, 1999). Because of uncertainties regarding the earth's climate system, there is much public debate over the extent to which increased concentrations of greenhouse gases have caused or will cause climate change, and over potential actions to limit and/or respond to climate change (AGU, 1999). Nevertheless, through the Kyoto Protocol, the developed world has already committed to reduce, by the year 2012, the amount of anthropogenic CO_2 released into the atmosphere to levels on average lower by 5.2% than 1990 levels.

Mitigation of human-induced climate change involves basically three approaches. The first approach is to increase the efficiency of primary energy conversion and end use, so that fewer units of primary fossil energy are required to provide the same energy service. The second approach is to substitute lower-carbon or carbon-free energy sources for the current sources. The third approach is carbon sequestration, whose purpose is to keep anthropogenic carbon emissions from reaching the atmosphere by capturing them, isolating them, and diverting them to secure storage, or to remove CO_2 from the atmosphere by various means and store it. Any viable system for sequestering carbon must be safe, environmentally benign, effective, economical, and acceptable to the public.

Large CO_2 sinks are terrestrial ecosystems (soils and vegetation), oceans, and geological media, with retention times of the order of $10–10^5$ years, respectively (Gunter et al., 1998). Terrestrial ecosystems represent a diffuse natural carbon sink that captures CO_2 from the atmosphere after release from various sources. The natural, diffuse, and slow exchange of CO_2 between the atmosphere and oceans can be artificially enhanced at concentrated points by injecting CO_2 at great depths where it will form either hydrates or heavier-than-water plumes (Aya et al., 1999). However, the technology of disposing of CO_2 from either ships or deep pipelines is only in the development stage, and the effects of disposing of CO_2 in oceans are not well known. Sequestration of CO_2 in geological media is likely to provide the first large-scale opportunity for concentrated sequestration of CO_2, being immediately applicable as a result of the experience already gained in oil and gas production, storage of natural gas, and groundwater resource management. The main issues remaining to be addressed are uncertainties regarding the volumes available for sequestration; identification of sequestration sites and appropriate methods; long-term integrity of sequestration; liability; and cost associated with CO_2 capture, transport, and injection.

Given their inherent advantages, such as availability, competitive cost, ease of transport and storage, and large resources, fossil fuels, which today provide about 75% of the world's energy (84% in the United States and 73% in Canada), are likely to continue to remain a major component of the world's energy supply for at least the next century. Fossil fuels are serendipitously linked with sedimentary basins in which CO_2 can be sequestered (Hitchon et al., 1999). In the near term, most anthropogenic CO_2 is likely to come from large point sources associated with power generation from fossil fuels and large industrial processes such as refineries and iron, steel, cement, and petrochemical plants. Carbon dioxide separated from flue gases and effluents and during fuel-decarbonization processes could be captured and concentrated into a liquid or gas stream that could be transported and injected into deep geological formations. The method of geological sequestration, available volumes, and retention time depend on media characteristics and CO_2 properties at in-situ conditions. These factors play an intrinsic role in the identification of potential sites for CO_2 sequestration. Surface considerations, such as CO_2 availability, infrastructure for capture and transport, and cost, further narrow the candidate sites for CO_2 sequestration in geological media.

RELEVANT CO_2 PROPERTIES

At normal atmospheric conditions, CO_2 is a gas heavier than air, having a density of 1.872 kg/m^3 at 15°C and 101.325 kilopascals (kPa). It is thermodynamically very stable, which makes its utilization very difficult at normal temperatures. At low temperatures (<10°C) and high pressures (>2–5 [megapascals, or MPa], depending on temperature), CO_2 forms a solid hydrate (CO_2 5.7H$_2$O) heavier than water (Baklid et al., 1996). This property leads to considering sequestering CO_2 as a hydrate at the bottom of oceans (Loken and Austvik, 1993), in sediments under the deep seabed (Koide et al., 1996), or under thick permafrost in northern sedimentary basins in the United States (Alaska's North Slope), Canada (Beaufort-Mackenzie), and in Siberia. For temperatures greater than 31.1°C and pressures greater than 7.38 MPa (critical point), CO_2 is in supercritical state (Figure 1a), behaving like a gas by filling all the available volume, but having a "liquid" density that increases with pressure to values greater than that of water (Figure 1b). Subcritical CO_2 is either a gas or a liquid, depending on temperature and pressure (Figure 1a). Carbon dioxide is soluble in water; its solubility increases with increasing pressure and decreases with increased temperature and water salinity (Figure 1c). Carbon dioxide is also soluble in oil. In oil reservoirs, dissolved CO_2 lowers the viscosity of the oil, reduces its interfacial tension (capillary pressure), and increases its mobility. This property is used for miscible flooding of light oil (>25° API) reservoirs in a process called enhanced oil recovery (EOR). Although subcritical CO_2 is not very reactive at normal temperatures, in aqueous solution it forms carbonic acid (H_2CO_3) that undergoes typical slow reactions of a weak acid. Finally, CO_2 adsorbs onto coal at a rate twice as high as that of methane, a gas commonly found in coal beds (Figure 1d). These CO_2 properties, the in-situ temperature and pressure conditions, and specific characteristics of the geological media determine the method used for the geological sequestration of CO_2 and its physical state (gas, liquid, or supercritical), with a corresponding impact on site selection and on surface methods for capture, transport, and injection.

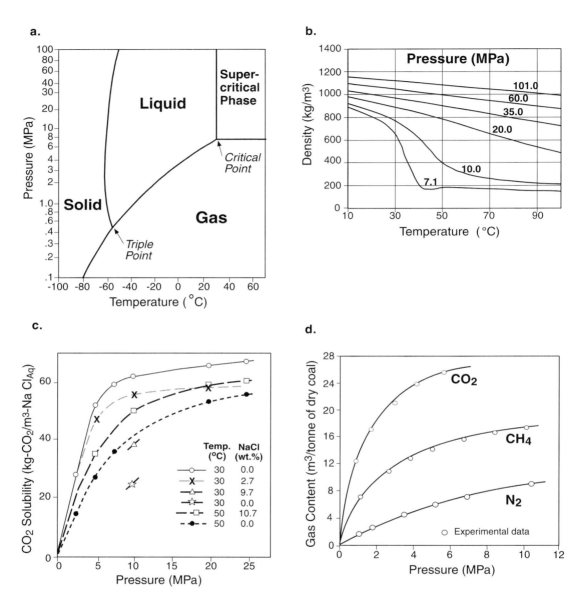

Figure 1 Relevant properties of carbon dioxide: (a) phase diagram, (b) variation of the density of liquid and supercritical CO_2 with temperature and pressure, (c) solubility variation with water salinity and temperature,[1] and (d) adsorption on coal compared with other gases.[2]

---

[1]Reprinted from *Engineering Geology*, v. 34, Koide et al., Underground storage of carbon dioxide in depleted natural gas reservoirs and useless aquifiers, p. 175–179, ©1996, with permission from Elsevier Science.

[2]©Society of Petroleum Engineers (SPE). Reprinted from Arri et al., "Modeling coalbed methane production with binary gas sorption," SPE Paper No. 24363, presented at the SPE Rocky Mountain Regional Meeting, Casper, Wyoming, 1992.

SEQUESTRATION MECHANISMS

Carbon dioxide can be sequestered in geological media by several mechanisms: 1) geological trapping, 2) solubility trapping, 3) hydrodynamic trapping, 4) mineral trapping, 5) adsorption trapping, and 6) cavern trapping. Carbon dioxide can be sequestered in any physical state in a **geological trap** (structural or stratigraphic) under static conditions, similar to hydrocarbons trapped in reservoirs. The same seals (caprock) that impeded the escape and migration of hydrocarbons over geological time should retain CO_2 "permanently," as long as pathways to the surface or adjacent formations are not created by fracturing the reservoir as a result of overpressuring, or by the presence of improperly completed or abandoned wells. Closed, depleted gas reservoirs are primary candidates as geological traps for CO_2 because primary recovery removes as much as 95% of the original gas in place. Closed, underpressured oil reservoirs that have not been invaded by water should also have good sequestration capacity. Oil and gas reservoirs in contact with underlying formation water (pressured by the water drive) are invaded by water as the reservoir is depleted. These reservoirs have less potential for CO_2 sequestration because CO_2 will have to displace (push back) the formation water. In this respect, these reservoirs behave more like deep aquifers.

Solubility trapping refers to CO_2 dissolved either in oil or in formation waters. Solubility trapping in oil still requires a geological trap (the oil reservoir). Unlike the CO_2 in geological traps, CO_2 dissolved in either oil or water is in a dynamic state. In an oil reservoir, the dissolved CO_2 flows toward the production well and is produced at the well bore within several months to one or two years, after which it is usually recirculated into the system. In the long term, the volume of CO_2 sequestered by this method is comparatively small (Gunter et al., 1998). Nevertheless, the technology is commercially proven and applied already in EOR operations at some 70 sites around the world, of which about 40 operations are in the Permian Basin of west Texas and New Mexico where natural CO_2 sources are being used. Large-scale use of CO_2 for EOR from anthropogenic sources is in an advanced stage of design and implementation for the Schrader Bluff and Weyburn oil fields in the Alaska North Slope and Williston Basin, respectively (McKean et al., 1999; Hattenbach et al., 1999). Up to 30% of subcritical CO_2 injected in deep saline aquifers dissolves over time in the formation water (Law and Bachu, 1996). The amount of dissolved CO_2 normally decreases with depth as a result of increasing temperature and formation-water salinity **(Figure 1c)** characteristic of many sedimentary basins. Once outside the radius of influence of the injection well, the dissolved CO_2 travels with the formation waters at their natural flow rate, which in sedimentary basins is about cm/yr (km/million years).

Injection and dissolution of subcritical CO_2 in regional-scale flow systems driven by compaction, tectonic compression, or topography **(Figures 2a, 2b, and 2c)** leads to **hydrodynamic trapping** because of the very small flow velocity and hydrodynamic dispersion that result in an extremely long retention time (10^4–10^7 years) (Bachu et al., 1994; Lindeberg and Holloway, 1999). This method of CO_2 sequestration is already used in the Sleipner West oil field in the North Sea offshore Norway, where ~1 million tonnes of carbon dioxide per year ($MtCO_2$/yr) are injected in the Utsira aquifer some 1000 m below the sea bottom (Korbol and Kaddour, 1995). Much smaller amounts of CO_2 (for a total of 0.25 $MtCO_2$/yr) are injected as acid gas (CO_2 and H_2S) in deep saline aquifers and depleted gas reservoirs at 23 sites in the Alberta Basin (Wichert and Royan, 1997). Another form of **hydrodynamic trapping** occurs when CO_2 is injected and dissolves in formation water that flows inward in the basin, driven by erosional rebound **(Figure 2d)**. In this case, CO_2 is practically sequestered permanently because the flow direction is downdip and into adjacent shales (Neuzill and Pollock, 1983; Bachu, 1995). The undissolved portion of the injected subcritical CO_2 segregates and forms a plume at the top of the aquifer as a result of density differences (Law and Bachu, 1996; Gupta et al., 1999). This plume will migrate updip along the bedding, driven by

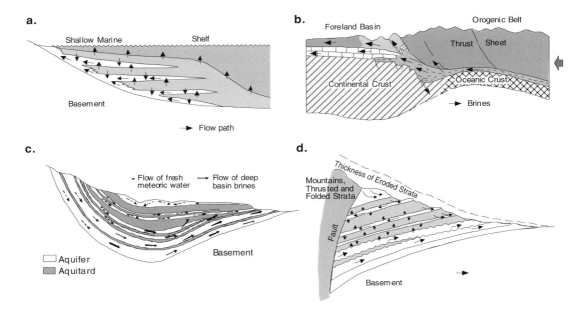

Figure 2 Mechanisms and type of flow in sedimentary basins: (a) driven by compaction in shelf basins, (b) driven by tectonic compression near active orogenic belts, (c) driven by topography in stable continental basins, and (d) driven by erosional or postglaciation rebound in continental basins. After Oliver (1986).

buoyancy in a process similar to oil and gas migration. Supercritical CO_2 injected in a deep saline aquifer will form a plume that, depending on its density relative to that of formation water **(Figure 1b)**, will segregate either at the top or at the bottom of the aquifer. Buoyancy will enhance the hydrodynamic entrapment of a heavier plume that will slowly migrate downdip, and will reduce the entrapment of a lighter plume that will migrate updip, similarly to subcritical CO_2. Hydrodynamic trapping of CO_2 does not need geological traps, but rather a competent regional-scale sealing aquitard (Bachu et al., 1994) that is not naturally or artificially fractured, or perforated by improperly completed or abandoned wells. However, the existence of geological traps along the natural-migration pathway would enhance the trapping efficiency of the overall system.

Given the extremely long retention time of CO_2 in deep saline aquifers, a considerable increase in permanently sequestered CO_2 would be achieved through precipitation of carbonate minerals, i.e., **mineral trapping** (Bachu et al., 1994), even though some chemical reactions between CO_2 and the rock of the host aquifer are slow. Solubility and mineral trapping are not independent, stand-alone sequestration mechanisms. Rather, they are linked intrinsically with geological trapping, in the case of CO_2 injected in oil reservoirs, or with hydrodynamic trapping, in the case of CO_2 injected in deep saline aquifers.

Adsorption trapping is achieved by preferential adsorption of gaseous CO_2 onto the coal matrix because of its higher affinity to coal than that of methane **(Figure 1d)**. Injection of CO_2 into deep uneconomic coal beds has the added advantage that it releases from the coal matrix methane that can be produced (Gunter et al., 1997a; Byrer and Guthrie, 1999) in a process called enhanced coalbed methane recovery (ECBMR) that is currently used in the San Juan Basin (Stevens et al., 1999). Although methane is a more potent greenhouse gas than CO_2, it is the cleanest fossil fuel

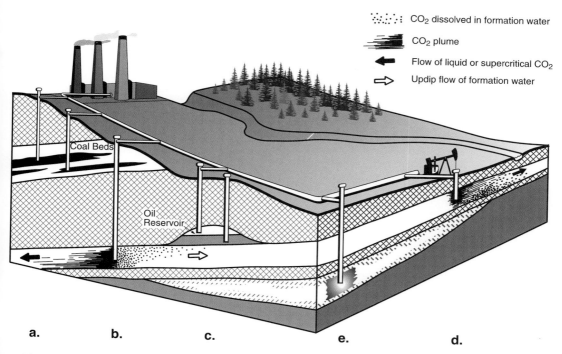

Figure 3 Diagrammatic representation of various mechanisms for CO_2 trapping in geological media: (a) adsorption trapping onto coal beds, (b) hydrodynamic and solubility trapping in descending plumes in saline aquifers, (c) geological and solubility trapping in oil reservoirs, (d) hydrodynamic and solubility trapping in ascending plumes in saline aquifers, and (e) cavern trapping in salt beds.

available and can be used instead of coal or oil for power generation. If CO_2 sequestration in coal beds is used in conjunction with methane production, the retention time is on the order of months to years, similar to EOR operations.

Finally, **cavern trapping** refers to CO_2 injection and permanent sequestration in liquid or supercritical phase in mined caverns in salt beds and domes, similar to the storage of natural gas (Tek, 1989; Crossley, 1998). Although, theoretically, salt caverns have a large storage capacity, the associated costs are high and the environmental problems related to brine disposal are significant. Thus, CO_2 sequestration in salt caverns is likely to be used only under special circumstances, such as in the case of caverns created by salt mining for public and industrial use, or in areas lacking other sequestration options. **Figure 3** shows diagrammatically the various types of CO_2 trapping mechanisms in geological media.

BASIN-SCALE CRITERIA FOR CO_2 SEQUESTRATION IN GEOLOGICAL MEDIA

The selection of the method, strata, and site for CO_2 sequestration in geological media depends on specific criteria being met to satisfy the general requirements of safety, benign environmental impact, and public acceptance. These criteria are scale dependent, some applying at the basin scale,

W   E

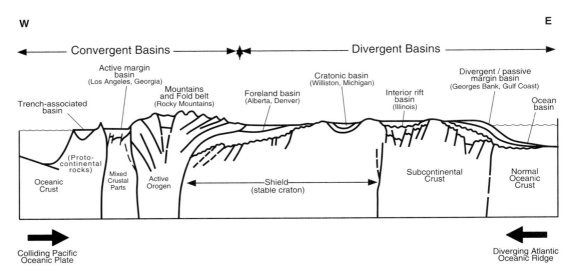

Figure 4   Main types of sedimentary basins, diagrammatically illustrated along a west-east cross section of the North American continent (adapted from Hitchon et al., 1999).

others being site specific. The basin-scale criteria (Bachu and Gunter, 1999) relate to a) tectonic setting, b) hydrodynamic and geothermal regimes, c) hydrocarbon potential and basin maturity, d) surface infrastructure, and e) sociopolitical conditions in the jurisdiction covering the basin. The last criterion will not be discussed in the following.

Tectonic Setting

Cratonic platforms, such as the Canadian, Guyana, and Brazilian shields, are generally unsuitable for geological sequestration of CO_2 because their crystalline or metamorphic rocks lack the porosity needed for storage space and the permeability needed for injection. Neither are orogenic belts suitable, because they lack continuous seals as a result of extensive faulting and fracturing during mountain forming. Sedimentary basins are the loci of geological sequestration of CO_2 because they possess the right type of porous and permeable rocks, and they are located where energy resources are found and produced and where most fossil-fuel-based power generation takes place (Hitchon et al., 1999). However, not all sedimentary basins are suitable for CO_2 sequestration. Convergent basins **(Figure 4)** along active tectonic margins (St. John et al., 1984), such as the circum-Pacific basins, are usually subject to volcanism, earthquakes, and active faulting. Thus, CO_2 sequestration in basins such as the Los Angeles Basin should be avoided because they pose a safety problem in case of accidental release of large quantities of CO_2 along open faults and fractures as a result of local catastrophic events such as earthquakes. Divergent basins **(Figure 4)** are located in tectonically stable areas (St. John et al., 1984):

- along the margins of older continental plates adjacent to an active orogenic belt, such as the foreland basins east of the Rocky Mountain (e.g., Alberta, Denver, Permian);
- on stable continental crust (rigid lithosphere) in areas of divergence or rifting (e.g., Williston, Michigan, Illinois, North Sea); and
- along divergent (passive) continental margins, such as the circum-Atlantic and circum-Arctic basins (e.g., Gulf Coast, Florida-Bahama, Scotian Shelf, Beaufort-Mackenzie).

Geological sequestration of CO_2 in divergent sedimentary basins, such as those in western Canada, the midwestern United States, and along the Gulf Coast and the eastern seaboard of North America, is preferable because of tectonic stability and general lack of significant hazardous earth events.

Hydrodynamic and Geothermal Regimes

Hydrodynamic criteria are very important when assessing the suitability of a basin, or parts thereof, for CO_2 sequestration, because water is a pervasive presence in all cases of geological sequestration except for salt caverns. It occurs within and adjacent to oil and gas reservoirs, is obviously present in aquifers, and is found in coal beds and adjacent strata. There is a close link between the type of sedimentary basin and the flow and pressure regime of formation waters. In basins located on the marine shelf, such as Beaufort-Mackenzie, Scotian Shelf, Gulf Coast, and the North Sea, the flow of formation water is driven by compaction vertically out of shales (aquitards) and laterally outward, toward the basin margin, in aquifers (**Figure 2a**). These basins are usually significantly overpressured at depths greater than 2000 m (e.g., Bredehoeft et al., 1988). Thus, CO_2 injection in such basins is recommended only in the shallower, normally pressured aquifers, such as the Utsira in the North Sea (Korbol and Kaddour, 1995), because the deeper ones pose technological and safety challenges. In basins adjacent to currently active orogenic belts, such as the Himalayas, hot and very saline water is expelled by lateral tectonic compression from underneath the mountains (**Figure 2b**) (Oliver, 1986). Because of the decreasing CO_2 solubility with increasing temperature and salinity, these deep aquifers are not well suited for CO_2 sequestration. Most flow systems in continental basins (cratonic and foreland), such as in midcontinent North America, are driven by topography from recharge areas at high topographic elevations to discharge areas at low elevations (**Figure 2c**). Reservoir and aquifer pressures are usually close to hydrostatic. The most suitable areas for CO_2 injection are in the upstream portion of aquifers confined by regional-scale aquitards, to increase the length of the flow path and residence time (hydrodynamic and mineral trapping). In foreland and intracratonic basins that underwent recent significant erosional and postglacial unloading, such as the Alberta and Williston Basins, flow is driven by erosional rebound vertically into thick shales and laterally inward, downdip, in thin adjacent aquifers (**Figure 2d**) (Neuzill and Pollock, 1983; Bachu, 1995; Gupta and Bair, 1997). These aquifers and the associated hydrocarbon reservoirs are the most suitable targets for CO_2 sequestration because of the enhanced hydrodynamic trapping. Zones of active hydrocarbon generation in sedimentary basins are overpressured (Spencer, 1987) and may pose a safety risk for CO_2 sequestration, besides the fact that the injected CO_2 may contaminate these resources. From the point of view of sequestration safety, the risk of leakage is lower in compacted continental basins where the flow is driven by topography or erosional rebound than in shelf basins undergoing active compaction (Hitchon et al., 1999).

Previous studies assumed, implicitly, that the pressure distribution in a sedimentary basin is hydrostatic, increasing linearly with depth at a rate of 1 MPa per 100 m (Holloway and Savage, 1993). Assuming also an average geothermal gradient of 25°C/km, it was determined that the conditions for CO_2 to be in supercritical state (T > 31.1°C, p > 7.38 MPa; **Figure 1a**) will be roughly met for depths greater than 800 m (Holloway and Savage, 1993; van der Meer, 1993). However, depending on basin type and flow-driving mechanism, the pressure distribution is different from hydrostatic in most cases, sometimes significantly so. Also, the temperature of 31.1°C is found at varying depths, depending on the geothermal regime, which is controlled by:

- basin type, age, and tectonism;
- basement heat flow;
- proximity to crustal heat sources, such as magma chambers, intrusives, and volcanoes;

- thermal conductivity and heat production in the sedimentary succession, which depend in turn on the lithology and porosity of the sedimentary rocks; and
- temperature at the top of the sedimentary succession.

For continental basins located in tropical-to-temperate regions, the ground-surface temperature depends on climatic conditions (geographic position, altitude, and local climate), varying from 25–27°C in a tropical basin such as the Llanos, in Colombia, to 4–7°C in a temperate basin such as the Alberta Basin, in Canada. For marine shelf basins such as the Gulf Coast and North Sea, the temperature at the top of the sedimentary succession (sea bottom) is about 3–4°C. In tropical and subtropical low-altitude basins, CO_2 can be injected only as a gas or in supercritical state because the 31.1°C isotherm is reached at shallow depths, varying between 150 and 500 m, where pressures are less than 7.38 MPa. In temperate and marine-shelf basins, such as Illinois, Michigan, Alberta, and the North Sea, CO_2 can be injected and sequestered as a gas, a liquid, or in supercritical state, depending on pressure and temperature distributions (relative depths of the 7.38 MPa isobar and the 31.1°C isotherm, respectively). For continental sedimentary basins located in arctic regions, such as Alaska North Slope and Beaufort-Mackenzie, the temperature at the top of the sedimentary succession (bottom of the permafrost zone) is about –2°C. Carbon dioxide can be safely sequestered as a solid hydrate in shallow ocean sediments (Koide et al., 1996) or below the permafrost in arctic basins, at depths greater than several hundred meters where the conditions for forming and maintaining CO_2 hydrates are met. Thus, the geothermal regime in the basin impacts the type and depth of CO_2 injection and sequestration.

Hydrocarbon Potential and Basin Maturity

Deep aquifers are found in all sedimentary basins around the world, hence their large potential for CO_2 sequestration (Gunter et al., 1998). However, hydrocarbons, coal, or salt beds are not found in all basins because of basin evolution, depositional history, lack of potential, or discoveries. Of 800 hydrocarbon "provinces" (basins) in the world, 82 produce or have discoveries of giant-size oil and/or gas fields, 177 produce from subgiant fields, and 540 are nonproductive, of which 208 have a fair potential and 332 are rated as having poor potential (St. John et al., 1984). Several reasons work against the use of young, high-hydrocarbon-potential basins as prime targets for CO_2 sequestration. First, most of the hydrocarbon resources are still to be discovered; therefore, there is a concern about their possible contamination. Second, there are no abandoned or depleted oil or gas reservoirs because the basin is immature with respect to development. Third, the geology and hydrogeology of the basin are not well known. This situation is the opposite for continental basins in a mature stage of exploration and exploitation, such as all the North American basins between the Rocky and Appalachian Mountains, from Alberta to the Gulf Coast. The geology, hydrogeology, and geothermal regime of a mature basin are usually well understood. Most of the hydrocarbon pools and/or coal beds have already been discovered and usually are in production, and some reservoirs might be already depleted, nearing depletion, or even abandoned as uneconomic.

Surface Infrastructure

Basin maturity in the geological sequestration of CO_2 is concerned with the degree of hydrocarbon production and the existing or needed infrastructure. In mature continental basins, such as in Texas and Alberta, the infrastructure is already in place (access roads, pipelines, and wells) relatively close to major CO_2 point sources, and injection sites are easy to access and inexpensive to develop. In immature basins, the infrastructure is usually either nonexistent or very rudimentary, such as the Tarim Basin in China. In the case of basins such as the Gulf Coast, North Sea, and

Beaufort-Mackenzie, developing the necessary infrastructure for CO_2 injection in geological media is very expensive, particularly under harsh climatic conditions. Nevertheless, when there are no other alternatives, these basins could be used, as in the case of the Sleipner West field in the Norwegian sector of the North Sea.

The basin-scale criteria for CO_2 sequestration in geological media are not completely independent. There is a casual link, long known in hydrocarbon exploration, among a basin origin and evolution, its geothermal regime and dominant type of fluid flow, and hydrocarbon generation, migration, and accumulation. They all ultimately impact the presence and distribution of confining strata (aquitards or caprock), hydrocarbon reservoirs, salt beds and domes, coal beds, and aquifer characteristics, i.e., the method of CO_2 sequestration. In addition, they affect the pressure and temperature distributions in the basin, thus impacting the physical state, and therefore the method of sequestration, of the injected CO_2. The maturity of a basin and the degree of development of the surface infrastructure, including power generation, affect the applicability of CO_2 sequestration in geological media and the selection of sites that need to meet local-scale criteria.

SITE-SPECIFIC CRITERIA FOR CO_2 SEQUESTRATION IN GEOLOGICAL MEDIA

Once a basin has been assessed as being suitable for geological sequestration of CO_2, site-specific criteria need to be applied for the selection of the site and method of sequestration. The site-specific criteria fall into several categories:

- surface;
- geological media;
- in-situ conditions;
- storage capacity;
- injectivity and flow dynamics; and
- sequestration efficiency (confinement safety).

Because of technological, safety, and cost considerations, a site for geological sequestration of CO_2 must be chosen within an acceptable distance from a major CO_2 point source, such as a thermal power plant. The surface infrastructure for CO_2 separation, capture, transport, and injection into the ground, including land and right-of-way, should be, as much as possible, already in place or easy and economical to implement.

The geological media in the vicinity of the CO_2 source that would be potential candidates for CO_2 sequestration need to be identified. Depending on basin geology, energy resources, and maturity, it could be any combination of active or depleted oil and gas reservoirs, uneconomic coal beds, deep saline aquifers, and salt beds or domes. Generally, geographic locations around a basin depocentre have more possibilities for CO_2 sequestration in terms of methods and targets (deep aquifers, oil and gas reservoirs, salt structures, coals), and geographic locations around a basin margin have less potential because of depth limitations, shorter flow paths, and fewer or shallower hydrocarbon reservoirs.

In-situ conditions such as temperature, pressure, stress, rock lithology, salinity of formation water, in the case of aquifers; oil density and viscosity, in the case of oil reservoirs; and coal rank and gas content, in the case of coal beds, are essential in establishing various parameters for CO_2 sequestration in geological media. They determine the physical state (gas, liquid, or supercritical) and solubility of the sequestered CO_2 and the applicability of the method of sequestration (EOR, ECBMR, hydrodynamic, mineral), provided that other conditions are satisfied, with a corresponding impact on surface facilities (e.g., pipelines, compressors, well completion). Knowledge of the in-situ stress conditions is essential in establishing the injection pressure so that the reservoir or aquifer rock is not fractured, thus compromising the sequestration efficiency and safety. Temperature, pressure, and stress are strongly related to depth, so that the depth of injection, which affects the cost, is a secondary parameter that results from geological and in-situ conditions.

The storage capacity (mass of CO_2) is a critical element in the selection of sequestration sites. The storage capacity depends mainly on the available pore space in the reservoir, aquifer, or coal seam. It also depends on the thickness, geometry, and heterogeneity of the reservoir, aquifer, or coal bed. In the case of hydrocarbon and methane reservoirs, the storage capacity depends also on their stage of depletion, and in the case of deep aquifers, it depends on the solubility of CO_2 in formation water. The amount of sequestered mass-CO_2 depends on the available volume and in-situ conditions.

Carbon-dioxide injectivity is the most critical element in the assessment of a specific site for CO_2 sequestration. It depends first and foremost on near- and far-field permeability (Law and Bachu, 1996). High near-well permeability is needed to achieve high rates of injection without high injection pressures that may cause local fracturing, and low far-field permeability is desirable to slow the flow rate in the reservoir or aquifer, thereby increasing the residence time. In the case of reservoirs, this would increase the sweeping efficiency of the EOR or ECBMR operation, with corresponding recovery of more oil or gas before CO_2 breakthrough at the production well. In the case of deep saline aquifers, the sweep efficiency will increase the amount of CO_2 dissolved in the formation water, and the capacity for mineral trapping. The flow dynamics (gravity segregation, overriding, buoyancy flow, fingering) depend on the density and viscosity difference between formation fluids and the liquid or supercritical CO_2 (Law and Bachu, 1996). The greater the density and viscosity differences between CO_2 and the formation fluid, the faster the undissolved CO_2 will separate and flow updip in the aquifer. The reservoir or aquifer heterogeneity also has a significant impact because it may either retard or enhance the CO_2 flow along preferred flow paths, thereby increasing or decreasing, respectively, the CO_2 residence time, dispersion, solubility, and mineral trapping capacity.

Sequestration efficiency is related to the residence, or retention time, which varies from several months in the case of EOR operations to millions of years for hydrodynamic and geological trapping (Lindeberg and Holloway, 1999), and to the safety of sequestration. The latter refers to the existence of thick, competent, confining caprock or aquitards in the case of reservoirs and aquifers, respectively, of low-permeability confining strata in the case of coal beds, and of high-purity salt structures that would prevent the vertical escape and migration of CO_2. The confining media must not be intersected by faults, fractured, brecciated (naturally or induced), or penetrated by improperly completed or abandoned wells. In the case of aquifer sequestration, the hydrostratigraphy and flow directions in the larger-scale basin flow system also have implications for CO_2 sequestration efficiency. Injection at depths close to temperature and pressure conditions corresponding to CO_2 phase change will induce the transition to the gaseous phase if CO_2 reaches slightly shallower depths. If this occurs, in the absence of stratigraphic or structural traps, gas bouyancy will lead to "rapid" CO_2 rise or flow through the sedimentary column and escape to the surface, similar to gas migration in hydrocarbon systems.

APPLICABILITY

Geological sequestration of CO_2 is immediately applicable from a technological point of view, based on the experience already gained by the upstream petroleum industry. A series of regional-scale and site-specific criteria needs to be applied for the selection of the appropriate method and site, and for implementation. These criteria can be applied as a general inventory (checklist), using databases and geographic information systems adapted specifically for this task, for each prospective sedimentary basin and sequestration site. The sites that meet most criteria, both as point characteristics and as a spatial intersection of the desired attributes, are prime candidates (targets) for geological sequestration of CO_2.

Considering that only Russia and the countries in the industrialized world are signatories of the Kyoto Protocol, application of CO_2 sequestration in geological media will likely occur in the very near future in the United States, Canada, and western Europe (under the terms of the protocol, Australia is allowed to actually increase its CO_2 emissions). These are the same countries that already have the necessary infrastructure and technological know-how for immediate application. However, CO_2 sequestration in oil reservoirs (EOR) or coal beds (ECBMR) may occur in the developing world in countries such as China, India, and Poland that need to maximize their oil production or have significant resources of low-quality coal and that would benefit economically from cleaner and more efficient fuels. Carbon-dioxide sequestration will likely not be implemented in countries that are politically unstable or lack the appropriate economic resources and infrastructure, or whose priority is economic development at any cost (Bachu and Gunter, 1999).

In North America and Europe, the usually small, convergent basins along the Pacific coast (from Alaska to California) and along the Mediterranean (St. John et al., 1984), respectively, which are prone to earth hazards, should be excluded from consideration for geological sequestration of CO_2. Circum-Arctic and north Atlantic basins, such as Beaufort-Mackenzie, Hudson Bay, Labrador, and Barents, or northern foreland basins such as Alaska's North Slope, have extremely harsh climatic conditions, lack fully developed infrastructure and major (or any) sources of CO_2, and pose significant technological challenges. Stable shelf basins along North America's Atlantic coast, from the Scotian Shelf to Florida-Bahama and the Gulf Coast, and along continental Europe's northern shores (North and Baltic Seas) could be used for CO_2 geological sequestration if continental basins are not available, as in the case of the Sleipner West field in the North Sea offshore Norway (Korbol and Kaddour, 1995; Baklid et al., 1996). Continental foreland, cratonic, and interior-rift basins are best suited for CO_2 sequestration in geological media based on regional-scale criteria. Such basins include all the North American basins east of the Rocky Mountains and west of the Canadian Precambrian shield and the Appalachian Mountains, the Paris Basin in France, and the European basins north of the Alps and Carpathian Mountains. Among these basins, the prime candidates for early implementation of CO_2 sequestration in geological media are the ones with the most hydrocarbon resources and a long production history (reservoir depletion), coupled with significant power generation from fossil fuels. In North America, such basins are the Alberta Basin in Canada (Bachu and Gunter, 1999; Bachu, 1999); Texas and Appalachian basins and Michigan and Illinois Basins in the United States (Winter and Bergman, 1993; Bergman and Winter, 1995); and the Williston Basin, shared by Canada and the United States. If one considers the location and amount of major CO_2 sources, surface infrastructure, stage of depletion of hydrocarbon reservoirs and their capacity, and the existence of other potential geological media for CO_2 sequestration such as deep aquifers, salt structures, and coal beds, then the southern part of the Alberta Basin and the onshore part of the Gulf Coast Basin in eastern Texas are the foremost targets for pilot and large-scale implementation of geological sequestration of CO_2.

After application of the regional-scale criteria for assessment of the potential for CO_2 sequestration in a basin, site-specific criteria have to be used to identify the methods and media for CO_2 sequestration. As an illustration, **Figure 5** shows in plan view and along a dip cross section various options for CO_2 geological sequestration for major CO_2 producers in the Alberta Basin. Along the

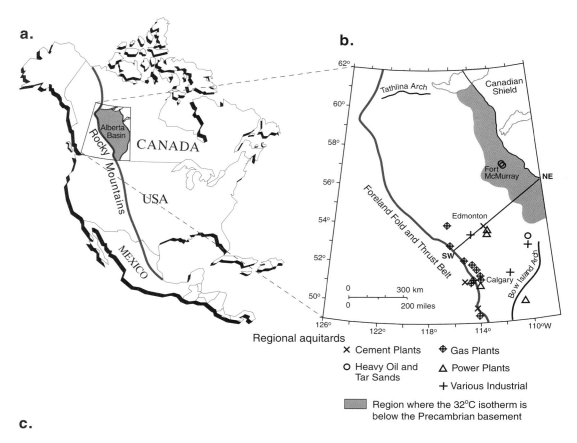

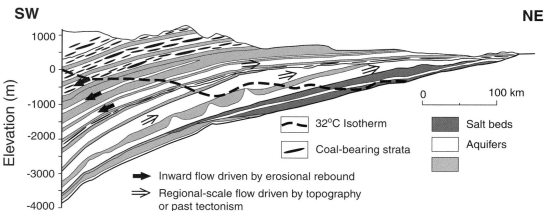

Figure 5 Characteristics of the Alberta Basin with respect to its potential for CO_2 sequestration in geological media: (a) basin location, (b) location of major CO_2 producers, and (c) cross section showing main saline aquifers, coal and salt beds, and hydrodynamic regime of formation waters.

eastern shallow edge of the basin, the 31.1°C isotherm is found at depths below the buried crystalline Precambrian basement **(Figure 5b)**. Significant coal beds are absent, and the saline aquifers are shallow and close to the discharge area **(Figure 5c)**. The oil sands plants near Fort McMurray, in northeastern Alberta **(Figure 5b)**, have limited options of disposing of CO_2, either as a gas in depleted shallow bitumen or gas reservoirs, or as a liquid compressed to high pressures in caverns mined in relatively thin Middle Devonian salt beds **(Figure 5c)**. The salt caverns will have to be mined specifically for this purpose, posing the problem of brine disposal. On the other hand, major CO_2 producers in the Edmonton area, such as power and cement plants and refineries, have a multitude of options for CO_2 sequestration in geological media:

- in Upper Cretaceous coal beds, producing methane at the same time;
- as a gas, liquid, or in supercritical state in active and depleted Cretaceous and Upper Devonian oil and gas reservoirs;
- as a liquid or in supercritical state in Upper Devonian saline aquifers, in regional-scale flow systems driven by topography from northern Montana to northern Alberta (Bachu, 1995); and
- in supercritical state in salt caverns mined for the petrochemical industry east of Edmonton.

Major CO_2 producers in southwestern Alberta, such as power, gas, and cement plants, have similar options for CO_2 sequestration, except for salt caverns. The hydrodynamic trapping is provided both by regional-scale flow in very deep Paleozoic aquifers and by inward, downdip flow driven by erosional rebound in Cretaceous aquifers **(Figure 5c)** (Bachu, 1995). In all these cases in which a multitude of options is available at a site, economic and safety considerations should play the decisive factor in making the final choice.

ISSUES

Geological sequestration of CO_2 is technologically feasible and is being practiced today on a very limited scale. Although the main mechanisms for CO_2 have been identified and a series of criteria for site assessment and selection has been developed, there are still many issues that must be addressed before full-scale implementation can occur. It is not yet possible to predict with confidence storage volumes, sequestration integrity, and the fate of injected CO_2 over long periods of time. Many important issues must be addressed to ensure safety, reduce costs, and gain full public acceptance. The primary uncertainty in large-scale implementation of CO_2 sequestration in geological media is the effectiveness of the operation—how easily it can be injected and how long it will be sequestered. The following is a nonexhaustive identification of issues that require attention in order to advance CO_2 sequestration in geological media from the conceptual and testing stages to full implementation.

Identification of Location and Capacity of Geological Traps for CO_2

Although general knowledge gained by the upstream energy industry can be applied for large-scale identification of sedimentary basins and areas most suitable for the geological sequestration of CO_2, specific information is severely missing. A concerted, well-coordinated effort should be directed toward the systematic identification (inventory) of specific sites and traps for CO_2 sequestration in geological media in the vicinity of major CO_2 producers, and the evaluation of these sites in terms of capacity and retention time.

Characterization of the Sequestration Medium

The sequestration medium (hydrocarbon reservoir, aquifer, coal bed, or salt structure) requires full characterization in terms of depth, geometry, internal architecture, lithology and mineralogy, porosity and permeability, and heterogeneity. The sealing unit requires characterization in terms of thickness, areal extent, permeability in relation to the permeability of the sequestration medium, degree of fracturing and heterogeneity, and overall integrity (e.g., possibility of leakage along natural and artificial vertical conduits). In mature sedimentary basins, such as in continental North America, there is a wealth of data collected by the upstream energy industry during exploration for and production of hydrocarbons. However, these data concentrate on hydrocarbon reservoirs, and they are very sparse or completely missing for nonprospective and nonproductive strata and areas. The data are often scarce or completely missing for aquifers, coal beds, and salt structures. In young basins, such as in the North Sea, the resolution of the data distribution is very low. Thus, a concerted effort must be directed toward data collection and integration for the characterization, representation, and modeling of the sequestration medium and confining unit.

In-Situ Conditions

The in-situ conditions of the sequestration medium and contained fluids need adequate characterization. The most important are stress, pressure, temperature, flow direction, and water salinity in the case of aquifers; oil and gas properties in the case of hydrocarbon reservoirs; and coal rank, quality, and gas content in the case of coal beds. These conditions which, again, are reasonably well known for hydrocarbon reservoirs and less or not at all known for the other media, have a significant impact on the characteristics of the CO_2 injection stream (wellhead pressure, temperature, and purity) and on the local-scale fate of the injected CO_2 (physical state, dissolved amount, gravity segregation, viscous fingering).

Predicting and Monitoring the Fate of the Injected CO_2

There is no practical knowledge about the behavior and long-term fate of the CO_2 injected in geological media. To date, besides conceptual models, no numerical models have been specifically developed to simulate the multiphase flow of CO_2 in various physical states, oil, gas, methane, and/or water, in reservoirs and/or aquifers. Numerical simulations of flow so far have used models adapted from reservoir simulators developed by the energy industry (e.g., Law and Bachu, 1996), and geochemical reactions have been addressed in isolation of flow processes (e.g., Gunter et al., 1997b). Adequate numerical models need developing, and monitoring programs put in place, to predict and determine the long-term fate of the injected CO_2 outside the immediate vicinity of the injection well (flow paths and rate, dissolution, mineral reactions), and for the detection of potential leakage.

Injection, Drilling, and Well-Completion Technology

Although CO_2 injection in aquifers, reservoirs, and coal beds is already taking place in a few EOR, ECBMR, and disposal operations, there is no experience yet regarding long-term special requirements for these operations to ensure their safety and long-term stability.

Energy Requirements for CO_2 Sequestration

The capture, purification, transport, and injection of a CO_2 stream require use of energy whose production by itself contributes to increased CO_2 emissions. Thus, improving the energy efficiency of the surface operations for CO_2 sequestration in geological media will improve the overall efficiency of the entire operation.

Performance Assessment

Because the issue of reducing CO_2 emissions to the atmosphere is very recent, and the concept of geological sequestration of CO_2 is in an incipient stage of development and application, there are no methods to design and optimize these operations. Such methods need to combine the identification of site-specific CO_2 geological traps ("inventory") with the cost of CO_2 capture, transport, and injection, the cost of monitoring, and other tangible and intangible economic and societal benefits. These comprehensive methods should help in selecting sites and assessing the overall performance of CO_2 sequestration in geological media.

Public Acceptance

Gaining public confidence and acceptance is critical for the large-scale implementation of geological sequestration of CO_2. To achieve this, the public must be credibly convinced primarily that it is a safe operation, with no risks for environment, property, and life. Addressing all the previous issues regarding CO_2 sequestration in geological media should lead, by themselves, to a high degree of public acceptance if all the aspects are properly communicated.

CONCLUSIONS

Atmospheric concentrations of CO_2 have risen since the beginning of the industrial revolution, primarily as a result of fossil-fuel use for energy production. Developed countries that signed the Kyoto Protocol have committed to reducing by 2012 the release into the atmosphere of anthropogenic CO_2 to levels lower by 5.2% than those of 1990. Reduction of CO_2 emissions can be achieved through increased energy efficiency and conservation, and through CO_2 management, of which CO_2 sequestration is an important option. Using the technology and experience already gained by the oil and gas industry and in groundwater resource management, sequestration of CO_2 in geological media is an immediately applicable option for the near- to medium-term mitigation of climate-change effects resulting from the release of anthropogenic CO_2 into the atmosphere. Based on its properties and in-situ conditions, CO_2 can be sequestered as a gas, a liquid, or in supercritical state in depleted oil and gas reservoirs, uneconomic coal beds, deep saline aquifers, and salt caverns. Use of CO_2 for miscible flooding of oil reservoirs or for methane production from coal beds has an added economic benefit. The main mechanisms responsible for CO_2 sequestration in geological media are geological, solubility, hydrodynamic, mineral, adsorption, and cavern trapping. Basin-scale criteria for CO_2 sequestration, such as tectonic setting, hydrodynamic and geothermal regimes, hydrocarbon potential and basin maturity, and surface infrastructure, should be used for the identification of sedimentary basins in the world that are suitable for CO_2 sequestration in geological media. Site-specific criteria, such as particular geological media, in-situ conditions, storage capacity, injectivity and flow dynamics, and sequestration efficiency, need to be applied to identify sites, methods, and capacity for CO_2 sequestration. Continental sedimentary basins in North America, foremost in Texas and Alberta, are the prime candidates for CO_2 sequestration in geological media, followed by circum-Atlantic shelf basins. However, a series of major issues still needs to be addressed before proceeding with full-scale implementation, such as identification of specific sites and their capacity; proper characterization of the sequestration medium and of in-situ conditions; prediction and monitoring the fate of the injected CO_2; surface CO_2 capture, transport, and injection; performance assessment; and, finally, general public acceptance. Nevertheless, CO_2 sequestration in geological media is very promising and is in advanced stages of development prior to application.

REFERENCES CITED

American Geophysical Union, 1999, Position Statement: EOS, Transactions of AGU, v. 80, p. 49.

Arri, L. E., D. Yee, W. D. Morgan and M. W. Jeansonne, 1992, Modeling coalbed methane production with binary gas sorption: Society of Petroleum Engineers Paper No. 24363, SPE Rocky Mountain Regional Meeting, Casper, Wyoming.

Aya, I., K. Yamane, and K. Shiozaki, 1999, Proposal of self sinking CO_2 sending system: COSMOS: in Greenhouse Gas Control Technologies (Eliasson, B., P. W. F. Riemer, and A. Wokaun, eds.), Pergamon, Elsevier Science Ltd., p. 269–274.

Bachu, S., 1995, Synthesis and model of formation-water flow, Alberta Basin, Canada: AAPG Bulletin, v. 79, p. 1159–1178.

Bachu, S., 1999, The potential for carbon dioxide sequestration in geological media in Alberta: in Proceedings (CD-ROM) of Combustion Canada '99 Conference on "Combustion and Global Climate Change: Canada's Challenges and Solutions," Canadian Environmental Industry Association, Calgary, Alberta, May 26–28.

Bachu, S., and W. D. Gunter, 1999, Storage capacity of CO_2 in geological media in sedimentary basins, with application to the Alberta basin: in Greenhouse Gas Control Technologies (Eliasson, B., P. W. F. Riemer, and A. Wokaun, eds.), Pergamon, Elsevier Science Ltd., p. 195–200.

Bachu, S., W. D. Gunter, and E. H. Perkins, 1994, Aquifer disposal of CO_2: Hydrodynamic and mineral trapping: Energy Conversion and Management, v. 35, p. 269–279.

Baklid, A., R. Korbol, and G. Owren, 1996, Sleipner west CO_2 disposal, CO_2 injection into a shallow underground aquifer: Society of Petroleum Engineers Paper No. 36600, 9 p.

Bergman, P. D., and E. M. Winter, 1995, Disposal of carbon dioxide in aquifers in the United States: Energy Conversion and Management, v. 36, p. 523–526.

Bredehoeft, J. D., R. D. Devanshir, and K. R. Belitz, 1988, Lateral fluid flow in a compacting sand-shale sequence: South Caspian basin: AAPG Bulletin, v. 72, p. 416–424.

Bryant, E., 1997, Climate Process and Change: Cambridge University Press, 209 p.

Byrer, C. W., and H. G. Guthrie, 1999, Coal deposits: Potential geological sink for sequestering carbon dioxide emissions from power plants: in Greenhouse Gas Control Technologies (Eliasson, B., P. W. F. Riemer, and A. Wokaun, eds.), Pergamon, Elsevier Science Ltd., p. 181–187.

Crossley, N. G., 1998, Conversion of LPG salt caverns to natural gas storage: "A Transgas experience": Journal of Canadian Petroleum Technology, v. 37, no. 12, p. 37–47.

Gunter, W. D., T. Gentzis, B. A. Rottenfusser, and R. J. H. Richardson, 1997a, Deep coalbed methane in Alberta, Canada: A fuel resource with the potential of zero greenhouse emissions: Energy Conversion and Management, v. 38S, p. S217–S222.

Gunter, W. D., B. Wiwchar, and E. H. Perkins, 1997b, Aquifer disposal of CO_2-rich greenhouse gases: Extension of the time scale of experiment for CO_2 sequestering reactions by geochemical modeling: Mineralogy and Petrology, v. 59, p. 121–140.

Gunter, W. D., S. Wong, D. B. Cheel, and G. Sjostrom, 1998, Large CO_2 sinks: Their role in the mitigation of greenhouse gases from an international, national (Canadian) and provincial (Alberta) perspective: Applied Energy, v. 61, p. 209–227.

Gupta, N., and E. S. Bair, 1997, Variable-density flow in the midcontinent basins and arches region of the United States: Water Resources Research, v. 33, p. 1785–1802.

Gupta, N., B. Sass, J. Sminchak, T. Naymik, and P. Bergman, 1999, Hydrodynamics of CO_2 disposal in a deep saline formation in the midwestern United States: in Greenhouse Gas Control Technologies (Eliasson, B., P. W. F. Riemer, and A. Wokaun, eds.), Pergamon, Elsevier Science Ltd., p. 157–162.

IPCC (Intergovernmental Panel on Climate Change), 1996, Climate Change 1995: The Science of Climate Change: Cambridge University Press, Cambridge, U.K., 572 p.

Hattenbach, R. P., M. Wilson, and K. R. Brown, 1999, Capture of carbon dioxide from coal combustion and its utilization for enhanced oil recovery: in Greenhouse Gas Control Technologies (Eliasson, B., P. W. F. Riemer, and A. Wokaun, eds.), Pergamon, Elsevier Science Ltd., p. 217–221.

Hitchon, B., W. D. Gunter, T. Gentzis, and R. T. Bailey, 1999, Sedimentary basins and greenhouse gases: A serendipitous association: Energy Conversion and Management, v. 40, p. 825–843.

Holloway, S., and D. Savage, 1993, The potential for aquifer disposal of carbon dioxide in the U.K.: Energy Conversion and Management, v. 34, p. 925–932.

Koide, H. G., Y. Tazaki, Y. Noguchi, M. Iijima, K. Ito, and Y. Shindo, 1993, Underground storage of carbon dioxide in depleted natural gas reservoirs and useless aquifers: Engineering Geology, v. 34, p. 175–179.

Koide, H. G., M. Takahashi, Y. Shindo, Y. Tazaki, M. Iijima, K. Ito, N. Kimura, and K. Omata, 1996, Hydrate formation in sediments in the sub-seabed disposal of CO_2: Energy—The International Journal, v. 22, p. 279–283.

Korbol, R., and A. Kaddour, 1995, Sleipner West CO_2 disposal—injection of removed CO_2 into the Utsira Formation: Energy Conversion and Management, v. 36, p. 509–512.

Law, D. H-S., and S. Bachu, 1996, Hydrogeological and numerical analysis of CO_2 disposal in deep aquifers in the Alberta sedimentary basin: Energy Conversion and Management, v. 37, p. 1167–1174.

Lindeberg, E., and S. Holloway, 1999, The next steps in geo-storage of carbon dioxide: in Greenhouse Gas Control Technologies (Eliasson, B., P. W. F. Riemer, and A. Wokaun, eds.), Pergamon, Elsevier Science Ltd., p. 145–150.

Loken, K. P., and T. Austvik, 1993, Deposition of CO_2 on the seabed in the form of hydrates, Part II: Energy Conversion and Management, v. 34, p. 1081–1087.

McKean, T. A. M., R. M. Wall, and A. A. Espie, 1999, Conceptual evaluation of using CO_2 extracted from flue gas for enhance oil recovery, Schraer Bluff field, North Slope, Alaska: in Greenhouse Gas Control Technologies (Eliasson, B., P. W. F. Riemer, and A. Wokaun, eds.), Pergamon, Elsevier Science Ltd., p. 207–215.

Meer, L. G. H. van der, 1993, The conditions limiting CO_2 storage in aquifers: Energy Conversion and Management, v. 34, p. 959–966.

Neuzill, C. E., and D. W. Pollock, 1983, Erosional unloading and fluid pressures in hydraulically "tight" rocks: Journal of Geology, v. 91, p. 179–193.

Oliver, J., 1986, Fluids expelled tectonically from orogenic belts: Their role in hydrocarbon migration and other geologic phenomena: Geology, v. 14, p. 99–102.

Spencer, C. W., 1987, Hydrocarbon generation as a mechanism for overpressuring in Rocky Mountain Region: AAPG Bulletin, v. 71, p. 368–388.

St. John, B., A. W. Bally, and H. D. Klemme, 1984, Sedimentary provinces of the world—hydrocarbon productive and nonproductive: AAPG, Tulsa, Oklahoma, 35 p.

Stevens, H. S., V. A. Kuuskra, D. Spector, and P. Riemer, 1999, CO_2 sequestration in deep coal seams: pilot results and worldwide potential: in Greenhouse Gas Control Technologies (Eliasson, B., P. W. F. Riemer, and A. Wokaun, eds.), Pergamon, Elsevier Science Ltd., p. 175–180.

Tek, M. R., ed., 1989, Underground storage of natural gas: Theory and practice: NATO ASI Series E, Applied Sciences, 171, Kluwer Academic, Boston, Maine, 458 p.

Wichert, E., and T. Royan, 1997, Acid gas injection eliminates sulfur recovery expense: Oil and Gas Journal, v. 95, p. 67–72.

Winter, E. M., and P. D. Bergman, 1993, Availability of depleted oil and gas reservoirs for disposal of carbon dioxide in the United States: Energy Conversion and Management, v. 34, p. 1177–1187.

16 | Near-Term Climate Prediction Using Ice-Core Data from Greenland

Sergey R. Kotov

*Institute of Precambrian Geology
and Geochronology RAS*
Saint Petersburg, Russia

ABSTRACT

Records from the Greenland Ice Sheet Project II (GISP2) Greenland ice core are considered in terms
of dynamical systems theory and nonlinear prediction. Dynamical systems theory allows us to reconstruct some properties of a phenomenon based only on past behavior without any mechanistic
assumptions or deterministic models. A short-term prediction of temperature, including a mean
estimate and confidence interval, is made for 800 years into the future. The prediction suggests that
the present short-time global warming trend will continue for at least 200 years and will be followed by a reversal in the temperature trend.

INTRODUCTION

Predicting the global climate is one of the major unresolved challenges facing twenty-first-century
applied science. Most previous efforts in climate prediction have required constructing deterministic models of global climate systems based on equations from mathematical physics. These
models consist of a complex of relationships in the form of diffusion equations, mass transfer equations, and mass balance conditions (Hansen et al., 1983; Russell et al., 1995). To be successful, such
a deterministic approach must be founded on strict conceptual grounds and the resulting models
must be implemented on extremely powerful computers.

Over the past 20 years, fundamentally new approaches to the prediction of time series have
been developed, based on dynamical systems theory (general discussions of this approach are
given in recent texts on time-series analysis such as Weigend and Gershenfeld, 1994). These new

procedures allow us to estimate certain fundamental properties that are required in theoretical models of nonlinear phenomena (such as the number of degrees of freedom in a system and its fundamental dimensions). More importantly from the viewpoint of climate modeling, these new procedures also provide a way to predict the near-term state of a complex system that exhibits chaotic behavior. Such approaches have already proven useful in scientific fields as diverse as physics, psychology, economy, and medicine, and seem sufficiently general and powerful enough to be useful for global climate modeling as well.

RECORDS OF CLIMATE CHANGE FROM GREENLAND ICE CORE

Here, we will apply procedures of dynamical systems theory to two natural climate records and, for illustrative purposes, to one artificial record. The first natural record was produced by the Greenland Ice Sheet Project II (GISP2), which investigated climatic and environmental changes over the past 250,000 years by analyzing a core drilled completely through the ice in the central part of the Greenland continental glacier (NSIDC, 1997).

The ratio of the two isotopes of oxygen, ^{16}O and ^{18}O, varies according to temperature in water because the lighter isotope is more volatile. This ratio can be used as a proxy measure of atmospheric temperature because the relative amount of ^{18}O is less in snow that has precipitated in colder air. The oxygen isotope ratio is conventionally expressed as $\delta^{18}O$, which is standardized relative to the standard mean oxygen isotope ratio in seawater (SMOW). The relationship between atmospheric temperature and $\delta^{18}O$ in Greenland has been empirically determined (Johnsen et al., 1992) to be $\delta^{18}O = 0.67T$ (°C) – 13.7. **Figure 1** shows the distribution of $\delta^{18}O$ averaged over 20-year intervals, back to 10,000 years B.P. (For conventions used in the GISP2 ice-core record, see Davis and Bohling, this volume).

On cursory examination, the temperature record throughout the Holocene appears chaotic, but closer examination shows trends of increasing or decreasing average temperatures over specific intervals of time, such as during the last 200 years. Davis and Bohling (this volume) have characterized the GISP2 record of 20-year average $\delta^{18}O$ values from a stochastic viewpoint. In contrast, we will examine the same record, regarding it as the output from a dynamic, nonlinear system complicated by random influences.

The climate dynamics of the Pleistocene prior to the start of the Holocene differ greatly from the dynamics that have been in operation since the collapse of continental ice sheets in the Northern Hemisphere. This is apparent in the dramatic change in the $\delta^{18}O$ record that occurs about 10,000 years B.P. **(Figure 2)**. The possible causes of this change are beyond the scope of this paper; interested readers can find an extensive discussion in Lowe and Walker (1998).

The GISP2 record of 20-year average $\delta^{18}O$ extends back only a short time into the Pleistocene (to 16,490 years B.P.) and consequently is not long enough to allow us to assess climate dynamics of the pre-Holocene interval. However, there are more extensive records of other constituents extracted from the GISP2 core, including Na, NH_4, K, Mg, Ca, Cl, NO_3, and SO_4. These variables can be combined into a single composite variable by principal component analysis (Gorsuch, 1983), yielding a new composite variable that is highly correlated with most of the measured constituents and that expresses more than 76% of the variation in all of the original variables. The record over time of this component is shown in **Figure 3**. By combining the different variables measured on the GISP2 ice core into a single dominant component, we not only avoid the problem of choosing the most appropriate variable for analysis, but we also suppress superfluous noise that is relegated to other, lesser components. An analysis by Mayewski et al. (1997) yielded an essentially identical composite variable they called the polar circulation index (PCI), which was interpreted as

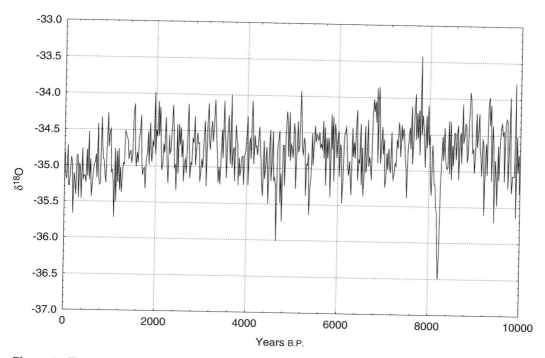

Figure 1 Twenty-year average record of $\delta^{18}O$ for the period of time 0–10 kyr B.P. (thousand years before the present) from the GISP2 ice core.

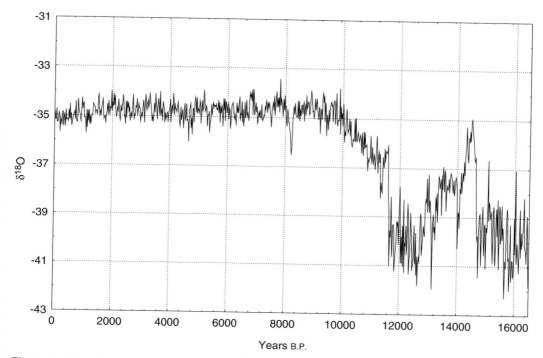

Figure 2 Twenty-year average record of $\delta^{18}O$ for the period of time 0–16 kyr B.P from the GISP2 ice core.

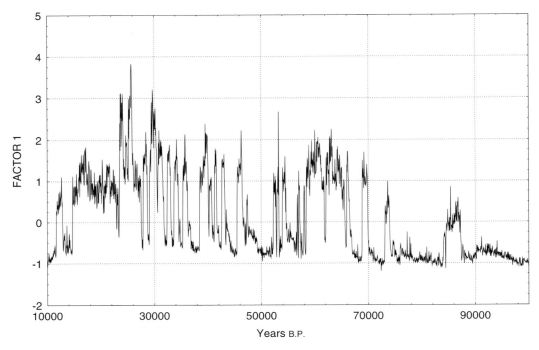

Figure 3 Distribution of first principal component of chemical constituents from GISP2 ice core for the period of time 10–100 kyr B.P.

reflecting increased continental dust and marine aerosols during cold intervals. We note that the behavior of the first principal component strongly reflects climatic conditions, because it is approximately inversely proportional to the temperature at which ice forms. This has been determined by correlating the component to $\delta^{18}O$. The linear correlation coefficient over the last 16,000 years is $r = -0.91$, which is significantly different from zero. In any case, strict dependence between temperature and component is not important for our purpose, which is to examine the dynamics of climatic behavior.

The main characteristic of the GISP2 component record is its intermittent behavior, with intervals representing periods that were relatively warm (on average) but that changed abruptly to periods that were cold. Within this general pattern, the record is characterized by high-frequency, low-amplitude oscillations. Causes of the major episodic alterations from relatively warm to cold and vice versa remain to be established. Most likely, these changes are a consequence of both external and internal influences that operated at a planetary scale. Possibly these long-term variations in climatic temperature were due to orbital forcing (Imbrie et al., 1993) modulated by changes in circulation within the oceanic and atmospheric covers of the earth (Lowe and Walker, 1998).

The third record to be considered in this paper is artificial, the *x*-coordinate of the realization of mathematical equations that describe the Lorenz system. This system of equations was developed by Edward Lorenz by simplifying and linearizing hydrodynamic equations as part of his research into weather patterns (Lorenz, 1963). The Lorenz system operates in three-dimensional phase space—the space in which variables describing the behavior of a dynamic system are entirely confined. For some values of the controlling parameters, the Lorenz system exhibits chaotic

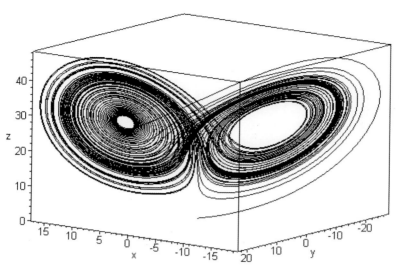

Figure 4 Trajectory of the Lorenz system.

behavior and any trajectory is attracted to a subset of phase space having a fractal dimension (Mandelbrot, 1977). The moving path described by this system **(Figure 4)** is wandering; that is, the trajectory follows several right-hand coils, then abruptly switches and follows several left-hand coils, then switches again, and so on. The trajectory is very sensitive to small variations in the initial parameters, making it extremely difficult to predict how many successive coils will be completed during some period of time before abruptly switching to the alternative state. This behavior is now popularly known as the "butterfly effect"—the idea that a butterfly flapping its wings in Saint Petersburg can set in motion a complicated chain of events that ultimately affects the weather in Kansas City.

When viewed in its full three dimensions, as in **Figure 4**, the Lorenz system seems to have no resemblance to the measures of climate recorded in the ice cores. However, if the wandering locus of the system of equations is projected onto a single dimension, its record appears quite different, as can be seen in **Figure 5**. This one-dimensional section of the Lorenz trajectory has, in a certain sense, features similar to those that appear in **Figure 3**: The trace of the Lorenz system consists of high-frequency, low-amplitude oscillations centered around local averages, with unpredictable "snaps" from one average state to the other. We will now consider how closely the dynamics of the Lorenz system match records from the Pleistocene ice core, and how this resemblance can be used for climate prediction.

SIMPLE NONLINEAR PREDICTION

Approaches stemming from dynamical systems theory allow us to make predictions both in strictly deterministic systems that exhibit chaotic behavior (such as the Lorenz system) and in systems which contain superimposed random noise. Different methods of prediction are used in such systems (Weigend and Gershenfeld, 1994), but most of them are based on the idea of the time decomposition of a single time series followed by phase space reconstruction (Grassberger and Procassia, 1983).

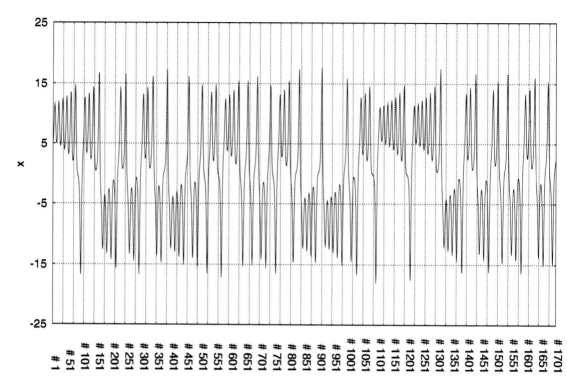

Figure 5 The *x*-coordinate of the Lorenz system. Horizontal axis is a discrete nondimensional time ($\delta t = 0.05$).

First, following this approach, we create a sequence of state vectors $X(i)$ from the available one-dimensional sequence, $x(i)$:

$$X(i) = \{x(i), x(i-L), \ldots, x[i-L(M-1)]\}.$$

Here, L is the "lag," or number of sampling intervals between successive components of the delay vectors, and M is the dimension of the delay vector. In other words, from a one-dimensional sequence of measured values, we construct a new sequence of M-dimensional vectors $X(i)$ which define some trajectory in M-dimensional space. A theorem by Takens (1981) and by Sauer et al. (1991) states that if the sequence $x(i)$ consists of a scalar measurement of the state of a dynamical system, then under certain assumptions, the time-delay procedure provides a one-to-one image of the original sequence, provided M is sufficiently large.

Next, we must determine M, the dimension of the phase space. This dimensional parameter is very important because it specifies the number of degrees of freedom in the system. Recall that the trajectory of the Lorenz system lies within a three-dimensional space **(Figure 4)**. Using the Grassberger-Procassia algorithm (Grassberger and Procassia, 1983), it is possible to estimate this dimension using only the record of the observed *x*-coordinate shown in **Figure 5**. So, for an arbitrary one-dimensional sequence it is possible to estimate the dimensionality of an entire dynamical system, if such a system exists. The estimate of the dimension of the phase space is also useful as a measure of the complexity of the system.

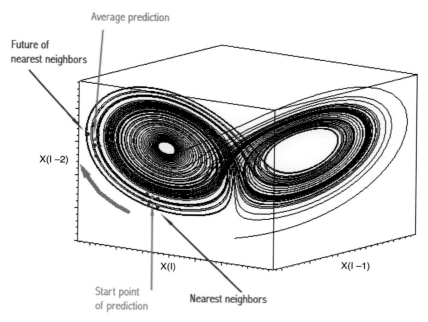

Figure 6   Prediction scheme in reconstructed space for the Lorenz system.

The last step is prediction itself. **Figure 6** shows, for illustrative purposes, a record of the Lorenz system. To predict $x(i+j)$ from the record at point i, we first impose a metric on the M-dimensional phase space (in this instance, we have used simple Euclidean distance as the metric) and find the k nearest neighbors of $X(i)$ from the past $X(l):l < i, l \in S$, where S is the set of indices of the k nearest neighbors. The prediction is simply the average over the "future" $X(l + j)$ of the neighbors $X(l), l = l_1, l_2, \ldots, l_k$. In other words, we must consider that part of the phase space around the predicted point, and see what happens within this domain during the evolution of the system **(Figure 6)**. To obtain a prediction in the one-dimensional space of the original data, we need only consider the first components of the delay-time vectors. The procedure is described in detail in Farmer and Sidorovich (1987) and Hegger et al. (1999).

These predictions can be made more useful by enclosing them in estimated confidence intervals for a specified level of probability. We may presume that inevitably, a theoretical natural dynamic system is confounded with independent random processes (random noise). Thus, we can consider the observed system trajectory to represent a random cloud of points surrounding an imaginary theoretical trajectory in multidimensional space. It is well known that the sums of a large number of independent random processes will form a normal (Gaussian) distribution of values. So we may assume that a normal distribution (representing random noise) has been superimposed on the response of the dynamic system. Student's criterion, a standard method for estimating confidence intervals around the average of a small number of random values, can be used to construct intervals around the prediction within which the true estimate will fall with specified probability.

It is important to emphasize that this methodology is intended for prediction only over a short time into the future. If a phenomenon truly exhibits chaotic behavior, the actual and predicted trajectories will diverge with time in an unpredictable manner because of the sensitivity of the system to the initial parameters. Long-term predictions should be made only with extreme caution.

EXAMPLES OF PREDICTIONS

The first illustration of the method uses artificial data, the x-coordinate of the Lorenz system, shown in **Figure 5**. The results of applying the short-time prediction technique are shown in **Figure 7**. Note that the Lorenz trace progresses from the "past" on the right side of the illustration to the "present" on the left, corresponding in orientation with the ice-core records. Predictions begin at and are based only on the characteristics of the prior record. Predictions and their confidence intervals have been made up to and can be compared with the actual values of the Lorenz system over this interval. There is very good correspondence between observations and predictions over the short time, and a gradual divergence between observations and predictions at longer times. Note that this artificial record is purely deterministic and free of random noise.

The approach is not restricted to deterministic systems, as we can see in an application to data from the GISP2 core for the interval prior to the Holocene (10,000-100,000 years B.P.), shown in **Figure 3**. Previous analyses have determined that the record of the first component for this period can be regarded as a dynamic system having a phase space dimension of $M = 3$. **Figure 8** shows a portion of the distribution of the first component as a function of time as well as the mean prediction and the confidence interval for the prediction ($M = 3, L = 1, k = 25$, 99% confidence interval). In this illustration, the record goes from the distant past (40,000 years B.P.) on the right side to the start of the Holocene (10,000 years B.P.) on the left side. Predictions and their confidence intervals begin at 25,000 years B.P. and extend to the start of the Holocene. As in the previous example, the predictions are based only on the part of the record prior to 25,000 years B.P.

A good correspondence between prediction and real data is apparent; although the confidence interval does not always cover the data, the tendencies in data and prediction are coincident. This result can be taken as supporting evidence for the notion that the Northern Hemisphere Pleistocene climate can be regarded as a type of low-dimensional dynamic system.

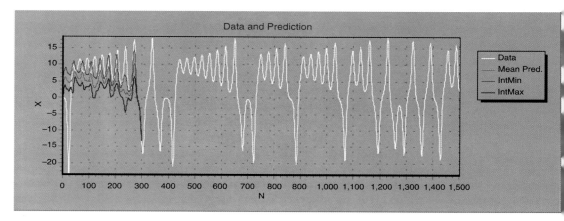

Figure 7   Prediction for trace of Lorenz system: White curve is x-coordinate of the Lorenz system (discrete time is on horizontal axis, $\delta t = 0.25$), red curve is average prediction, blue curves represent 99% confidence interval. $M = 3, L = 1, k = 25$.

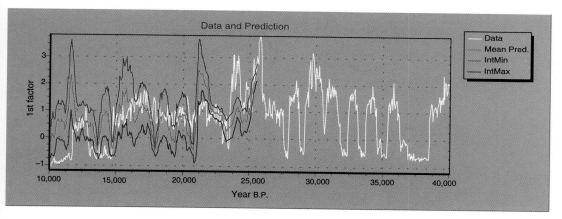

Figure 8   Distribution of first principal component for the period 10–40 kyr B.P. and prediction for the period 10,000–25,000 years B.P.

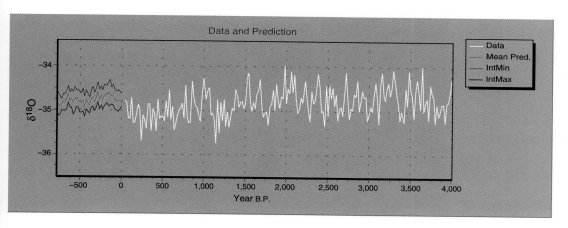

Figure 9   Distribution of $\delta^{18}O$ for the period 0–4,000 years B.P. and prediction for the period 1140 years B.P. to A.D. 2800.

Next, we will examine the results from analyzing the $\delta^{18}O$ ice-core record for the last 10,000 years (Holocene). In dynamic systems terms, the behavior of the trajectory for this record is more complicated—the estimated phase space dimension M is 8. The results of testing the predictions are shown on **Figure 9** ($M = 8$, $L = 1$, $k = 25$, 99% confidence interval). The prediction begins at $N = 54$ (1040 years B.P.) and extends through the end of the ice-core record (–30 years B.P., or A.D. 1980) and beyond, into the future. There is good correspondence between prediction and observation for a short period of time (approximately 300 years) and a general correspondence between prediction and data for the rest of the time interval.

Figure 10 shows the mean prediction and the confidence interval for prediction over the next 800 years into the future ($M = 8$, $L = 1$, $k = 25$, 99% confidence interval). This prediction is based on characteristics of the entire ice-core record since the start of the Holocene (10,000 years B.P.); predictions begin at the latest date in the 20-year average $\delta^{18}O$ record (A.D. 1980) and extend into

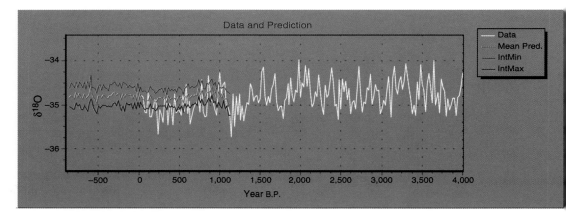

Figure 10 Distribution of $\delta^{18}O$ for the period 0–4,000 years B.P. and prediction for 800 years into the future.

the future. Note that the most recent 200-year time interval is characterized by a pattern of increasing temperatures. The onset of such a pattern has been noted over the past decades and often is ascribed to global warming caused by a greenhouse effect associated with the industrial revolution. It is true that this pattern coincides with the duration of the industrial revolution, which began approximately 150 years ago. On the other hand, if the analytical method has been chosen appropriately, then from these results, it follows that this positive trend in temperature is not an extraordinary one for the Holocene. Moreover, this short-time warming is predicted to continue for about 200 years and to be followed by a reversal in the temperature trend.

CONCLUSIONS

Traditional models of global climate change are extremely complicated, rely on a number of fundamental assumptions, and are difficult and costly to operate. Perhaps most troublesome is the brevity of the historical record of climate on which the models are conditioned. However, records of various constituents measured in Greenland ice cores provide information about past variations in climate over thousands of years before the present, which can be used to help understand better the processes of global climate change. Methods for reconstructing the entire multidimensional trajectory of a process from its one-dimensional sequence of measured values allows us to estimate important characteristics of dynamic systems such as the number of degrees of freedom and the true dimensionality of the system. The dimensionality reflects the number of independent variables in a system and must be taken into account in the construction of a global climate model. In addition, this approach permits us to make short-term predictions of important climatic variables.

The main characteristic of the 90,000 years that preceded the Holocene is the intermittent behavior of the climate. Periods that were relatively warm (on average) were abruptly followed by periods of intense cold. Within these relatively warm or cold intervals, temperatures varied through high-frequency, low-amplitude oscillations. Described in system terms, the climate behaved as a dynamical system with three degrees of freedom. We obtain a good correspondence between predicted behavior and reality by modeling climate as a low-dimensional dynamical system over this period of time. The same characteristics are exhibited by a Lorenz system—the simplest nonlinear deterministic model that can be applied. Such a system has three degrees of freedom and demonstrates high-frequency oscillations with unpredictable "snaps" from one space domain to another.

In terms of dynamical systems, the behavior of the climatic record in the Holocene is more complicated—the dimensionality of the phase space M is 8. Estimates of temperature based on a dynamical system model indicate that a positive trend in temperature over a 200-year period is not unexpected for the Holocene. Predictions based on the characteristics of the Holocene and extending into the future indicate that the present short-term warming trend may continue for at least 200 years and may be followed by a reversal in the temperature trend.

ACKNOWLEDGMENTS

The author would like to acknowledge the help of Dr. G. Bohling with the editing of text. Especially, the author expresses his gratitude to Dr. J. Davis for valuable remarks and proofreading of the article. The work described in this article was done while the author was a visiting scientist in the Mathematical Geology Section at Kansas Geological Society, based at the University of Kansas.

REFERENCES CITED

Davis, J. C., and Bohling, G., 2000, The search for patterns in ice-core temperature curves: this volume, Chapter 11, p. 213–229.

Farmer, J. D., and Sidorowich, J. J., 1987, Predicting chaotic time series: Physical review letters, v. 59, no. 8, p. 845–848.

Gorsuch, R., 1983, Factor analysis: L. Erlbaum Associated, Hillsdale, New Jersey, 452 p.

Grassberger, P., and Procassia, I., 1983, Characterization of strange attractors: Physical review letters, v. 50, no. 5, p. 346–349.

Hansen, J. G., et al., 1983, Efficient three-dimensional global models for climate studies: Models I and II: Monthly Weather Review, v. 11, p. 609–662.

Hegger, R., Kantz, H., and Schreiber, T., 1999, Practical implementation of nonlinear time series methods: The TISEAN package: Chaos, v. 9, no. 2, p. 413–435.

Imbrie, J., Berger, A., and Shackleton, N., 1993, Role of orbital forcing: A two-million-years perspective, in Eddy, J., and Oeschger, H., eds., Global Changes in the Perspective of the Past: John Wiley & Sons, Ltd., Chichester, U.K., p. 263–277.

Johnsen, S., Dansgaard, W., and White, J., 1992, The origin of arctic precipitation under present and glacial conditions: Tellus, v. B41, p. 452–468.

Lorenz, E., (1963), Deterministic non-periodic flow: Journal of the Atmospheric Sciences 20, p. 130–141.

Lowe, J. J, and Walker, M. J., 1998, Reconstructing quaternary environments, second ed: Longman Inc., White Plains, New York, 446 p.

Mandelbrot, B., 1977, Fractals: Form, chance and dimension: Freeman Publ. Co., San Francisco, California, 346 p.

Mayewski, P. A., et al., 1997, Major features and forcing of high-latitude northern hemisphere atmospheric circulation using a 110,000-year-long glaciochemical series: Journal of Geophysical Research, v. 102, no. C12, p. 26, 345–26, 366.

NSIDC User Services, 1997, The Greenland Summit Ice Cores CD-ROM, GISP-2/GRIP: World Data Center A for Glaciology, CIRES, Univ. Colorado, Boulder, CD-ROM.

Russell, G. L., Miller, J. R., and Rind, D., 1995, A coupled atmosphere-ocean model for transient climate change studies: Atmosphere-Ocean, v. 33, p. 683–730.

Sauer, T., Yorke, J., and Casdagli, M., 1991, Embedology: Journal of Statistical Physics, v. 65, p. 579–616.

Takens, F., 1981, Detecting strange attractor in turbulence: Lecture Notes in Mathematics, v. 898, p. 366–381.

Weigend, A. S., and Gershenfeld, N. A., eds., 1994, Time series prediction: Addison-Wesley, Reading, Massachusetts, 643 p.

Idso, S. B., Carbon-dioxide-induced global warming: A skeptic's view of potential climate change, 2001, *in* L. C. Gerhard, W. E. Harrison, and B. M. Hanson, eds., Geological perspectives of global climate change, p. 317–336.

17 | Carbon-Dioxide-Induced Global Warming: A Skeptic's View of Potential Climate Change[1]

Sherwood B. Idso

U.S. Water Conservation Laboratory
Phoenix, Arizona, U.S.A.

ABSTRACT

Over the course of the past two decades, I have analyzed a number of natural phenomena that reveal how earth's near-surface air temperature responds to surface radiative perturbations. These studies all suggest that a 300 to 600 parts per million (ppm) doubling of the atmosphere's carbon dioxide (CO_2) concentration could raise the planet's mean surface air temperature by only about 0.4°C. Even this modicum of warming may never be realized, however, for it could be negated by a number of planetary cooling forces that are intensified by warmer temperatures and by the strengthening of biological processes that are enhanced by the same rise in atmospheric CO_2 concentration that drives the warming. Several of these cooling forces have individually been estimated to be of equivalent magnitude, but of opposite sign, to the typically predicted greenhouse effect of a doubling of the air's CO_2 content, which suggests to me that little net temperature change will ultimately result from the ongoing buildup of CO_2 in earth's atmosphere. Consequently, I am skeptical of the predictions of significant CO_2-induced global warming that are being made by state-of-the-art climate models and believe that much more work on a wide variety of research fronts will be required to properly resolve the issue.

INTRODUCTION

Twenty years ago I was heavily involved in the measurement of solar and thermal radiation fluxes at the surface of the earth, concentrating on their responses to changes in atmospheric composition

---

[1]Reprinted, with permission, from *Climate Research*, 1998, v. 10, pp. 69–82.

that were produced by unique local weather phenomena. About the same time I also became interested in carbon-dioxide-induced global warming, and I decided to see if I could learn something about the subject from the natural experiments provided by the special meteorological situations I was investigating.

My idea was to determine the magnitudes of radiative perturbations created by various climatic events and observe how the near-surface air temperature responded to the resultant changes in the surface radiation balance. From this information I sought to develop a surface air temperature sensitivity factor, defined as the rise in surface air temperature divided by the increase in surface-absorbed radiation that prompted the temperature rise. Then, by multiplying this factor by the increase in downwelling thermal radiation expected to be received at the surface of the earth as a result of a doubling of the atmosphere's CO_2 concentration, I hoped to obtain a rough estimate of the likely magnitude of future CO_2-induced global warming.

DECIPHERING NATURE'S "EXPERIMENTS"

Natural Experiment 1

The first of the unique meteorological situations I investigated was the change in atmospheric water vapor that typically occurs at Phoenix, Arizona, U.S.A., with the advent of the summer monsoon (Hales, 1972, 1974; Douglas et al., 1993). During the initial phases of the establishment of this more humid weather regime, atmospheric vapor pressure exhibits large day-to-day fluctuations, creating significant variations in the solar and thermal radiation fluxes received at the earth's surface. Consequently, for all cloudless days of the 45-day period centered on the summer monsoon's mean date of arrival, I plotted the prior 30 years' daily maximum and minimum air temperatures as functions of surface vapor pressure to see to what degree these specific temperatures were influenced by fluctuations in atmospheric water vapor.

In the case of the maximum air temperature, which typically occurs in the middle of the afternoon, there was no dependence on the surface vapor pressure, due to the opposing effects of atmospheric water vapor on the fluxes of solar and thermal radiation received at the earth's surface at that time of day. But in the case of the minimum air temperature, which typically occurs just prior to sunrise after several hours' absence of solar radiation, there was a strong relationship, since the effect of water vapor on the flux of solar radiation is absent at that time and is thus unable to mask the effect of atmospheric moisture on the downwelling flux of thermal radiation.

Using an equation I had developed previously (Idso, 1981a), which specifies the downward-directed flux of thermal radiation at the earth's surface as a function of surface vapor pressure (e_0) and air temperature (T_0), I calculated that in going from the low-end values of this relationship (e_0 = 0.4 kilopascal [kPa], T_0 = 18.3°C) to its high-end values (e_0 = 2.0 kPa, T_0 = 29.4°C), the flux of thermal radiation to the earth's surface would rise by approximately 64.1 watts per square meter (Wm^{-2}). Consequently, the surface air-temperature sensitivity factor I obtained from this natural experiment was (29.4°C − 18.3°C)/64.1 Wm^{-2}, or 0.173°C/(Wm^{-2}) (Idso, 1982).

At this stage, however, I had no reason to believe that this result applied anywhere beyond the bounds of Phoenix, Arizona, or that it was a true measure of any portion of the planet's climatic sensitivity. What bothered me most with respect to this latter point was the large size of the experiment's radiative perturbation and the short time period over which the temperature response was determined. I thought it unlikely that these experimental features would yield a surface air temperature sensitivity factor equivalent to that produced by a much smaller radiative perturbation introduced over a considerably longer time span, such as is occurring in response to the ongoing rise in the air's CO_2 content. Hence, I initiated a search for a set of meteorological circumstances that would more closely approximate this latter situation.

Natural Experiment 2

I found what I sought in the naturally occurring vertical redistribution of dust that occurs at Phoenix, Arizona, each year between summer and winter (Idso, S. B. and Kangieser, 1970). With respect to this phenomenon, I had previously demonstrated that the restriction of airborne dust to a much shallower depth of atmosphere during winter does not alter the transmittance of the atmosphere for the total flux of solar radiation, but that it increases the atmosphere's downwelling flux of thermal radiation at the earth's surface by 13.9 Wm^{-2} (Idso, S. B., 1981b). In another study, a colleague and I had also determined that winter surface air temperatures were 2.4°C warmer than what would be expected for the vertical distribution of dust that exists in summer (Idso, S. B., and Brazel, 1978). Assuming that this temperature increase was a consequence of the extra thermal radiation produced by the seasonal redistribution of atmospheric dust, I divided the latter of these two numbers by the former to obtain a surface air-temperature sensitivity factor that was identical to the result derived from my first natural experiment: 0.173°C/(Wm^{-2}).

The perfect agreement of these two results was totally unexpected. The radiative perturbation of the second experiment was significantly less than that of the first—13.9 versus 64.1 Wm^{-2}—and it unfolded over a time span of months, as opposed to a period of days. Both of these differences, I had thought, should have increased the opportunity for slower-acting secondary and tertiary feedback processes (which I had initially assumed to be predominantly positive) to manifest themselves in the second experiment, leading to a possibly very different (and presumably larger) result from that of the first experiment. Such was not the case, however, and in analyzing the situation in retrospect, I could think of no compelling reason why the net effect of all of the feedback processes that play significant roles in earth's climatic system should be strongly positive. Indeed, it seemed more logical to me that the opposite would be true; for if strong positive feedbacks existed, the earth would likely exhibit a radically unstable climate, significantly different from what has characterized the planet over the eons (Walker, 1986).

Be that as it may, I felt that I needed more real-world evidence for the value of the surface air-temperature sensitivity factor I had derived from my first and second natural experiments, before I made too much of what I had found. For one thing, the perfect agreement of the two results could well have been coincidental; and they both were derived from data pertaining to but a single place on the planet. Consequently, I began to look for a third set of meteorological circumstances that would broaden my geographical database and allow me to make yet another evaluation of surface air-temperature response to radiative forcing.

Natural Experiment 3

The next phenomenon to attract my attention was the annual cycle of surface air temperature that is caused by the annual cycle of solar radiation absorption at the earth's surface. To derive what I sought from this set of circumstances, I obtained values for the annual range of solar radiation reception at 81 locations within the United States (Bennett, 1975), multiplied them by 1 minus the mean global albedo (Ellis et al., 1978), and plotted against the resulting values of this parameter the annual air temperature ranges of the 81 locations.

The results fell into two distinct groups: one for the interior of the country and one for the extreme West Coast, which is greatly influenced by weather systems originating over the Pacific Ocean. Each of these data sets was thus treated separately; and the linear regressions that were run on them were forced to pass through the origin, since for no cycle of solar radiation absorption there should be no cycle of air temperature. The slopes of the two regressions then yielded surface air-temperature sensitivity factors of $0.171°C/(Wm^{-2})$ for the interior of the United States—which result was essentially identical to the results of my earlier natural experiments at Phoenix—and $0.087°C/(Wm^{-2})$ for the extreme West Coast.

Since this latter sensitivity factor was only half as great as the sensitivity factor of the rest of the country, it was clear that it was largely determined by the surface energy balance of the adjacent sea, the effects of which are advected inland and gradually diminish with distance from the coast. It was also clear, however, that this predominantly ocean-based sensitivity factor had to be somewhat elevated, due to the influence of the land over which it was determined. Consequently, assuming that it represented an upper limit for the water surfaces of the globe, which cover approximately 70% of the earth, and assuming that the result for the interior of the United States applied to all of earth's land surfaces, I combined the two results to obtain an upper-limiting global average of $0.113°C/(Wm^{-2})$ for the surface air-temperature sensitivity factor of the entire planet (Idso, S. B., 1982), recognizing, of course, that this result was probably still no more than a rough approximation of reality, for there were many climatically significant real-world phenomena not explicitly included in the natural experiments I had analyzed.

Initial Implications

Although there was no way to prove that the results I had obtained were applicable to long-term climatic change, their consistency gave me enough confidence to pursue their implications with respect to the ongoing rise in the air's CO_2 content. Consequently, I derived a value of $2.28 \ Wm^{-2}$ for the radiative perturbation likely to be expected from a 300 to 600 ppm doubling of the atmospheric CO_2 concentration, and multiplying this result by the upper-limiting surface air-temperature sensitivity factor I had derived for the globe as a whole, I calculated that the earth could warm by no more than $0.26°C$ for a doubling of the air's CO_2 content (Idso, 1980). Others, however, have calculated by more sophisticated methods that the radiative perturbation for this change in CO_2 is probably closer to $4 \ Wm^{-2}$ (Smagorinsky et al., 1982; Nierenberg et al., 1983; Shine et al., 1990), and this presumably more realistic result suggests that the earth could warm by no more than $0.45°C$ for a 300 to 600 ppm doubling of the atmosphere's CO_2 concentration.

Soon after the publication of these findings, and primarily on the basis of the concerns expressed above, my approaches and conclusions were both questioned (Leovy, 1980; Schneider et al., 1980), as well they should have been for there was no apparent reason why the surface air-temperature sensitivity factor I had derived should necessarily describe the long-term climatic response of the earth to the impetus for warming produced by the atmosphere's rising CO_2 concentration. Hence, I began to look for a situation where the surface air temperature of the entire planet had achieved unquestioned equilibrium in response to a change in surface-absorbed radiant energy in a natural experiment that implicitly included all processes that combine to determine the climatic state of the earth, and I ultimately identified two such situations.

Natural Experiment 4

The first of these two global-equilibrium natural experiments consisted of simply identifying the mean global warming effect of the entire atmosphere and dividing it by the mean flux of thermal radiation received at the surface of the earth that originates with the atmosphere and which would be nonexistent in its absence (Idso, S. B., 1984). Calculation of both of these numbers is straightforward (Idso, S. B., 1980, 1982), and there is no controversy surrounding either of the results: a total greenhouse warming of approximately $33.6°C$ sustained by a thermal radiative flux of approximately 348

Wm^{-2}. Hence, the equilibrium surface air-temperature sensitivity factor of the entire planet, as defined by this fourth approach to the problem, is $0.097°C/(Wm^{-2})$.

Natural Experiment 5

My second global-equilibrium experiment made use of the annually averaged equator-to-pole air temperature gradient that is sustained by the annually averaged equator-to-pole gradient of total surface-absorbed radiant energy (Idso, S. B., 1984). Mean surface air temperatures (Warren and Schneider, 1979) and water-vapor pressures (Haurwitz and Austin, 1944) for this situation were obtained for each 5° latitude increment stretching from 90°N to 90°S. From these data, I calculated values of clear-sky atmospheric thermal radiation (Idso, S. B., 1981a) incident upon the surface of the earth at the midpoints of each of the specified latitude belts. Then, from information about the latitudinal distribution of cloud cover (Sellers, 1965) and the ways in which clouds modify the clear-sky flux of downwelling thermal radiation at the earth's surface (Kimball et al., 1982), I appropriately modified the clear-sky thermal radiation fluxes and averaged the results over both hemispheres. Similarly averaged fluxes of surface-absorbed solar radiation (Sellers, 1965) were then added to the thermal radiation results to produce 18 annually averaged total surface-absorbed radiant energy fluxes stretching from the equator to 90°NS, against which I plotted corresponding average values of surface air temperature.

This operation produced two distinct linear relationships—one of slope $0.196°C/(Wm^{-2})$, which extended from 90°NS to approximately 63°NS, and one of slope $0.090°C/(Wm^{-2})$, which extended from 63° NS to the equator. Hence, I weighted the two results according to the percentages of earth's surface area to which they pertained (12% and 88%, respectively) and combined them to obtain a mean global value of $0.103°C/(Wm^{-2})$. Averaging this result with the preceding analogous result of $0.097°C/(Wm^{-2})$ then produced a value of $0.100°C/(Wm^{-2})$ for what I truly believe is earth's long-term climatic sensitivity to radiative perturbations of the surface energy balance. And this value is just slightly less than the upper-limiting value I had obtained from my first three experiments, as indeed it should be.

Intermediate Implications

In light of the exceptional agreement of the two preceding results, as well as their appropriate relationship to the results of my first three natural experiments (which relationship may still be argued to be largely coincidental, however), I began to think that my observationally derived value of $0.1°C/(Wm^{-2})$ may be much more than a rough approximation of earth's climatic sensitivity to surface radiative forcing. Indeed, I began to think that it may be a very good representation of it, and to assess its implications I multiplied it by $4\ Wm^{-2}$, which is the radiative perturbation likely to be produced by a 300- to 600-ppm doubling of the atmosphere's CO_2 concentration, to obtain a mean global temperature increase of 0.4°C (Idso, S. B., 1984), which is but a tenth to a third of the warming that has historically been predicted for this scenario by most general circulation models of the atmosphere (Kacholia and Reck, 1997).

To resolve the discrepancy between the predictions of these two approaches to the CO_2-climate problem, major changes need to be made to one or both of them. If the discrepancy were to be totally resolved in the realm of my natural experiments, for example, it is clear that my estimate of earth's climatic sensitivity would have to be increased by a factor of 3.3 to 10. However, I can conceive of no way that an adjustment of this magnitude could be made to the analyses of my two global-equilibrium natural experiments, which I believe are the best-founded analyses of the five situations discussed thus far. In addition, the results of still other natural experiments I have identified and analyzed suggest—at least to me—that the value I have derived for earth's climatic sensitivity is indeed correct.

Natural Experiment 6

Consider what we can learn from our nearest planetary neighbors, Mars and Venus. In spite of the tremendous differences that exist between them, and between them and the earth, their observed surface temperatures have been said to confirm "the existence, nature, and magnitude of the greenhouse effect" by two select committees of the U.S. National Research Council (Smagorinsky et al., 1982; Nierenberg et al., 1983), which conclusion appears also to be accepted by the Intergovernmental Panel on Climate Change (Trenberth et al., 1996). So what can these planets tell us about CO_2-induced warming on earth?

Venus exhibits a greenhouse warming of approximately 500°C (Oyama et al., 1979; Pollack et al., 1980) that is produced by a 93-bar atmosphere of approximately 96% CO_2 (Kasting et al., 1988). Mars exhibits a greenhouse warming of 5° to 6°C (Pollack, 1979; Kasting et al., 1988) that is produced by an almost pure CO_2 atmosphere that fluctuates over the Martian year between 0.007 and 0.010 bar (McKay, 1983). Plotting the two points defined by these data on a log-log coordinate system of CO_2-induced global warming versus atmospheric CO_2 partial pressure and connecting them by a straight line produces a relationship that, when extrapolated to CO_2 partial pressures characteristic of present-day earth, once again yields a mean global warming of only 0.4°C for a 300- to 600-ppm doubling of the air's CO_2 content (Idso, S. B., 1988a).

Natural Experiment 7

The same result may also be obtained from the standard resolution of the paradox of the faint early sun (Sagan and Mullen, 1972; Owen et al., 1979; Kasting, 1997), which dilemma (Longdoz and Francois, 1997; Sagan and Chyba, 1997) is most often posed by the following question. How could earth have supported life nearly 4 billion years ago when, according to well-established concepts of stellar evolution (Schwarzchild et al., 1957; Ezer and Cameron, 1965; Bahcall and Shaviv, 1968; Iben, 1969), the luminosity of the sun was probably 20% to 30% less at that time than it is now (Newman and Rood, 1977; Gough, 1981), so that, all else being equal, nearly all of earth's water should have been frozen and unavailable for sustaining life?

Most of the people who have studied the problem feel that the answer to this question resides primarily in the large greenhouse effect of earth's early atmosphere—which is believed to have contained much more CO_2 than it does today (Hart, 1978; Wigley and Brimblecombe, 1981; Holland, 1984; Walker, 1985)—with a secondary contribution coming from the near-global extent of the early ocean (Henderson-Sellers and Cogley, 1982; Henderson-Sellers and Henderson-Sellers, 1988; Jenkins, 1995). Consequently, based on the standard assumption of a 25% reduction in solar luminosity 4.5 billion years ago, I calculated the strength of the greenhouse effect required to compensate for the effects of reduced solar luminosity at half-billion-year intervals from 3.5 billion years ago—when we are confident of the widespread existence of life (Schopf and Barghoorn, 1967; Schopf, 1978; Schidlowski, 1988; Mojzsis et al., 1996; Eiler et al., 1997)—to the present, and I plotted the results as a function of atmospheric CO_2 concentration derived from a widely accepted atmospheric CO_2 history for that period of time (Lovelock and Whitfield, 1982). The relationship derived from that exercise (Idso, S. B., 1988a) is nearly identical to the one derived from the preceding comparative planetary climatology study, once again implying a mean global warming of only 0.4°C for a 300- to 600-ppm doubling of the atmosphere's CO_2 concentration. And it is the essentially perfect agreement of the results of these last four global equilibrium natural experiments that leads me to believe that I have indeed obtained the proper answer to the question of potential CO_2-induced climate change. There is, nevertheless, yet another natural phenomenon that warrants consideration within this context.

Natural Experiment 8

A final set of empirical evidence that may be brought to bear on the issue of CO_2-induced climate change pertains to the greenhouse effect of water vapor over the tropical oceans (Raval and Ramanathan, 1989; Ramanathan and Collins, 1991; Lubin, 1994). This phenomenon has recently been quantified by Valero et al. (1997), who used airborne radiometric measurements and sea-surface temperature data to evaluate its magnitude over the equatorial Pacific. Their direct measurements reveal that a 14.0 Wm^{-2} increase in downward-directed thermal radiation at the surface of the sea increases surface water temperatures by 1.0°C, and dividing the latter of these two numbers by the former yields a surface water-temperature sensitivity factor of 0.071°C/(Wm^{-2}), which would imply a similar surface air-temperature sensitivity factor at equilibrium. By comparison, if I equate my best estimate of the surface air-temperature sensitivity factor of the world as a whole [0.100°C/(Wm^{-2})] with the sum of the appropriately weighted land and water-surface factors [0.3 × 0.172°C/(Wm^{-2}) + 0.7 × W, where W is the surface air-temperature sensitivity factor over the open ocean], I obtain a value of 0.069°C/(Wm^{-2}) for the ocean-based component of the whole-earth surface air-temperature sensitivity factor, in close agreement with the results of Valero et al.

Of course, there is no compelling reason why the results of these two evaluations should necessarily agree so well, for different portions of the planet can clearly exhibit different surface air temperature sensitivity factors, as I have shown herein to be the case for land versus water (Natural Experiment 3) and for high latitudes versus low latitudes (Natural Experiment 5). Indeed, it is only when the planet is treated in its entirety that one should expect the same result, such as is obtained from Natural Experiments 4, 6, and 7. Nevertheless, the good agreement between the results of this last natural experiment and all of my prior evaluations is still gratifying, although a significant disagreement would not have been all that discouraging.

RISING CO_2 AND THE GLOBAL WARMING OF THE PAST CENTURY

As demonstrated by the results of the several natural experiments described above, a large body of real-world evidence points to the likelihood of a future CO_2-induced global warming of but a tenth to a third of what is currently predicted by theoretical numerical models of the earth-ocean-atmosphere system. However, the observed global warming of the past century, which has occurred in concert with a 75-ppm rise in the air's CO_2 content, has already exceeded the 0.4°C increase in temperature that my analyses suggest would require an atmospheric CO_2 increase of fully 300 ppm. It is only natural to wonder if this relatively large warming of the last hundred years was produced by the relatively small concurrent rise in the air's CO_2 content. This question is of crucial importance, for if the global warming of the past century was wholly the result of the concurrent rise in atmospheric CO_2, it would imply that the primary conclusion derived from my natural experiments is incorrect.

Although the question cannot be unequivocally resolved at the present time, it is possible that the warming of the earth over the last hundred years may well have been wholly unrelated to the concurrent rise in atmospheric CO_2, for the observed temperature increase may have been produced by changes in a number of other climatically important factors, such as the energy output of the sun, for example, which is looking more and more like a major determinant of earth's climate each year (Baliunas and Jastrow, 1990; Foukal and Lean, 1990; Friis-Christensen and Lassen, 1991;

Lockwood et al., 1992; Scuderi, 1993; Charvatova and Strestik, 1995; Lean et al., 1995; Baliunas and Soon, 1996, 1998; Soon et al., 1996; Hoyt and Schatten, 1997). Indeed, it is even possible that the global warming of the past century may have been nothing more than a random climatic fluctuation.

That some alternative explanation of the observed warming is, in fact, quite plausible is readily evident when the temperature increase of the past century is viewed from the broader perspective of the past millennium. From this improved vantage point, the warming of the last hundred years is seen to be basically a recovery (Idso, S. B., 1988b; Reid 1993) from the global chill of the Little Ice Age, a several-hundred-year period of significantly cooler temperatures than those of the present that persisted until the end of the nineteenth century (Grove, 1988; Whyte, 1995). And as ice-core data give no indication of any drop in atmospheric CO_2 over the period of the Little Ice Age's induction (Friedli et al., 1984, 1986), something other than CO_2 had to have initiated it, implying that the inverse of that something—or even something else (or nothing at all, in the case of a random climatic fluctuation)—is likely to have been the cause of its demise.

But what if temperatures were to rise even higher in the future? Here, again, the long historical perspective proves invaluable, for it reveals that the Little Ice Age was preceded by a several-centuries-long period of significantly warmer temperatures than those of the present (Le Roy Ladurie, 1971; Lamb, 1977, 1984, 1988; Keigwin, 1996). And while the earth was traversing the entire temperature range from the maximum warmth of this Little Climatic Optimum (Dean, 1994; Petersen, 1994; Serre-Bachet, 1994; Villalba, 1994) to the coolest point of the Little Ice Age, the CO_2 content of the atmosphere, as inferred from ice-core data, varied not at all (Idso, S. B., 1988b). Consequently, the earth can clearly warm even more than it has already warmed over the last century without any change in atmospheric CO_2, suggesting that even continued global warming—which appears to have peaked (Hurrell and Trenberth, 1997; Spencer, 1997)—would imply very little (and possibly nothing at all) about the potential for future CO_2-induced climatic change.

COOLING THE GLOBAL GREENHOUSE

Although the evidence I have presented suggests that a doubling of the air's CO_2 content could raise earth's mean surface air temperature by only about 0.4°C, there are a number of reasons to question whether even this minor warming will ever occur. There is, for example, a variety of ways by which rising temperatures may strengthen the cooling properties of clouds and thereby retard global warming. In addition, several biological processes that are enhanced by the aerial fertilization effect of atmospheric CO_2 enrichment can directly intensify these climate-cooling forces.

With respect to the first of these subjects, it has long been recognized that the presence of clouds has a strong cooling effect on earth's climate (Barkstrom, 1984; ERBE Science Team, 1986; Nullet, 1987; Nullet and Ekern, 1988; Ramanathan et al., 1989). In fact, it has been calculated that a mere 1% increase in planetary albedo would be sufficient to totally counter the entire greenhouse warming typically predicted to result from a doubling of the atmosphere's CO_2 concentration (Ramanathan, 1988). And as the typically predicted warming may be three to 10 times larger than what could actually occur, according to my interpretation of the real-world evidence I have presented herein, it is possible that but a tenth to a third of a 1% increase in planetary albedo may be sufficient to accomplish this feat.

Within this context, it has been shown that a 10% increase in the amount of low-level clouds could completely cancel the typically predicted warming of a doubling of the air's CO_2 content by reflecting more solar radiation back to space (Webster and Stephens, 1984). In addition, Ramanathan and Collins (1991), by the use of their own natural experiments, have shown how the warming-induced production of high-level clouds over the equatorial oceans totally nullifies the greenhouse effect of water vapor there, with high clouds dramatically increasing from close to 0% coverage at sea surface temperatures of 26°C to fully 30% coverage at 29°C (Kiehl, 1994). And in describing the implications of this strong negative feedback mechanism, Ramanathan and Collins (1991) state that "it would take more than an order-of-magnitude increase in atmospheric CO_2 to increase the maximum sea surface temperature by a few degrees," which they acknowledge is a considerable departure from the predictions of most general circulation models of the atmosphere.

In addition to increasing their coverage of the planet, as they appear to do in response to an increase in temperature (Henderson-Sellers, 1986a, b; McGuffie and Henderson-Sellers, 1988; Dai et al., 1997), clouds in a warmer world would also have greater liquid-water contents than they do now (Paltridge, 1980; Charlock, 1981, 1982; Roeckner, 1988). And because the heat-conserving greenhouse properties of low- to midlevel clouds are already close to the maximum they can attain (Betts and Harshvardhan, 1987), although their reflectances for solar radiation may yet rise substantially (Roeckner et al., 1987), an increase in cloud liquid-water content would tend to counteract an initial impetus for warming, even in the absence of an increase in cloud cover. By incorporating just this one negative feedback mechanism into a radiative-convective climate model, for example, the warming predicted to result from a doubling of the air's CO_2 content has been shown to fall by fully 50% (Somerville and Remer, 1984). However, a 20% to 25% increase in cloud liquid water path has been shown to totally negate the typically predicted warming of a doubling of the air's CO_2 content in a three-dimensional general circulation model of the atmosphere (Slingo, 1990).

Another negative feedback mechanism involving clouds, which is estimated to be of the same strength as the typically predicted greenhouse effect of CO_2 (Lovelock, 1988; Turner et al., 1996), has been described by Charlson et al. (1987). They suggest that the productivity of oceanic phytoplankton will increase in response to an initial impetus for warming, with the result that one of the ultimate by-products of the enhanced algal metabolism—dimethyl sulfide, or DMS—will be produced in more copious quantities. Diffusing into the atmosphere, where it is oxidized and converted into particles that function as cloud condensation nuclei, this augmented flux of DMS is projected to create additional and/or higher-albedo clouds, which will thus reflect more solar radiation back to space, thereby cooling the earth and countering the initial impetus for warming (Shaw, 1983, 1987).

There is much evidence—700 papers in the past 10 years (Andreae and Crutzen, 1997)—to support the validity of each link in this conceptual chain of events. First, there is the demonstrated propensity for oceanic phytoplankton to increase their productivity in response to an increase in temperature (Eppley, 1972; Goldman and Carpenter, 1974; Rhea and Gotham, 1981), which is clearly evident in latitudinal distributions of marine productivity (Platt and Sathyendranath, 1988; Sakshaug, 1988). Second, as oceanic phytoplankton photosynthesize, they are known to produce a substance called dimethylsulfonio propionate (Vairavamurthy et al., 1985), which disperses throughout the surface waters of the oceans when the phytoplankton either die or are eaten by zooplankton (Dacey and Wakeham, 1988; Nguyen et al., 1988) and which decomposes to produce DMS (Turner et al., 1988). Third, it has been shown that part of the DMS thus released to the earth's oceans diffuses into the atmosphere, where it is oxidized and converted into sulfuric and methanesulfonic acid particles (Bonsang et al., 1980; Hatakeyama et al., 1982; Saltzman et al., 1983; Andreae et al., 1988; Kreidenweis and Seinfeld, 1988) that function as cloud condensation nuclei or CCN (Saxena, 1983; Bates et al., 1987). And more CCN can clearly stimulate the production of new clouds and dramatically increase the albedos of preexistent clouds by decreasing the sizes of the clouds' component droplets (Twomey and Warner, 1967; Warner and Twomey, 1967; Hudson, 1983; Coakley et al., 1987; Charlson and Bates, 1988; Durkee, 1988), which phenomenon tends to cool the planet by enabling clouds to reflect more solar radiation back to space (Idso, S. B., 1992; Saxena et al.,

1996). In fact, it has been calculated that a 15% to 20% reduction in the mean droplet radius of earth's boundary-layer clouds would produce a cooling influence that could completely cancel the typically predicted warming influence of a doubling of the air's CO_2 content (Slingo, 1990).

Another way in which the enhanced production of CCN may retard warming via a decrease in cloud droplet size is by reducing drizzle from low-level marine clouds, which lengthens their life-span and thereby expands their coverage of the planet (Albrecht, 1988). In addition, because drizzle from stratus clouds tends to stabilize the atmospheric boundary layer by cooling the sub-cloud layer as a portion of the drizzle evaporates (Brost et al., 1982; Nicholls, 1984), a CCN-induced reduction in drizzle tends to weaken the stable stratification of the boundary layer, enhancing the transport of water vapor from ocean to cloud. As a result, clouds containing extra CCN tend to per-sist longer and perform their cooling function for a longer period of time.

The greater numbers of CCN needed to enhance these several cooling phenomena are also produced by biological processes on land (Went, 1966; Duce et al., 1983; Roosen and Angione, 1984; Meszaros, 1988), and in the terrestrial environment, the volatilization of reduced sulfur gases from soils is particularly important in this regard (Idso, S. B., 1990). Here, too, one of the ways in which the ultimate cooling effect is set in motion is by an initial impetus for warming. It has been re-ported, for example, that soil DMS emissions rise by a factor of 2 for each 5°C increase in tempera-ture between 10° and 25°C (Staubes et al., 1989). As a result of the enhanced microbial activity produced by increasing warmth (Hill et al., 1978; MacTaggart et al., 1987), there is a 25-fold in-crease in soil-to-air sulfur flux between 25°N and the equator (Adams et al., 1981). Of even greater importance, however, is the fact that atmospheric CO_2 enrichment alone can initiate the chain of events that leads to cooling.

Consider the fact, impressively supported by literally hundreds of laboratory and field experi-ments (Lemon, 1983; Cure and Acock, 1986; Mortensen, 1987; Lawlor and Mitchell, 1991; Drake, 1992; Poorter, 1993; Idso, K. E. and Idso, S. B., 1994; Strain and Cure, 1994), that nearly all plants are better adapted to higher atmospheric CO_2 concentrations than those of the present, and that the pro-ductivity of most herbaceous plants rises by 30% to 50% for a 300- to 600-ppm doubling of the air's CO_2 content (Kimball, 1983; Idso, K. E., 1992), while the growth of many woody plants rises even more dramatically (Idso, S. B., and Kimball, 1993; Ceulemans and Mousseau, 1994; Wullschleger et al., 1995, 1997). Because of this stimulatory effect of elevated CO_2 on plant growth and development, the productivity of the biosphere has been rising hand in hand with the recent historical rise in the air's CO_2 content (Idso, S. B., 1995), as is evident in (1) the ever-increasing amplitude of the seasonal cycle of the air's CO_2 concentration (Pearman and Hyson, 1981; Cleveland et al., 1983; Bacastow et al., 1985; Keeling et al., 1985, 1995, 1996; Myneni et al., 1997), (2) the upward trends in a number of long tree-ring records that mirror the progression of the industrial revolution (LaMarche et al., 1984; Graybill and Idso, 1993), and (3) the accelerating growth rates of numerous forests on nearly every continent of the globe over the past several decades (Kauppi et al., 1992; Phillips and Gentry, 1994; Pimm and Sugden, 1994; Idso, 1995).

In consequence of this CO_2-induced increase in plant productivity, more organic matter is re-turned to the soil (Leavitt et al., 1994; Jongen et al., 1995; Batjes and Sombroek, 1997), where it stim-ulates biological activity (Curtis et al., 1990; Zak et al., 1993; O'Neill, 1994; Rogers et al., 1994; Godbold and Berntson, 1997; Ineichen et al., 1997; Ringelberg et al., 1997) that results in the en-hanced emission of various sulfur gases to the atmosphere (Staubes et al., 1989). Thereupon, more CCN are created (as described above), which tends to cool the planet by altering cloud properties in ways that result in the reflection of more solar radiation back to space. In addition, many non-sulfur biogenic materials of the terrestrial environment play major roles as both water- and ice-nucleating aerosols (Schnell and Vali, 1976; Vali et al., 1976; Bigg, 1990; Novakov and Penner, 1993;

Saxena et al., 1995; Baker, 1997); and the airborne presence of these materials should also be enhanced by atmospheric CO_2 enrichment.

That analogous CO_2-induced cooling processes operate at sea is implied by the facts that (1) atmospheric CO_2 enrichment stimulates the growth of both macro- (Titus et al., 1990; Sand-Jensen et al., 1992; Titus, 1992; Madsen, 1993; Madsen and Sand-Jensen, 1994) and micro- (Raven, 1991, 1993; Riebesell, 1993; Shapiro, 1997) aquatic plants, and (2) experimental iron-induced (Coale et al., 1996) increases (acting as surrogates for CO_2-induced increases) in the productivity of oceanic phytoplankton in high-nitrate low-chlorophyll waters of the equatorial Pacific (Behrenfeld et al., 1996) have been observed to greatly increase surface-water DMS concentrations (Turner et al., 1996). There is also evidence to suggest that a significant fraction of the ice-forming nuclei of maritime origin are composed of organic matter (Rosinski et al., 1986, 1987), and the distribution of these nuclei over the oceans (Bigg, 1973) has been shown to be strongly correlated with surface patterns of biological productivity (Bigg, 1996; Szyrmer and Zawadzki, 1997). Hence, there may well exist an entire suite of powerful planetary cooling forces that can respond directly to the rising carbon-dioxide content of the atmosphere over both land and sea. And in view of the relative weakness of the CO_2 greenhouse effect at current atmospheric CO_2 partial pressures, as revealed by the natural experiments I have described herein—a likely warming of only 0.4°C for a 300- to 600-ppm doubling of the air's CO_2 content—these CO_2-induced cooling forces could potentially negate a large portion (or even all) of the primary warming effect of a rise in atmospheric CO_2, leading to little net change in mean global air temperature.

SUMMARY AND CONCLUSIONS

There is no controversy surrounding the claim that atmospheric CO_2 concentrations are on the rise; direct measurements demonstrate that fact. The basic concept of the greenhouse effect is also not in question; rising CO_2 concentrations, in and of themselves, clearly enhance the thermal blanketing properties of the atmosphere. What is debatable, however, is the magnitude of any warming that might result from a rise in the air's CO_2 concentration. Although admittedly incomplete and highly approximate general circulation models of the atmosphere predict that a 300- to 600-ppm doubling of the air's CO_2 content will raise mean global air temperature a few degrees Celsius, natural experiments based on real-world observations suggest that a global warming of no more than a few tenths of a degree could result from such a CO_2 increase. Which conclusion is correct?

Several complexities of earth's climate system make accurate predictions of global climate change very difficult for general circulation models of the atmosphere and probably account for the deviations of their predictions from those of the natural experiments I have described herein. First, a number of planetary cooling forces are intensified by increases in temperature and therefore tend to dampen any impetus for warming. Many of these phenomena are only now beginning to be fully appreciated, much less adequately incorporated into the models. Second, nearly all of these cooling forces can be amplified by increases in biological processes that are directly enhanced by the aerial fertilization effect of atmospheric CO_2 enrichment, and most of these phenomena are also not included in general circulation model studies of potential CO_2-induced climate change. Third, many of these cooling forces have individually been estimated to have the capacity to totally thwart the typically predicted (and likely overestimated) warming of a doubling of the atmosphere's CO_2 concentration. Fourth, real-world measurements have revealed that contemporary climate models have long significantly underestimated the cooling power of clouds (Cess et al., 1995; Pilewskie and Valero, 1995; Ramanathan et al., 1995; Heymsfield and McFarquhar, 1996), even when demonstrating their ability to completely negate the likely overestimated global warming typically predicted to result from a doubling of the air's CO_2 content.

In light of these observations, it is my belief that it will still be a very long time before any general circulation model of the atmosphere will be able to accurately determine the ultimate consequences of the many opposing climatic forces that are both directly and indirectly affected by the rising CO_2 content of earth's atmosphere. Consequently, although many equally sincere and thoughtful scientists may feel otherwise, I believe that these models do not yet constitute an adequate basis for developing rational real-world policies related to potential climate change.

REFERENCES CITED

Adams, D. F., Farwell, S. O., Robinson, E., Pack, M. R., Bamesberger, W. L., 1981. Biogenic sulfur source strengths. Environmental Science and Technology 15:1493–1498.

Albrecht, B. A., 1988. Modulation of boundary layer cloudiness by precipitation processes. In Proceedings: Symposium on the role of clouds in atmospheric chemistry and global climate. American Meteorological Society, Boston, p. 9–13.

Andreae, M. O., Crutzen, P. J., 1997. Atmospheric aerosols: Biogeochemical sources and role in atmospheric chemistry. Science 276:1052–1058.

Andreae, M. O., Berresheim, H., Andreae, T. W., Kritz, M. A., Bates, T. S., Merril, J. T., 1988. Vertical distribution of dimethylsulfide, sulfur dioxide, aerosol ions and radon over the northeast Pacific Ocean. Journal of Atmospheric Chemistry 6:149–173.

Bacastow, R. B., Keeling, C. D., Whorf, T. P., 1985. Seasonal amplitude increase in atmospheric CO_2 concentration at Mauna Loa, Hawaii, 1959–1982. Journal of Geophysical Research 90: 10540–10592.

Bahcall, J. N., Shaviv, G., 1968. Solar models and neutrino fluxes. The Astrophysical Journal 153:113–126.

Baker, M. B., 1997. Cloud microphysics and climate. Science 276:1072–1078.

Baliunas, S., Jastrow, R., 1990. Evidence for long-term brightness changes of solar-type stars. Nature 348:520–522.

Baliunas, S. L., and Soon, W. H., 1996. The sun-climate connection. Sky and Telescope 92(6):38–41.

Baliunas, S. L., and Soon, W. H., 1998. An assessment of the sun-climate relation on time scales of decades to centuries: The possibility of total irradiance variations. In Pap, J. M., Frohlich, C., Ulrich, R. (eds.). Proceedings of the SOLERS22 1996 Workshop. Kluwer Academic Publishers, Dordrecht, p. 401–411.

Barkstrom, B. R., 1984. The Earth Radiation Budget Experiment (ERBE). Bulletin of the American Meteorologist Society 65:1170–1185.

Bates, T. S., Charlson, R. J., Gammon, R. H., 1987. Evidence for the climatic role of marine biogenic sulphur. Nature 329:319–321.

Batjes, N. H., Sombroek, W. G., 1997. Possibilities for carbon sequestration in tropical and subtropical soils. Global Change Biology 3:161–173.

Behrenfeld, M. J., Bale, A. J., Kolber, Z. S., Aiken, J., Falkowski, P., 1996. Confirmation of iron limitation of phytoplankton photosynthesis in the equatorial Pacific Ocean. Nature 383:508–511.

Bennett, I., 1975. Variation of daily solar radiation in North America during the extreme months. Geophysics and Bioclimatology Series B 23:31–57.

Betts, A. K., Harshvardhan, 1987. Thermodynamic constraint on the cloud liquid water feedback in climate models. Journal of Geophysical Research 92:8483–8485.

Bigg, E. K., 1973. Ice nucleus concentrations in remote areas. Journal of the Atmospheric Sciences 30:1153–1157.

Bigg, E. K., 1990. Measurement of concentrations of natural ice nuclei. Atmospheric Research 25:397–408.

Bigg, E. K., 1996. Ice forming nuclei in the high Arctic. Tellus 48B:223–233.

Bonsang, B., Nguyen, B. C., Gaudry, A., Lambert, G., 1980. Sulfate enhancement in marine aerosols owing to biogenic sulfur compounds. Journal of Geophysical Research 85:7410–7416.

Brost, R. A., Lenschow, D. H., Wyngaard, J. C., 1982. Marine stratocumulus layers. Part II. Turbulence budgets. Journal of the Atmospheric Sciences 39:818–836.

Cess R. D., Zhang, M. H., Minnis, P., Corsetti, L., Dutton, E. G., Forgan, B. W., Garber, D. P., Gates, W. L., Hack, J. J., Harrison, E. F., Jing, X., Kiehl, J. T., Long, C. N., Morcrette, J. J., Potter, G. L., Ramanathan, V., Sub-asilar, B., Whitlock, C. H., Young, D. F., Zhou, Y., 1995. Absorption of solar radiation by clouds: observations versus models. Science 267:496–499.

Ceulemans, R., Mousseau, M., 1994. Effects of elevated atmospheric CO_2 on woody plants. New Phytologist 127:425–446.

Charlock, T. P., 1981. Cloud optics as a possible stabilizing factor in climate change. Journal of the Atmospheric Sciences 38:661–663.

Charlock, T. P., 1982. Cloud optical feedback and climate stability in a radiative-convective model. Tellus 34:245–254.

Charlson, R. J., Bates, T. S., 1988. The role of the sulfur cycle in cloud microphysics, cloud albedo, and climate. In Proceedings: Symposium on the role of clouds in atmospheric chemistry and global climate. American Meteorological Society, Boston, p. 1–3.

Charlson, R. J., Lovelock, J. E., Andreae, M. O., Warren, S. G., 1987. Oceanic phytoplankton, atmospheric sulfur, cloud albedo and climate. Nature 326:655–661.

Charvatova, I., Strestik, J., 1995. Long-term changes of the surface air temperature in relation to solar internal motion. Climate Change 29:333–352.

Cleveland, W. S., Frenny, A. E., Graedel, T. E., 1983. The seasonal component of atmospheric CO_2: information from new approaches to the decomposition of seasonal time-series. Journal of Geophysical Research 88:10934–10940.

Coakley, J. A., Bernstein, R. L., Durkee, P. A., 1987. Effect of ship-stack effluents on cloud reflectivity. Science 237:1020–1022.

Coale, K. H., Johnson, K. S., Fitzwater, S. E., Gordon, R. M., Tanner, S., Chavez, F. P., Ferioli, L., Sakamoto, C., Rogers, P., Millero, F., Steinberg, P., Nightingale, P., Cooper, D., Cochlan, W. P., Landry, M. R., Constantinou, J., Rollwagen, G., Trasvina, A., Kudela, R., 1996. A massive phytoplankton bloom induced by an ecosystem-scale iron fertilization experiment in the equatorial Pacific Ocean. Nature 383:495–501.

Cure, J. D., Acock, B., 1986. Crop responses to carbon dioxide doubling: A literature survey. Agricultural and Forest Meteorology 8: 127–145.

Curtis, P. S., Balduman, L. M., Drake, B. G., Whigham, D. F., 1990. Elevated atmospheric CO_2 effects on below ground processes in C_3 and C_4 estuarine marsh communities. Ecology 71:2001–2006.

Dacey, J. W. H., Wakeham, S. G., 1988. Oceanic dimethylsulfide: Production during zooplankton grazing on phytoplankton. Science 233:1314–1316.

Dai, A., Del Genio, A. D., Fung, I. Y., 1997. Clouds, precipitation and temperature range. Nature 386:665–666.

Dean, J. S., 1994. The medieval warm period on the southern Colorado Plateau. Climate Change 26:225–241.

Douglas, M. W., Maddox, R. A., Howard, K., 1993. The Mexican monsoon. Journal of Climate 6:1665–1677.

Drake, B. G., 1992. The impact of rising CO_2 on ecosystem production. Water, Air, and Soil Pollution 64:25–44.

Duce, R. A., Mohnen, V. A., Zimmerman, P. R., Grosjean, D., Cautreels, W., Chatfield, R., Jaenicke, R., Ogsen, J. A., Pillizzari, E. D., Wallace, G. T., 1983. Organic material in the global troposphere. Reviews of Geophysics and Space Physics 21:921–952.

Durkee, P. A., 1988. Observations of aerosol-cloud interactions in satellite-detected visible and near-infrared radiance. In Proceedings: Symposium on the role of clouds in atmospheric chemistry and global climate. American Meteorological Society, Boston, p. 157–160.

Eiler, J. M., Mojzsis, S. J., Arrhenius, G., 1997. Carbon isotope evidence for early life. Nature 386:665.

Ellis, J. S., Vonder Haar, T. H., Levitus, S., Oort, A. H., 1978. The annual variation in the global heat balance of the earth. Journal of Geophysical Research 83:1958–1962.

Eppley, R. W., 1972. Temperature and phytoplankton growth in the sea. Fish Bull 70:1063–1085.

ERBE Science Team, 1986. First data from the Earth Radiation Budget Experiment (ERBE). Bulletin of the American Meteorologist Society 67: 818–824.

Ezer, D., Cameron, A. G. W., 1965. A study of solar evolution. Canadian Journal of Physics 43:1497–1517.

Foukal, P., Lean, J., 1990. An empirical model of total solar irradiance variation between 1874 and 1988. Science 247:556–558.

Friedli, H., Lotscher, H., Oeschger, H., Siegenthaler, U., Stauffer, B., 1986. Ice core record of the $^{13}C/^{12}C$ ratio of atmospheric CO_2 in the past two centuries. Nature 324:237–238.

Friedli, H., Moor, E., Oeschger, H., Siegenthaler, U., Stauffer, B., 1984. Ice core record of the $^{13}C/^{12}C$ ratios in CO_2 extracted from Antarctic ice. Geophysical Research Letters 11: 1145–1148.

Friis-Christensen, E., Lassen, K., 1991. Length of the solar cycle: An indicator of solar activity closely associated with climate. Science 254:698–700.

Godbold, D. L., Berntson, G. M., 1997. Elevated atmospheric CO_2 concentration changes ectomycorrhizal morphotype assemblages in *Betula papyrifera*. Tree Physiology 17: 347–350.

Goldman, J. C., Carpenter, E. J., 1974. A kinetic approach to the effect of temperature on algal growth. Limnology and Oceanography 19:756–766.

Gough, D. O., 1981. Solar interior structure and luminosity variations. Solar Physics 74:21–34.

Graybill, D. A., Idso, S. B., 1993. Detecting the aerial fertilization effect of atmospheric CO_2 enrichment in tree-ring chronologies. Global Biogeochemical Cycles 7:81–95.

Grove, J. M., 1988. The Little Ice Age. Routledge, London.

Hales, J. E. Jr. 1972. Surges of maritime tropical air northward over the Gulf of California. Monthly Weather Review 100:298–306.

Hales, J.E. Jr., 1974. Southwestern United States summer monsoon source—Gulf of Mexico or Pacific Ocean? Journal of Applied Meterology 13:331–342.

Hart, M. H., 1978. The evolution of the atmosphere of the Earth. Icarus 33:23–29.

Hatakeyama, S. D., Okuda, M., Akimoto, H., 1982. Formation of sulfur dioxide and methane sulfonic acid in the photo-oxidation of dimethylsulfide in the air. Geophysical Research Letters 9:583–586.

Haurwitz, B., Austin, J. M., 1944. Climatology. McGraw-Hill, New York.

Henderson-Sellers, A., 1986a. Cloud changes in a warmer Europe. Climate Change 8:25–52.

Henderson-Sellers, A., 1986b. Increasing cloud in a warming world. Climate Change 9:267–309.

Henderson-Sellers, A., Cogley, J. G., 1982. The Earth's early hydrosphere. Nature 298:832–835.

Henderson-Sellers, A., Henderson-Sellers, B., 1988. Equable climate in the early Archaean. Nature 336:117–118.

Heymsfield, A. J., McFarquhar, G. M., 1996. High albedos of cirrus in the tropical Pacific warm pool: Microphysical interpretations from CEPEX and from Kwajalein, Marshall Islands. Journal of the Atmospheric Sciences 53:2424–2451.

Hill, F. B., Aneja, V. P., Felder, R. M., 1978. A technique for measurement of biogenic sulfur emission fluxes. Environmental Science and Health 13:199–225.

Holland, H. D., 1984. The chemical evolution of the atmosphere and oceans. Princeton, New Jersey: Princeton University Press.

Hoyt, D. V., Schatten, K. H., 1997. The role of the sun in climate change. Oxford University Press, Oxford.

Hudson, J. D., 1983. Effects of CCN concentrations on stratus clouds. Journal of the Atmospheric Sciences 40:480–486.

Hurrell, J. W., Trenberth, K. E., 1997. Spurious trends in satellite MSU temperatures from merging different satellite records. Nature 386:164–167.

Iben, I., 1969. The Cl^{37} solar neutrino experiment and the solar helium abundance. Annuals of Physics 54:164–203.

Idso, K. E., 1992. Plant responses to rising levels of atmospheric carbon dioxide: A compilation and analysis of the results of a decade of international research into the direct biological effects of atmospheric CO_2 enrichment. Office of Climatology, Arizona State University, Tempe.

Idso, K. E., Idso, S. B., 1994. Plant responses to atmospheric CO_2 enrichment in the face of environmental constraints: A review of the past 10 years' research. Agricultural and Forest Meteorology 69:153–203.

Idso, S. B., 1980. The climatological significance of a doubling of earth's atmospheric carbon dioxide concentration. Science 207:1462–1463.

Idso, S. B., 1981a. A set of equations for full spectrum and 8–14 μm and 10.5–12.5 μm thermal radiation from cloudless skies. Water Resources Research 18:295–304.

Idso, S. B., 1981b. An experimental determination of the radiative properties and climatic consequences of atmospheric dust under non-duststorm conditions. Atmospheric Environment 15:1251–1259.

Idso, S. B., 1982. A surface air temperature response function for earth's atmosphere. Boundary-Layer Meteorology 22:227–232.

Idso, S. B., 1984. An empirical evaluation of earth's surface air temperature response to radiative forcing, including feedback, as applied to the CO_2-climate problem. Archives for Meteorology, Geophysics and Biclimatology Series B 34:1–19.

Idso, S.B., 1988a. The CO_2 greenhouse effect on Mars, Earth, and Venus. The Science of the Total Environment 77:291–294.

Idso, S. B., 1988b. Greenhouse warming or Little Ice Age demise: A critical problem for climatology. Theoretical and Applied Climatology 39:54–56.

Idso, S. B., 1990. A role for soil microbes in moderating the carbon dioxide greenhouse effect? Soil Sci ences 149:179–180.

Idso, S. B., 1992. The DMS-cloud albedo feedback effect: Greatly underestimated? Climate Change 21:429–433.

Idso, S. B., 1995. CO_2 and the biosphere: The incredible legacy of the Industrial Revolution. Department of Soil, Water and Climate, University of Minnesota, St. Paul.

Idso, S. B., Brazel, A. J., 1978. Climatological effects of atmospheric particulate pollution. Nature 274:781–782.

Idso, S. B., Kangieser, P. C., 1970. Seasonal changes in the vertical distribution of dust in the lower troposphere. Journal of Geophysical Research 75:2179–2184.

Idso, S. B., Kimball, B. A., 1993. Tree growth in carbon dioxide enriched air and its implications for global carbon cycling and maximum levels of atmospheric CO_2. Global Biogeochemical Cycles 7:537–555.

Ineichen, K., Wiemken, V., Wiemken, A., 1997. Shoots, roots and ectomycorrhiza formation of pine seedlings at elevated atmospheric carbon dioxide. Plant, Cell and Environment 18: 703–707.

Jenkins, G. S., 1995. Early Earth's climate: Cloud feedback from reduced land fraction and ozone concentrations. Geophysical Research Letters 22:1513–1516.

Jongen, M., Jones, M. B., Hebeisen, T., Blum, H., Hendrey, G., 1995. The effects of elevated CO_2 concentration on the root growth of Lolium perenne and Trifolium repens grown in a FACE system. Global Change Biology 1:361–371.

Kacholia, K., Reck, R. A., 1997. Comparison of global climate change simulations for $2 \times CO_2$-induced warming: An intercomparison of 108 temperature change predictions published between 1980 and 1995. Climate Change 35:53–69.

Kasting, J. F., 1997. Warming early Earth and Mars. Science 276:1213–1215.

Kasting J. F., Toon, O. B., Pollack, J. B., 1988. How climate evolved on the terrestrial planets. Scientific American 258(2):90–97.

Kauppi, P. E., Mielikainen, K., Kuusela, K., 1992. Biomass and carbon budget of European forests, 1971–1990. Science 256:70–74.

Keeling, C. D., Chin, J. F. S., Whorf, T. P., 1996. Increased activity of northern vegetation inferred from atmospheric CO_2 measurements. Nature 382:146–149.

Keeling, C. D., Whorf, T. P., Wahlen, M., van der Pilcht, J., 1995. Interannual extremes in the rate of rise of atmospheric carbon dioxide since 1980. Nature 375:666–670.

Keeling, C. D., Whorf, T. P., Wong, C. S., Bellagay, R. D., 1985. The concentration of carbon dioxide at ocean weather station P from 1969–1981. Journal of Geophysical Research 90:10511–10528.

Keigwin, L. D., 1996. Sedimentary record yields several centuries of data. Oceanus 39(2):16–18.

Kiehl, J. T., 1994. On the observed near cancellation between longwave and shortwave cloud forcing in tropical regions. Journal of Climate 7:559–565.

Kimball, B. A., 1983. Carbon dioxide and agricultural yield: an assemblage and analysis of 770 prior observations. U.S. Water Conservation Laboratory, Phoenix.

Kimball, B. A., Idso, S. B., Aase, J. K., 1982. A model of thermal radiation from partly cloudy and overcast skies. Water Resources Research 18:931–936.

Kreidenweis, S. M., Seinfeld, J. H., 1988. Nucleation of sulfuric acid-water and methanesulfonic acid-water solution particles: Implications for the atmospheric chemistry of organosulfur species. Atmospheric Environment 22:283–296.

LaMarche, V. C. Jr., Graybill, D. A., Fritts, H. C., Rose, M. R., 1984. Increasing atmospheric carbon dioxide: Tree ring evidence for growth enhancement in natural vegetation. Science 223:1019–1021.

Lamb, H. H., 1977. Climate history and the future. Methuen, London.

Lamb, H. H., 1984. Climate in the last thousand years: Natural climatic fluctuations and change. *In* Flohn, H., Fantechi, R. (eds.). The climate of Europe: Past, present and future. D. Reidel, Dordrecht, p. 25–64.

Lamb, H. H., 1988. Weather, climate and human affairs. Routledge, London.

Lawlor, D. W., Mitchell, R. A. C., 1991. The effects of increasing CO_2 on crop photosynthesis and productivity: A review of field studies. Plant, Cell and Environment 14:807–818.

Le Roy Ladurie, E., 1971. Times of feast, times of famine: A history of climate since the year 1000. Doubleday, New York.

Lean, J., Beer, J., Bradley, R., 1995. Reconstruction of solar irradiance since 1610: Implications for climate change. Geophysical Research Letters 22:3195–3198.

Leavitt, S. W., Paul, E. A., Kimball, B. A., Hendrey, G. R., Mauney, J. R., Rauschkolb, R., Rogers, H., Lewin, K. F., Nagy, J., Pinter, P. J. Jr., Johnson, H. B., 1994. Carbon isotope dynamics of free-air CO_2-enriched cotton and soils. Agricultural and Forest Meteorology 70:87–101.

Lemon, E. R., 1983. CO_2 and plants: the response of plants to rising levels of atmospheric carbon dioxide. Westview Press, Boulder, Colorado.

Leovy, C. B., 1980. Carbon dioxide and climate. Science 210: 6–8.

Lockwood, G. W., Skiff, B. A., Baliunas, S. L., Radick, R. R., 1992. Long-term solar brightness changes estimated from a survey of sun-like stars. Nature 360:653–655.

Longdoz, B., Francois, L. M., 1997. The faint young sun climatic paradox: Influence of the continental configuration and of the seasonal cycle on the climatic stability. Global Planet Change 14:97–112.

Lovelock, J. E., 1988. The ages of Gaia: A biography of our living Earth. Norton, New York.

Lovelock, J. E., Whitfield, M., 1982. Life span of the biosphere. Nature 296:561–563.

Lubin, D., 1994. The role of the tropical super greenhouse effect in heating the ocean surface. Science 265:224–227.

MacTaggart, D. L., Adams, D. F., Farwell, S. O., 1987. Measurement of biogenic sulfur emissions from soils and vegetation using dynamic enclosure methods: total sulfur gas emissions via MFC/FD/FPD determinations. Journal of Atmospheric Chemistry 5:417–437.

Madsen, T. V., 1993. Growth and photosynthetic acclimation by *Ranunculus aquatilis* L. in response to inorganic carbon availability. New Phytologist 125:707–715.

Madsen, T. V., Sand-Jensen, K., 1994. The interactive effects of light and inorganic carbon on aquatic plant growth. Plant, Cell and Environment 17:955–962.

McGuffie, K., Henderson-Sellers, A., 1988. Is Canadian cloudiness increasing? Atmosphere-Ocean 26:608–633.

McKay, C., 1983. Section 6. Mars. *In* Smith, R. E., West, G. S. (eds.). Space and planetary environment criteria guidelines for use in space vehicle development. Marshall Space Flight Center, Alabama.

Meszaros, E., 1988. On the possible role of the biosphere in the control of atmospheric clouds and precipitation. Atmospheric Environment 22:423–424.

Mojzsis, S. J., Arrhenius, G., McKeegan, K. D., Harrison, T. M., Nutman, A. P., Friend, C. R. L., 1996. Evidence for life on Earth before 3, 800 million years ago. Nature 384:55–59.

Mortensen, L. M., 1987. Review: CO_2 enrichment in greenhouses. Crop responses. Scientia Horticulturae 33:1–25.

Myneni, R. B., Keeling, C. D., Tucker, C. J., Asrar, G., Nemani, R. R., 1997. Increased plant growth in the northern high latitudes from 1981 to 1991. Nature 386:698–702.

Newman, M. J., Rood, R. T., 1977. Implication of the solar evolution for the Earth's early atmosphere. Science 198:1035–1037.

Nguyen, B. C., Belviso, S., Mihalopoulos, N., Gostan, J., Nival, P., 1988. Dimethyl sulfide production during natural phyto-planktonic blooms. Marine Chemistry 24:133–141.

Nicholls, S., 1984. The dynamics of stratocumulus: Aircraft observations and comparisons with a mixed layer model. Quarterly Journal of the Royal Meteorological Society 110:783–820.

Nierenberg, W. A., Brewer, P. G., Machta, L., Nordhaus, W. D., Revelle, R. R., Schelling, T. C., Smagorinsky, J., Waggoner, P. E., Woodwell, G. M., 1983. Synthesis. *In* Changing climate: Report of the carbon dioxide assessment committee. National Academy Press, Washington, D.C., p. 5–86.

Novakov, T., Penner, J. E., 1993. Large contribution of organic aerosols to cloud-condensation-nuclei concentrations. Nature 365:823–826.

Nullet, D., 1987. Sources of energy for evaporation on tropical islands. Physical Geography 8:36–45.

Nullet, D., and Ekern, P. C., 1988. Temperature and insolation trends in Hawaii. Theoretical and Applied Climatology 39:90–92.

O'Neill, E. G., 1994. Responses of soil biota to elevated atmospheric carbon dioxide. Plant Soil 165:55–65.

Owen, T., Cess, R. D., Ramanathan, V., 1979. Enhanced CO_2 greenhouse to compensate for reduced solar l uminosity on early earth. Nature 277:640–642.

Oyama, Y. I., Carle, G. C., Woeller, F., Pollack, J. B., 1979. Venus lower atmospheric composition: analysis by gas chromatography. Science 203:802–805.

Paltridge, G. W., 1980. Cloud-radiation feedback to climate. Quarterly Journal of the Royal Meteorological Society 106:895–899.

Pearman, G. I., Hyson, P., 1981. The annual variation of atmospheric CO_2 concentration observed in the northern hemisphere. Journal of Geophysical Research 86:9839–9843.

Petersen, K. L., 1994. A warm and wet little climatic optimum and a cold and dry little ice age in the southern Rocky Mountains, U.S.A. Climate Change 26:243–269.

Phillips, O. L., Gentry, A. H., 1994. Increasing turnover through time in tropical forests. Science 263:954–958.

Pilewskie, P., Valero, F. P. J., 1995. Direct observations of excess solar absorption by clouds. Science 267:1626–1629.

Pimm, S. L., Sugden, A. M., 1994. Tropical diversity and global change. Science 263:933–934.

Platt, T., Sathyendranath, S., 1988. Oceanic primary production: Estimation by remote sensing at local and regional scales. Science 241:1613–1620.

Pollack, J. B., 1979. Climate change on terrestrial planets. Icarus 37:479–553.

Pollack, J. B., Toon, O. B., Boese, R., 1980. Greenhouse models of Venus' high surface temperature, as constrained by Pioneer Venus measurements. Journal of Geophysical Research 85:8223–8231.

Poorter, H., 1993. Interspecific variation in the growth response of plants to an elevated ambient CO_2 concentration. Vegetatio 104–105:77–97.

Ramanathan, V., 1988. The greenhouse theory of climate change: A test by an inadvertent global experiment. Science 240:293–299.

Ramanathan, V., Collins, W., 1991. Thermodynamic regulation of ocean warming by cirrus clouds deduced from observations of the 1987 El Niño. Nature 351:27–32.

Ramanathan, V., Cess, R. D., Harrison, E. F., Minnis, P., Barkstrom, B. R., Ahmed, E., Hartmann, D., 1989. Cloud-radiative forcing and climate: Results from the Earth Radiation Budget Experiment. Science 243:57–63.

Ramanathan, V., Subasilar, B., Zhang, G. J., Conant, W., Cess, R. D., Kiehl, J. T., Grassl, H., Shi, L., 1995. Warm pool heat budget and shortwave cloud forcing: A missing physics? Science 267:499–503.

Raval, A., Ramanathan, V., 1989. Observational determination of the greenhouse effect. Nature 342:758–761.

Raven, J. A., 1991. Physiology of inorganic C acquisition and implications for resource use efficiency by marine phytoplankton: Relation to increased CO_2 and temperature. Plant, Cell and Environment 14:779–794.

Raven, J. A., 1993. Phytoplankton: Limits on growth rates. Nature 361:209–210.

Reid, G. C., 1993. Do solar variations change climate? EOS: Transactions: American Geophysical Union 74:23.

Rhea, G. Y., Gotham, I. J., 1981. The effect of environmental factors on phytoplankton growth: Temperature and the interactions of temperature with nutrient limitation. Limnology and Oceanography 26:635–648.

Riebesell, U., Wolf-Gladrow, D. A., Smetacek, V., 1993. Carbon dioxide limitation of marine phytoplankton growth rates. Nature 361:249–251.

Ringelberg, D. B., Stair, J. O., Almeida, J., Norby, R. J., O'Neill, E. G , White, D., 1997. Consequences of rising atmospheric carbon dioxide levels for the belowground microbiota associated with white oak. Journal of Environmental Quality 26:495–503.

Roeckner, E., 1988. A GCM analysis of the cloud optical depth feedback. In Proceedings: Symposium on the role of clouds in atmospheric chemistry and global climate. American Meteorological Society, Boston, Massachusetts, p. 67–68.

Roeckner, E., Schlese, U., Biercamp, J., Loewe, P., 1987. Cloud optical depth feedbacks and climate modeling. Nature 329:138–140.

Rogers, H. H., Runion, G. B., Krupa, S. V., 1994. Plant responses to atmospheric CO_2 enrichment with emphasis on roots and the rhizosphere. Environmental Pollution 83:155–189.

Roosen, R. G., Angione, R. J., 1984. Atmospheric transmission and climate: Results from Smithsonian measurements. Bulletin of the American Meteorologist Society 65:950–957.

Rosinski, J., Haagenson, P. L., Nagamoto, C. T., Parungo, F., 1986. Ice-forming nuclei of maritime origin. Journal of Aerosol Sciences 17:23–46.

Rosinski, J., Haagenson, P. L., Nagamoto, C. T., Parungo, F., 1987. Nature of ice-forming nuclei in marine air masses. Journal of Aerosol Sciences 18:291–309.

Sagan, C., Chyba, C., 1997. The early faint sun paradox: Organic shielding of ultraviolet-labile greenhouse gases. Science 276:1217–1221.

Sagan, C., Mullen, G., 1972. Earth and Mars: Evolution of atmospheres and surface temperatures. Science 177:52–56.

Sakshaug, E., 1988. Light and temperature as controlling factors of phytoplankton growth rate in temperate and polar regions. EOS: Transactions: American Geophysical Union 69:1081.

Saltzman, E. S., Savoie, D. L., Zika, R. G , Prospero, J. M., 1983. Methane-sulfonic acid in the marine atmosphere. Journal of Geophysical Research 88:10897–10902.

Sand-Jensen, K., Pedersen, M. F., Laurentius, S., 1992. Photosynthetic use of inorganic carbon among primary and secondary water plants in streams. Freshwater Biology 27:283–293.

Saxena, P., Hildemann, L. M., McMurry, P. H., Seinfeld, J. H., 1995. Organics alter hygroscopic behavior of atmospheric particles. Journal of Geophysical Research 100:18755–18770.

Saxena, V. K., 1983. Evidence of the biogenic nuclei involvement in Antarctic coastal clouds. Journal of Physical Chemistry 87:4130.

Saxena, V. K., Durkee, P. A., Menon, S., Anderson, J., Burns, K. L., Nielsen, K. E., 1996. Physico-chemical measurements to investigate regional cloud-climate feedback mechanisms. Atmospheric Environment 30:1573–1579.

Schidlowski, M., 1988. A 3,800-million-year isotopic record of life from carbon in sedimentary rocks. Nature 333:313–318.

Schneider, S. H., Kellogg, W. W., Ramanathan, V., 1980. Carbon dioxide and climate. Science 210:6–8.

Schnell, R. C., Vali, G., 1976. Biogenic ice nuclei. Part I. Terrestrial and marine sources. Journal of the Atmospheric Sciences 33:1554–1564.

Schopf, J. W., 1978. The evolution of the earliest cells. Scientific American 239(3):110–138.

Schopf, J. W., Barghourn, E. S., 1967. Alga-like fossils from the early Precambrian of South Africa. Science 156:507–512.

Schwarzchild, M., Howard, R., Harm, R., 1957. Inhomogeneous stellar models. V. A solar model with convective envelope and inhomogeneous interior. The Astrophysical Journal 125:233–241.

Scuderi, L. A., 1993. A 2000-year tree ring record of annual temperatures in the Sierra Nevada mountains. Science 259:1433–1436.

Sellers, W. D., 1965. Physical climatology. University of Chicago Press.

Serre-Bachet F., 1994. Middle Ages temperature reconstructions in Europe, a focus on Northeastern Italy. Climate Change 26:213–224.

Shapiro, J., 1997. The role of carbon dioxide in the initiation and maintenance of blue-green dominance in lakes. Freshwater Biology 37:307–323.

Shaw, G. E., 1983. Bio-controlled thermostasis involving the sulfur cycle. Climate Change 5:297–303.

Shaw, G. E., 1987. Aerosols as climate regulators: A climate-biosphere linkage? Atmospheric Environment 21:985–986.

Shine, K. P., Derwent, R. G., Wuebbles, D. J., Morcrette, J. J., 1990. Radiative forcing of climate. In Houghton, J. T., Jenkins, G. J., Ephraums, J. J., (eds.). Climate change: The IPCC scientific assessment. Cambridge University Press, p. 41–68.

Slingo, A., 1990. Sensitivity of the Earth's radiation budget to changes in low clouds. Nature 343:49–51.

Smagorinsky, J., Armi, L., Bretherton, F. P., Bryan, K., Cess, R. D., Gates, W. L., Hansen, J., Kutzbach, J. E., Manabe, S., 1982. Carbon dioxide and climate: A second assessment. National Academy Press, Washington, D.C.

Somerville, R. C. J., Remer, L. A., 1984. Cloud optical thickness feedbacks in the CO_2 climate problem. Journal of Geophysical Research 89:9668–9672.

Soon, W. H., Posmentier, E. S., Baliunas, S. L., 1996. Inference of solar irradiance variability from terrestrial temperature changes, 1880–1993: An astrophysical application of the sun-climate connection. The Astrophysical Journal 472:891–902.

Spencer, R. W., 1997. 1996: A preview of cooler days ahead. In Michaels, P. J. (ed.). State of the climate report: Essays on global climate change. New Hope Environmental Services, New Hope, p. 14–17.

Staubes, R., Georgii, H. W., Ockelmann, G., 1989. Flux of COS, DMS and CS_2 from various soils in Germany. Tellus 41B:305–313.

Strain, B. R., Cure, J. D., 1994. Direct effects of atmospheric CO_2 enrichment on plants and ecosystems: An updated bibliographic data base. Oak Ridge National Laboratory, Oak Ridge, Tennessee.

Szyrmer, W., Zawadzki, I., 1997. Biogenic and anthropogenic sources of ice-forming nuclei: A review. Bulletin of the American Meteorologist Society 78:209–228.

Titus, J. E., 1992. Submersed macrophyte growth at low pH. II. CO_2 sediment interactions. Oecologia 92:391–398.

Titus, J. E., Feldman, R. S., Grise, D., 1990. Submersed macrophyte growth at low pH. I. CO_2 enrichment effects with fertile sediment. Oecologia 84:307–313.

Trenberth, K. E., Houghton, J. T., Meira Filho, L. G., 1996. The climate system: An overview. In Houghton, J. T., Meira Filho, L. G., Callander, B. A., Harris, N., Kattenberg, A., Maskell, K. (eds.). Climate change 1995: the science of climate change. Cambridge University Press, p. 51–64.

Turner, S. M., Malin, G., Liss, P. S., Harbour, D. S., Holligan, P. M., 1988. The seasonal variation of dimethyl sulfide and dimethyl-sulfoniopropionate concentrations in nearshore waters. Limnology and Oceanography 33:364–375.

Turner, S. M., Nightingale, P. D., Spokes, L. J., Liddicoat, M. I., Liss, P. S., 1996. Increased dimethyl sulphide concentrations in sea water from in situ iron enrichment. Nature 383:513–517.

Twomey, S. A., Warner, J., 1967. Comparison of measurements of cloud droplets and cloud nuclei. Journal of the Atmospheric Sciences 24:702–703.

Vairavamurthy, A., Andreae, M. O., Iverson, R. L., 1985. Biosynthesis of dimethylsulfide and dimethylpropiothetin by Hymenomonas carterae in relation to sulfur source and salinity variations. Limnology and Oceanography 30:59–70.

Valero, F. P. J., Collins, W. D., Pilewskie, P., Bucholtz, A., Flatau, P. J., 1997. Direct radiometric observations of the water vapor greenhouse effect over the equatorial Pacific Ocean. Science 275:1773–1776.

Vali, G., Christensen, M., Fresh, R. W., Galyan, E. L., Maki, L. R., Schnell, R. C., 1976. Biogenic ice nuclei. Part II: Bacterial sources. Journal of the Atmospheric Sciences 33:1565–1570.

Villalba, R., 1994. Tree-ring and glacial evidence for the medieval warm epoch and the little ice age in southern South America. Climate Change 26:183–197.

Walker, J. C. G., 1985. Carbon dioxide on the early Earth. Origins of Life 16:117–127.

Walker, J. C. G., 1986. The Earth history: The several ages of the Earth. Jones and Bartlett, Boston.

Warner, J., Twomey, S. A., 1967. The production of cloud nuclei by cane fires and the effect on cloud droplet concentration. Journal of the Atmospheric Sciences 24:704–706.

Warren, S. G., Schneider, S. H., 1979. Seasonal simulation as a test for uncertainties in the parameterizations of a Budyko-Sellers zonal climate model. Journal of the Atmospheric Sciences 36:1377–1391.

Webster, P. J., Stephens, G. L., 1984. Cloud-radiation interaction and the climate problem. In Houghton, J. T. (ed.). The global climate. Cambridge University Press, p. 63–78.

Went, F. W., 1966. On the nature of Aitken condensation nuclei. Tellus 18:549–555.

Whyte, I. D., 1995. Climatic change and human society. Arnold, London.

Wigley, T. M. L., Brimblecombe, P., 1981. Carbon dioxide, ammonia and the origin of life. Nature 291:213–215.

Wullschleger, S. D., Post, W. M., King, A. W., 1995. On the potential for a CO_2 fertilization effect in forests: Estimates of the biotic growth factor based on 58 controlled-exposure studies. *In* Woodwell, G. M., Mackenzie, F. T. (eds.). Biotic feed-backs in the global climatic system. Oxford University Press, New York, p. 85–107.

Wullschleger, S. D., Norby, R. J., Gunderson, C. A., 1997. Forest trees and their response to atmospheric CO_2 enrichment: A compilation of results. *In* Allen, L. H. Jr., Kirkham, M. B., Olszyk, D. M., Whitman, C. E., (eds.). Advances in CO_2 effects research. American Society of Agronomy, Madison, p. 79–100.

Zak, D. R., Pregitzer, K. S., Curtis, P. S., Teeri, J. A , Fogel, R., Randlett, D. L., 1993. Elevated atmospheric CO_2 and feedback between carbon and nitrogen cycles. Plant Soil 151:105–117.

Jenkins, D. A. L., Potential impact and effects of climate change, 2001, *in* L. C. Gerhard, W. E. Harrison, and B. M. Hanson, eds., Geological perspectives of global climate change, p. 337–359.

18 | Potential Impact and Effects of Climate Change

David A. L. Jenkins
Chartwood Resources Ltd.
Weybridge, Surrey, U.K.

ABSTRACT

An awareness and understanding of the palaeoclimatological history of the planet allows a different construct for today's concerns about the impact of future climatic changes.

In particular, a better appreciation of the magnitude and rate of change during the past few hundred thousand years demonstrates that the changes anticipated during the next few hundred are well within the range experienced during the Pleistocene Era.

The planet is now in a period of gradual cooling from the time of the postglacial thermal optimum 6000–9000 years ago. Temperatures are now on an irregular downward path, comparable to the Eemian interglacial, although at present we are experiencing a minor temperature increase as a partial recovery from the "Little Ice Age," which ended 150 years ago.

Climate will always change. The planet is extremely resilient. As the most intelligent species that has colonised the surface, humans clearly have ample capability to adapt. However, our ability to do so will be limited by our political and behavioural patterns.

The well-documented evidence from the climate changes in the Northern Hemisphere during the past 1000 years indicates that changes in average global temperatures of 2°C will have significant regional impacts on precipitation and vegetation patterns and lead to further changes in sea level. From the viewpoint of the demands of an increasing global population, these could be managed, given appropriate levels of investment. Warming is definitely easier to cope with than cooling. Stability is not an option.

INTRODUCTION

Climate change has over the past decade appeared to be one of the most compelling politicised global concerns. It has become an issue often highlighted in the discussions around the fashionable topic of "sustainability," and in some quarters it has overtaken population increase, poverty, urban congestion, and pollution as the number one environmental worry.

The concerns are focused on the impacts that the magnitude of temperature change will have, together with the rate at which these projected changes will take place. The frequently expressed view is that both the projected magnitude and the rate of change over the next 100 years will have serious detrimental consequences for life on earth and that we must take action now to try to stop the changes occurring. This view is certainly held by the environmental nongovernmental organisations (NGOs). However, as enunciated at the Kyoto conference in December 1997 and Buenos Aires in November 1998, this viewpoint also now is widely embedded in political thinking.

Extremely comprehensive and very extensively referenced material forms part of reports compiled under the auspices of the International Panel on Climate Change (IPCC). The current definitive global work is the Second Assessment Report, a massive compilation completed in 1995 and published in the following year (Watson et al., 1996). This was later supplemented by another extensive report that examined the regional impacts (IPCC, 1998). It will be updated by the Third Assessment Report, due for publication in 2001. This will have the benefit of improved coupled global circulation models (GCMs) with better regional definition, a flavour of which was provided in the December 1997 booklet from the British Meteorological Office (U.K. Meteorological Office, 1997).

EQUILIBRIUM AND NONEQUILIBRIUM STATES

Before describing the conclusions of the IPCC analyses, it is important to set the philosophy and context for the subject. The context is particularly important for geologists, because we can readily comprehend the present state of the planet's surface conditions in a wide temporal frame of reference. We are also very conscious of both the magnitude of variability and the rate at which change has occurred in the past. Nevertheless, we have to remain mindful of the concerns, worries, and fears of our fellow citizens, whose beliefs are no less real simply because they lack a knowledge of earth history. In particular, we must not argue that since the Quaternary Period was characterised by very rapid climate change, there need not be a concern about possible future changes.

There is certainly a concern, but it is not actually a worry, about the climate's changing. The worry is that there is quite a widespread belief that climate would be stable were it not for human influence. The climatologists and atmospheric physicists recognise that the earth's surface is a highly dynamic system. Nevertheless, there is a feeling that whilst the natural rate and magnitude of change would be manageable, the postulated anthropogenic forcing from increases in concentration of CO_2 in the atmosphere moves us into a state that gives rise to considerable uncertainty. The language and perceptions that grow up and develop around these concepts then lead to the simplified political expression of climate stability and instability that we hear so often expressed today.

It is genuinely important that we recognise and become comfortable with the intellectual construct of a dynamic world, constantly changing at varying rates. That is reality. Striving after stability is a hopeless endeavour. Attempting to mitigate the rate of change is plausible, but highly uncertain. Accepting change and planning to use our capability to adapt is also a plausible approach. The debate around the relative effort to devote to these two divergent stances will certainly be a concern for mankind during the first quarter of this century.

The argument for change is put most cogently by Philip Stott in an editorial in the *Journal of Biogeography* (Stott, 1998). He points out how today's mode of referring to ecological systems (a "metalanguage") subtly reinforces in people's minds the concept of continuity and equilibrium as the "natural" state. Hence we speak about "sustainability," "balance," "optimal," "stable." We talk of nature being in equilibrium. We strive to preserve biodiversity, in the context of preventing any species from dying out. And the key words of the equilibrium metalanguage have been adjectivally cleverly modified to reinforce the concept. Hence development should now always be sustainable, whereas the nonequilibrium world is characterised in "multiple equilibrium" states. Humans then have to become the problem, for as ever-increasing teeming millions, they destroy the earth and upset balance and stability. It presents an interesting intellectual dilemma.

Stott (1998) feels we need a new metalanguage that better describes the world as it actually behaves, a nonequilibrium world in which change is occurring all the time, in different directions, at different scales, gradually, catastrophically, and unpredictably. In a world where means are "meaningless" there is chaotic variability and prediction is very difficult. The key words for a metalanguage that describes this viewpoint would be "adaptation," "migration," "movement," "flexibility," "resilience," "opportunism." They create a totally different image for the way an ecological system would behave. Under this concept, ecological stress simply becomes another opportunity, as does the rate of change.

The other aspect of this approach is that it considers the living world behaving autecologically, where changes in a population of species occur in a highly individualistic manner. The environment does not behave and change as a system, but as an evolving assemblage of individuals. This is referred to as abiotic change. As individuals their ability to deal with change is different and very variable.

These concepts of eternal nonequilibrium and taking change as the norm lead to a different perspective on the nature of our concerns when faced with potentially rapid and effectively unpredictable change. In effect they shape our attitude to change. Since predictions both on the degree of regional change and the capability of flora and fauna to cope are subject to a high level of uncertainty, then the interpretations will be influenced by one's ecological stance. For this reason, the concepts of equilibrium and nonequilibrium states are important. Those who yearn for equilibrium and stability will be more worried about change than those who view change as the norm.

CLIMATE IMPACT PARAMETERS

The starting point for a study of impacts is the projection for the magnitude and rates of change in key environmental parameters, in particular temperature, precipitation, and sea level. The main work in this field has been conducted under the auspices of the United Nations Framework Convention on Climate Change. The scientific studies to inform the parties to this convention are carried out by the International Panel on Climate Change (IPCC), which was established in 1988 by the World Meteorological Organisation and the United Nations Environment Programme. The IPCC has produced two assessment reports, in 1990 and 1995. The latter provides the predictions on which the work on the analyses of impacts has been based (Watson et al., 1996).

The key conclusion of the IPCC report is as follows. Based on the range of sensitivities of climate to changes in the atmospheric concentrations of greenhouse gases and plausible changes in emissions, climate models project that the mean annual global surface temperature will increase between 1.0° and 3.5°C by 2100. Global mean sea level will rise by between 15 and 95 cm. The report emphasises that this rate of change is faster than any seen in the past 10,000 years, but also points out that the actual annual to decadal rate will include considerable natural variability and that regional changes will differ substantially from the global mean value.

THE EARTH'S RECENT HISTORY

Any discussions of climatic changes and their impacts today benefit from an understanding of the earth's recent history. For notwithstanding that the current anthropogenic enhancement of greenhouse gas concentrations have introduced a new climatic driving agent, this has to operate in conjunction with the other drivers that have been influencing the world during the Quaternary. This period of earth history, as all geologists are aware, is one in which the climate has been, and in geological terms remains today, unusually cold.

A very neat "translation" of the earth's time-temperature chart **(Figure 1)** was produced by Webb (1991) from studies carried out for the Office of Interdisciplinary Earth Studies at the University Corporation for Atmospheric Research at Boulder, Colorado. The charts move through a series of time windows, enlarging in quantum steps from the Mesozoic Era to the present day.

During the decade since this compilation was constructed, there has been a major increase in the effort to study the earth's recent climatic history. It is very instructive to focus attention on three time windows, namely the past 150,000 years, which covers the last ice age and the preceding Eemian Interglacial, the past 10,000 years, which covers the Holocene, and the past millennium.

PLEISTOCENE CLIMATIC CHANGES

There are a number of ways to interpret temperatures during the Quaternary, including sensitive floral and faunal assemblages, tree-ring growth, lacustrine and peat bog layering, and isotope ratios. Some of the most graphic depictions of the last 150,000 years of the Quaternary Period come from cores recovered from the seabed of the northeast Atlantic. These have been collected as part of the North-East Atlantic Palaeogeography and Climate Change Project (NEAPACC), conducted from 1992 to 1998 under the stewardship of the U.K.'s Natural Environment Research Council (NERC). The sea surface temperatures are calculated from the oxygen isotope composition of the planktonic fauna. The results are shown in **Figure 2**, which is taken from Chapman and Shackleton (1998).

The core sampling is at 1-cm-depth intervals and the summer and winter estimates are derived from quantitative analysis of the planktonic foraminiferal assemblages. The location is at 40°N latitude, north of the Azores.

This record demonstrates a number of very interesting points.

- Today's sea-surface temperature broadly corresponds to an interglacial highstand.
- Looking at the length of the last period of high interglacial temperature from c. 120,000 to 110,000 years ago, we might expect that the earth is now due for a significant temperature drop.
- The past 150,000 years have been characterised by limited windows of warmth and extended periods of progressive cooling.
- Temperature increases are rapid and large.
- There is considerable variation of temperature with time.

The most striking variations are the rapid temperature fluctuations marked as H1 to H11. These are the Heinrich events (named after the individual who identified them) that represent floods (armadas) of icebergs entering the North Atlantic from Baffin Bay as a result of the periodic collapse of the Laurentian ice shield (Broecker, 1994). The melting of these icebergs led to a freshening of the surface waters of the northward-flowing Gulf Stream. This fresher water then became insufficiently dense to sink as it cooled toward Iceland, with the consequence that the ocean current system in the North Atlantic changed completely. The "North Atlantic Conveyor," the term

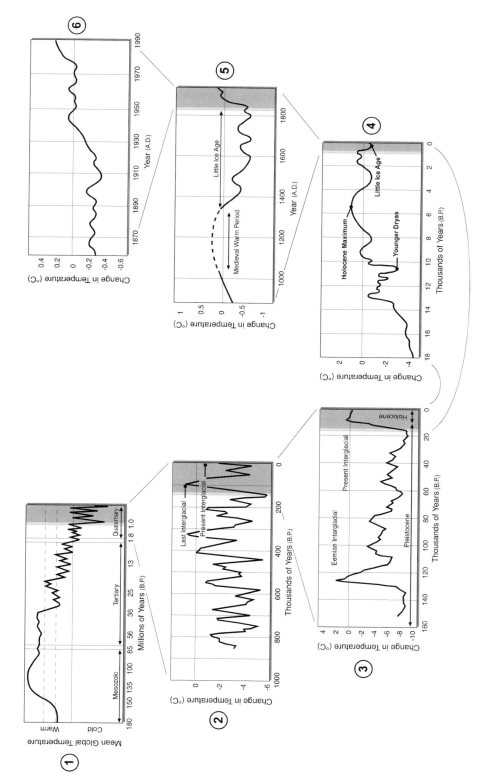

Figure 1 Changes through time in the temperature of the earth (from Webb, 1991).

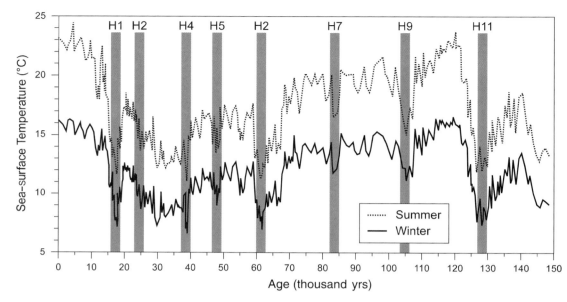

Figure 2 Summer and winter sea-surface temperatures from North Atlantic core SU90-03 at 40°N lat. 32°W long. (from Chapman and Shackleton, 1998).

used for the northward-flowing surface Gulf Stream and the returning deep southward-flowing current, was "turned off" and cold surface arctic currents moved rapidly south. The effect was dramatic and geologically instantaneous, occurring in only a few decades. Ironically, some of the recent modeling work indicates that changes in the precipitation over the Arctic from an increase in global temperatures could give us a similar effect today, with consequent dramatic cooling in western Europe. The possibility of the countries of western Europe limiting energy consumption to curb CO_2 emissions only to find the climate then cooling rapidly provides an interesting potential future challenge for political credibility.

THE HOLOCENE

Another set of cores taken during the NEAPACC project was from the Barra Fan off northwest Scotland (Kroon et al., 1997). These can be used to show the detail of the temperature changes at the end of the last glacial period **(Figure 3)**.

The beginning of the Holocene at 11,500 years before the present day (years B.P.) at the end of the 1500-year cold spell known as the Younger Dryas is well defined. What is also interesting at this location is the warmth at around 14,500 years B.P., the time of very rapid recovery from the extreme cold associated with the last Heinrich event (H1 on **Figure 2**).

The "10,000" years of the Holocene, now accurately dated as commencing 11,500 years B.P., have been subject to extensive recent research, with a wide range of publications. One of the best-referenced up-to-date books is by Roberts (1998). It charts the Holocene environments and their evolution during the climatic recovery from the last ice age, together with the impact on the fauna and flora and the effects on and adaptation displayed by our human ancestors.

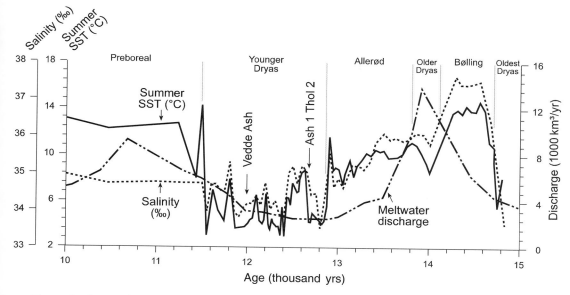

Figure 3 Sea-surface temperature (SST) changes during the past 15,000 years. Core 56/-10/36 at 56°N lat. 9°W long. (from Kroon et al., 1997).

Figure 4 shows the temperature changes across the transition to the current interglacial (Holocene) from the closing stages of the last glacial period. The temperature interpretation uses different parameters from the Barra Fan profile, which was based on plankton isotope data, but the correlation is good.

The mean July temperatures are for the U.K. and are based on changes in the assemblages of the beetle order Coleoptera (Atkinson et al., 1987). The Greenland Ice Sheet Project II (GISP2) summit ice-core data show rates of ice accumulation and concentrations of dust, indicative of variation in local precipitation and central Asian aridity, respectively (Alley et al., 1993, Mayewski et al., 1994). The rate of sea-level rise is from work on Barbados coral reefs by Fairbanks (1989) and Bard et al. (1990). The Greenland ice-core data demonstrate particularly rapid rates of change, with an implication of a warming at the ice-cap crest of 7°C in only 50 years at the end of the Younger Dryas (Roberts, 1998, p. 74).

The data and the research for the Holocene are predominantly for the Northern Hemisphere, but they allow us to build up a good picture of the broad changes in temperature, precipitation, and sea level that are the impacts of climate change which are of primary interest today.

Reflecting the same broad pattern as occurred in the last interglacial 120,000 to 110,000 years B.P., the time of Holocene thermal maximum was around 9000 to 5500 years B.P., following a rapid increase of global temperature from 11,500 to 9000 years B.P. During the past 5000 years, the general tendency has been toward global cooling in a series of steps, the most marked being at around 2600 years B.P.

One of the best guides to Holocene temperatures is the elevation of the tree line. **Figure 5** is reproduced from Lamb (1995), using data from Markgraf (1974) for temperate latitudes.

A second line of evidence is from the temperature profiles constructed from an analysis of the variations in oxygen isotope data in the ice cores from the Greenland ice sheet. Cuffey and Clow

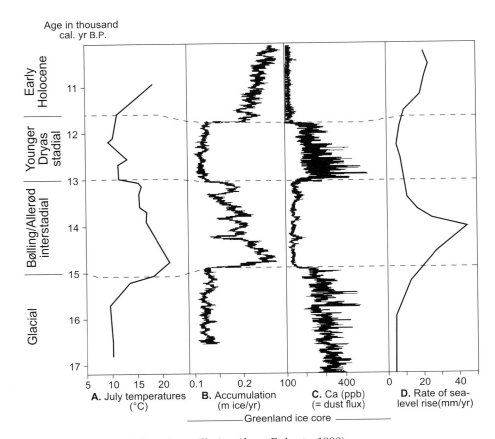

Figure 4 The late glacial climatic oscillation (from Roberts, 1998).

(1997) have calculated the temperature history from the GISP2 summit borehole and the record for the past 15,000 years is reproduced in **Figure 6**. It shows the same gradual decline as Markgraf's (1974) tree-line data and has quite a remarkable similarity to the Eemian interglacial temperature profile calculated from the oxygen isotope record in the seabed core shown in **Figure 2**.

The implication from these data sets is that the long-run temperature profile for the earth is now one of gradual cooling. However, the temperature calculation from the Azores seabed core at 40°N latitude (Chapman and Shackleton, 1998, and **Figure 2**) shows an opposing warming trend. Clearly, more data points are required for a valid global inference to be drawn.

The Holocene experienced wide variations in precipitation. Perhaps the most noteworthy difference during the first 5000 years was the greatly expanded area of the world's large lakes, whose shrunken remnants remain today, together with the existence of lakes across a large part of what is now the Sahara Desert. This contrasted with the glacial periods, when the wind circulation patterns led to greater areas of desert and tundra. During the second half of the Holocene, there was a change to moister (and cooler) conditions in northern temperate regions and a progressive desiccation of northern Africa and the Middle East. In North America, the midwestern prairie-forest boundary between the Rockies and the Mississippi moved eastwards until 8000 years B.P. and then subsequently shifted back to the west under the influence of changes in moisture (Webb et al., 1984).

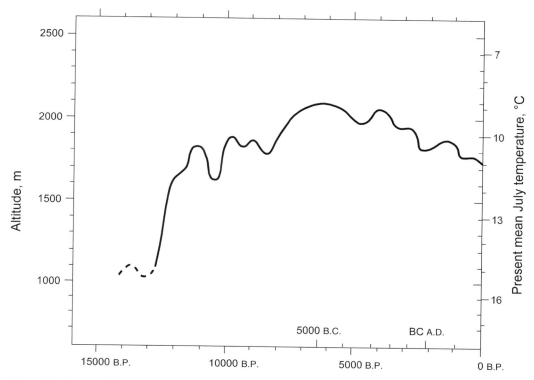

Figure 5 Average height of the upper tree line of mountains in temperate latitudes (from Lamb, 1995, Figure 43, after Markgraf, 1974).

At the close of the last glacial period, 15,000 years ago, today's midlatitude deciduous and boreal woodlands were located nearer to the equator and often restricted to isolated pockets. Similarly, the rain forests of Africa and Amazonia were reduced in extent and fragmented by climatic aridity. The return of the forests during the first part of the Holocene is well documented by palynology and reflects the impact of temperature and precipitation changes and the manner in which the ice sheets retreated. The process (as mentioned earlier, see Stott, 1998) was quite chaotic and may have taken 2000 years to reach a distribution equilibrium (Roberts, 1998, p. 106).

The general temperature pattern of a mid-Holocene maximum followed by cooling is reflected in the reversal of northward advance of the boreal forest and tundra in Canada (Ritchie, 1984) and similarly in the tree elevations reached in temperate latitudes, referred to earlier **(Figure 5)**.

However, two other aspects of the Holocene climatic changes are worth noting. One is the occasional "climatic crisis," recorded in the abrupt fall in African lake levels at 8200 and 5200 years B.P. (Gasse and van Campo, 1994). This is thought to reflect sudden and dramatic changes in levels of precipitation that persisted for several hundred years and must have been related to a significant climatic change. The second, during the last 5000 years, is the secular oscillation between cool-wet and warm-dry conditions, shown from the analysis of Danish raised bogs (Aaby, 1976). Here the frequency seems to indicate a change at 260-year intervals. The cause is not known.

One event that will definitely result in global cooling is a major volcanic eruption. These do leave a clear signature, both in the ice-core record (from anomalously high sulphate concentrations)

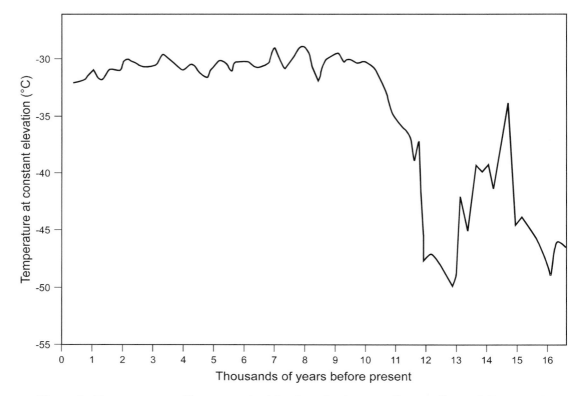

Figure 6 Temperature profile at summit of the Greenland ice cap (from Cuffey and Clow, 1997).

and also, in more recent time, from tree rings. Although important at the time, the climate perturbation is typically less than a decade. The Santorini eruption at 3578 years B.P. is one of the most spectacular of the latter part of the Holocene and is documented in the GISP2 ice-core record (Zielinski et al., 1994).

The third significant climatic-change impact during the Holocene was the rise in sea level. Although temperature at the end of the Younger Dryas rose rapidly, the immense bulk of the ice sheets meant that they melted gradually and sea level rose from about 55 m below today's level to reach modern elevations only around 6000 years B.P. (Bard et al., 1996).

Since then the data from the Tahitian and Barbados coral reefs (Shinn et al., 1965) indicate that sea level has fluctuated within a range of only a few metres. In large parts of the Northern Hemisphere, the coastline has been affected more by the rate of adjustment to eustatic response to the removal of the ice sheets than by changes in sea level.

THE PAST TWO MILLENNIA

The third time window that gives context for studying today's climate change impacts and effects is the last 1000 to 2000 years. The documentation for this period in northern Europe is extensive and serves as the reference record against which to place the more fragmentary evidence from

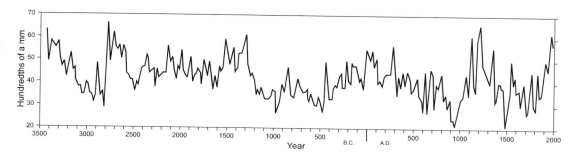

Figure 7 Tree-ring widths (20-yr average) of bristlecone pine (from Lamb, 1995, Figure 52) based on data supplied by Prof. V. C. La Marche at the Laboratory of Tree Ring Research, University of Arizona.

elsewhere in the world. The book by H. H. Lamb (1995) is unquestionably the best referenced compilation of information. What is particularly striking, as one in effect "drills down" to an order of magnitude increase in detail from the previous 9000 years of the Holocene, is that the distribution of variability is maintained.

In broad terms, the past 2000 years in the Northern Hemisphere have witnessed a warm period until approximately A.D. 300 to 400, followed by a cooling to c. A.D. 800, then the well-defined Medieval Warm Period to c. A.D. 1300, followed by the "Little Ice Age" from c. A.D. 1400 to 1850. We are presently in the recovery period of warming from the Little Ice Age.

There are many lines of evidence supporting the interpretation of these changes. One of the best records is from the yearly tree-ring widths of the bristlecone pines on the White Mountains of California, where there is a continuous annual record going back almost 5500 years. There are some timing differences with the European data, but the overall cyclicity is comparable. Although not shown on the summary plot in **Figure 7**, there is also marked variability on both decadal and annual frequency scales.

Using our knowledge of temperature/height lapse rate of 0.6 to 0.7°C per 100-m change in elevation, long-term changes in temperature can be gauged from the upper tree-line level. The deduction from the bristlecone pine data above is that by about 200 B.C., the average annual temperature was some 0.5°C lower than at the mid- Holocene (5000 years B.P.). In Europe, it is estimated to have been 1°C cooler (Lamb, 1995, p. 143). In fact, from the bristlecone data it could be deduced that the general slow cooling trend continues through to the present century. This is what would be expected based on the climatic history of the past 150,000 years using seabed and ice-core data as discussed in the preceding section.

One further "cameo" along this theme is the interpretation that judging from the distribution of vineyards in Europe during the Medieval Warm Period, average summer temperatures were approximately 1°C warmer than in the twentieth century. Based on all these long-run data, it is perhaps doubtful that we would ever return to this level without the impact of enhanced atmospheric CO_2 concentrations.

The record that allows us to take temperature granularity to its tightest definition is the central England temperature plot. This is a compilation of mean temperatures for each month of each year from 1659 to the present day, at a lowland site in the central part of the U.K. This unpublished record is maintained by the U.K. Meteorological Office at its Hadley Centre research headquarters. The data are reproduced in **Figure 8**.

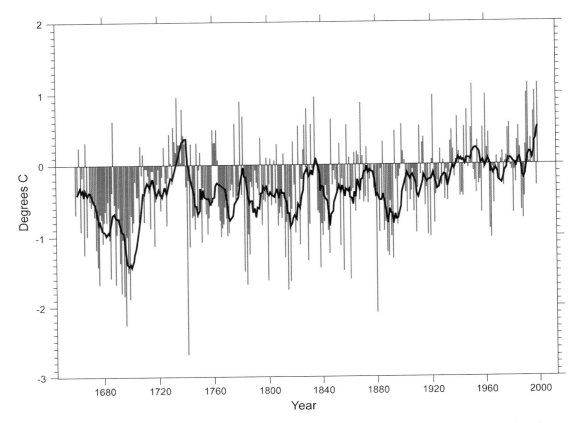

Figure 8 Annual central England temperature differences (from 1961–1990 average) for the years 1659 to 1997 (from the U.K. Meteorological Office, unpublished).

The global temperature chart from 1860 that is the standard reference for the IPCC and all of today's commentary on climate change **(Figure 9)** is reflected in the central England record. However, the recent global increases in temperature do not appear particularly unusual when viewed against the detailed U.K. evidence of 350 years of continuous and rapid change. What is compelling is the degree of this variability both in magnitude and rate.

The coldest period of the Little Ice Age occurred during the last few decades of the seventeenth century. This was followed, from 1700 to 1740, by 40 years when the temperature recovered at a rate of 5°C per 100 years (twice the current IPCC projection for the twenty-first century) before plunging abruptly. There is good evidence that the coldest years during this 350-year record were associated with major volcanic events, a correlation for which there are data throughout the Holocene. And there has been speculation that the cooling of the 1950s could be, at least in part, related to a doubling of volcanic material in the Northern Hemisphere atmosphere (Bryson and Goodman, 1980). However, the drivers for the decadal-scale variability remain uncertain and are the subject of intensive ongoing research today. Changes in solar flux would certainly be a component, but the precise relationship remains unclear.

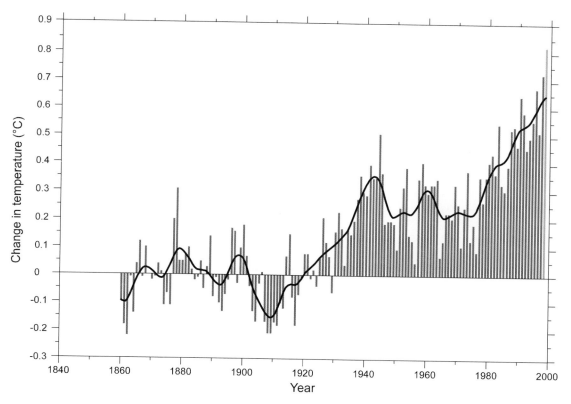

Figure 9 Changes in global average temperature, 1860–1998 (updated from the U.K. Meteorological Office, 1997).

There is another dimension, though, to temperature variability in the context of practical impacts, and that is the seasonal variations. **Figure 8** for the central U.K. shows that the average temperature difference between the last decade of the seventeenth and twentieth centuries was about 2°C. However, the monthly means for the winter (December to February) and summer (June to August) months show that in the 1690s the temperatures were up to 8°C colder in the winter and 4°C colder in the summer (tables from Lamb, 1995, p. 230). In terms of practical impact on the population, particularly on agriculture, these were very dramatic differences. They demonstrate that an apparently small change in an annual averaged temperature can mask big impacts when translated into monthly ranges and, even more so, monthly extremes.

Similarly, global averaging masks substantial local variation. The severe U.K. winter of 1962–1963 was colder than some of those in the seventeenth century when frost fairs were held on the Thames, but on the global chart the period ranked as warm as any prior to the most recent 20 years. The reason, incidentally, is that northwestern Europe came under the influence of increasing ice in the east Greenland Sea. This was related to a decrease in salinity from the Arctic Ocean that was in turn due to greatly increased runoff from the Mackenzie River system, caused by a change in the atmospheric circulation (Mysak et al., 1990; Mysak and Powers, 1991). More recently, it has been hypothesised (Driscoll and Haug, 1998) that increases in the Siberian river runoff into the Arctic Ocean could have been a contributory factor in the onset of the Pleistocene glacial epoch.

PERCEPTIONS OF THE IMPACT OF TEMPERATURE VARIATION

One of the more interesting results of the details of temperature variations over the past few hundred years that are so well documented by Lamb (1995) is that the study allows us to address the question of what a change of 1°C in the annual average temperature would actually feel like to people living at a particular location. This is an important general point, because what the public hears from the IPCC work is that the expected rise in temperature over the next 100 years will be between 1 and 3.5°C, with 2°C as the best estimate. These numbers are global averages, but in the absence of predicted regional variations, which the models are only now beginning to be able to address, the impression being created in the public's mind is that a 2°C change is what they individually will experience. Compounding the disconnect between the scientific/political process and public perception is the fact that everyone understands temperature, but people's awareness exists predominantly in the field of short time windows, rarely exceeding a month. Our images are those of a cold January, a hot summer. As pointed out, a 1°C annual change can smooth over substantial variations during the year.

We can then continue the analogy and bring the issue of climate perception into very recent times. The cooling from 1950 to 1975, which on the global 5-year moving-average data was only 0.15°C and in the U.K. was 0.5°C, nevertheless felt sufficiently pronounced to lead to genuine concern that we might be returning to a period of more long-run cooling. Given the historical temperature data that have been summarised here, this was a very logical conclusion.

Finally, there is the observation that we are all aware of the magnitude of the monthly and daily variations that we experience. So when we are told that a rise in temperature during the next 100 years is going to have serious consequences, the public remains puzzled. From the viewpoint of personal experience, it is not apparent why this should be the case. Two °C does not sound like much. However, 2°C is the predicted global average and there will be wide regional variations, with high latitudes on land showing a much greater increase. Furthermore, as was demonstrated above by data from the U.K., a 2°C annual change can lead to very marked seasonal variations. It is a safe prediction, therefore, that this level of change would have a very significant impact and would require substantial adaptation in many countries. It should by now be apparent that the degree of variability in the Quaternary climate system is so great that it is absurd for humans to think that it could be stabilised. From any point in time, the temperature will become either warmer or cooler. It will always be changing. As mentioned earlier, there is no equilibrium state.

Add to this the fact that the balance of evidence indicates that the long-run temperature trend is now one of global cooling, with the mid-Holocene 6000 years B.P. representing the peak of the current interglacial. As geologists, we have an interesting dilemma when faced with the now widespread view that 100 years of 2°C warming has to be prevented. If the public had to make a choice between that or 2°C global cooling, it is unlikely that many would choose the latter, certainly not when faced with the evidence of conditions in the Northern Hemisphere 300 years ago. Such a choice is not, of course, being offered. Nevertheless, given the planet's climate history, we could take the view that it is quite fortunate that the balance of prediction means there is a strong probability the world is going to warm quite rapidly, with the long-run cooling trend being deferred for a few hundred years. However, this is not an opinion one hears expressed today.

GLOBAL WARMING PREDICTIONS

The warming prediction is based on atmosphere modeling. Given the complexity of the subject, these predictions have a high degree of uncertainty. Nevertheless, the basic physics of the impact of greenhouse gas concentrations is well understood. Hoyle (1996) provides a succinct recent paper on this subject, which has been a topic of scientific study for 200 years. As concentration increases so does the temperature. The relationship is nonlinear (Augustsson and Ramanathan, 1977). Their study was focused on expected temperature increases from enhanced atmospheric CO_2 concentration, but their work also showed the very rapid drop in temperature as CO_2 concentration falls below 100 ppm. Doubling of atmospheric CO_2 concentration from the preindustrial level of 280 ppm would, absent any other changes, produces a 1°C average annual increase in the global surface temperature. Naturally, there are many consequent changes and feedback mechanisms, and the relationship of these lead to the uncertainty in ranges of temperature prediction that we have today.

In addition, we know there are a number of external drivers for the climate of the planet. The variability of the past few million years, which has been dramatic in both magnitude and rate of change, has been related to factors other than changes in atmospheric CO_2 concentration. This has also applied during the majority of the last millennium, including the recovery from the Little Ice Age 150 years B.P.

Although the direct impact of increasing CO_2 concentration cannot be disputed, the question of the degree of indirect effect and its magnitude relative to the other drivers remains an area of uncertainty. The decade of the 1990s proved to be unusually warm. The average annual temperatures in the U.K. were only just within the range of variability seen in the 350-year record **(Figure 8)**. Past patterns of natural variability (which mean changes due to all factors other than anthropogenic CO_2 emissions) would certainly lead us to expect cooler global temperatures in the first decade of this new millennium. Should these not occur, not only would this indicate that increased greenhouse gas concentrations have become the dominant climatic driver, but this would also suggest that their impact was in fact considerably greater than predicted by current GCMs.

CLIMATE CHANGE IMPACTS

The main references for the impacts of a 2°C global average temperature increase during the next 100 years come from the IPCC (IPCC Working Group II, 1998). These were compiled within the context of placing a positive value on the perceived current equilibrium state and hence expressing a high level of concern about change. The geological evidence from the Holocene demonstrates that the fauna and flora can handle high levels of change, certainly equivalent to those now being predicted, but the timescale is longer than is now viewed as acceptable. Given time, species populations do adjust, but in the short term the diminution in numbers at locations subject to high stress can lead to the inference that the species itself is at risk. In addition, the effect of mankind on the planet is now so great that not only is there little scope for natural adaptation to occur, but the pressure of concern for the impacts on large and ever-increasing population groupings means that the focus is now quite different.

As mentioned earlier, there are broadly three axes along which impacts need to be considered: temperature, precipitation, and sea level. Given the current coolness of global temperatures (in terms of geological history), a 2°C warming is certainly not a problem for the planet. The evidence from both historic and Quaternary change demonstrates that the predicted rate is within the range experienced, albeit at the top end. In addition, given the regional differences in the temperature changes, which mean little relative change where it is presently hot and much larger relative changes where it is presently cool, direct adaptation to temperature by humans should not be a major issue.

Since adaptation to change is easier when the organism is simple, insects will adapt much more quickly than vertebrates and bacteria most rapidly of all. This is quite a worrying prospect and will require a technological response.

It is perhaps worth emphasising at this juncture that based on the history of human innovation, when there is sufficient will and coherent effort, the range of technological solutions to address the need is very large. Nothing in any of the model projections is outside the planet's range of experience. However, the planet has never recognised the concepts of equilibrium and sustainability, which comprise today's fashions of concern. These will certainly be in conflict with the degree of change that may be a consequence of the more extreme predictions. Identifying technological solutions is easier than making them politically acceptable.

One aspect of climatic change that is certain to become a significant issue is the provision of adequate supplies of fresh water, particularly given the scale of present demand and the current inability of humans to migrate. Water resources per se are infinite, but the majority are saline and we rely on the sun's energy for free desalination. The question then is the change in distribution of the precipitation as the climate changes. Where this is significant, the cost of providing the right quantity of the right quality in the right place at the right rate will require very substantial investment with long lead times. This can certainly be done if we elect to make the commitment, but the political challenge is only just beginning to be recognised. The models show marked regional changes that are dealt with in outline form below.

The second major issue is rising sea levels. Sea level responds to global temperature changes because of thermal expansion in the oceans. It is quite difficult to get reliable data on recent changes, but the data sets shown in **Figure 10** from Warrick and Oerlemans (1990) are based on different assessments from a large number of observations. They both show a steady rise in sea level of about 120 mm over the past 100 years. Warrick and Oerlemans (1990) estimate the overall global rate to be within a range of 1 to 2 mm/year.

The question of the ice flux through the Antarctic and Greenland ice caps is the subject of current research using very accurate satellite radar measurements. The latest data (Wingham et al., 1998) show that the main Antarctic ice sheet is currently stable, with the height changing by less than 1 cm per year, although there are worrying signs of shrinkage in west Antarctica. The modeled prediction from the IPCC (1998) for sea-level rise over the next 100 years is 150 to 950 mm. Given the very slow rate of change, there is clearly time to put in place the means to provide appropriate additional protection. Most populated low-lying areas already need protection from high tides and storm surges. However, a destabilisation of the west Antarctic ice sheet would give a rise of several metres in a short space of time and would be very serious. Previous work (Vaughan and Doake, 1996) suggested that a temperature rise of 10°C would be required for large-scale destabilisation, but the uncertainties on this issue are one of the few genuinely worrying factors around the changes resulting from a warmer world.

RESULTS OF COMPUTER MODELING
AND REGIONAL IMPACT ASSESSMENTS

The IPCC Regional Impacts Special Report (IPCC Working Group II, 1998) draws from the sector impact assessments of the Climate Change Second Assessment Report (Watson et al., 1996). It is a lengthy treatise, comprehensively referenced, and for each of 10 regions of the globe it discusses

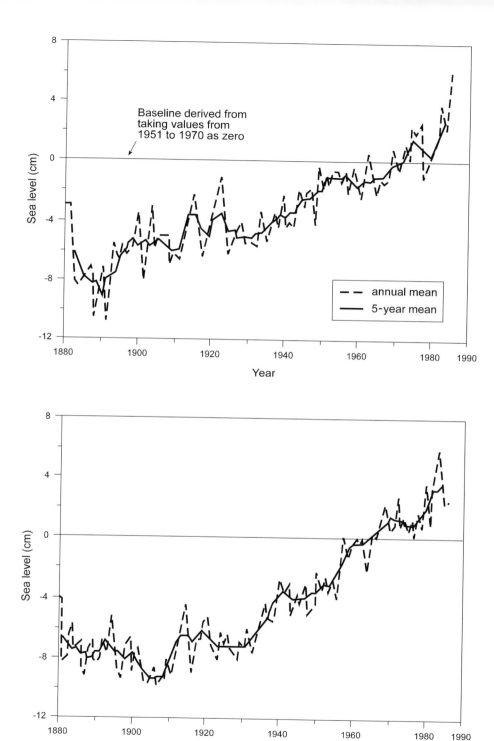

Figure 10 Composite global ocean sea-level curves from different data sets derived from a large number of tidal stations (from Warrick and Oerlemans, 1990).

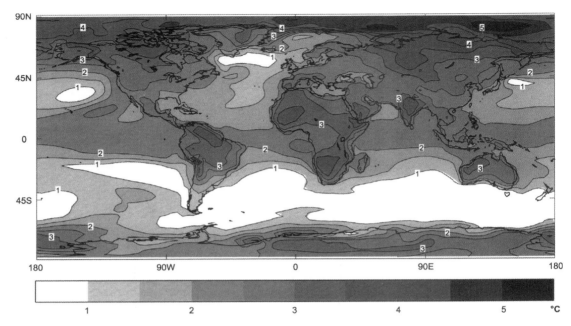

Figure 11 Change in annual temperatures for the 2050s compared with the present day. Model assumes a 1% increase per year in CO_2 concentration (from the U.K. Meteorological Office, 1997).

the impact of global warming models on a number of different sectors. These include ecosystems, water resources, food and fibre production, coastal systems, human settlements, and health. The nature of the impacts is directed toward the vulnerability of these sectors to the assumed changes, *vulnerability* being defined as the sensitivity of the sector and its ability to adapt.

Although the analyses are very comprehensive and detailed, the nature of the subject makes it difficult to capture the uncertainty range on the projected outcomes. In particular, the GCMs that were then available did not provide the level of definition required to give confidence to regional predictions (U.K. Meteorological Office, 1997). The current generation of atmosphere-ocean coupled GCMs gives a much higher confidence level for this capability. **Figures 11** and **12** are the predictions for temperature and precipitation changes from today to the middle of the twenty-first century. They are taken from the latest analyses by the U.K. Meteorological Office at its Hadley Research Centre (U.K. Meteorological Office, 1997).

The enormous regional variability is very noticeable. Notwithstanding the uncertainties, it is possible to provide an indication of the broad pattern of change that should be anticipated and for which the implementation of adaptive measures would be prudent. As has been mentioned before, the overall attitude should be one that is expecting change and then looks at the balance of probability for the direction which that change is most likely to take. We would then consider the options to deal with the anticipated outcomes. There is a practical difficulty, of course, in judging the level of precaution to adopt when evaluating the options, but that is inevitably a subjective judgment with a strong political flavour. The approach can only to a degree be informed by the scientific analysis, because of the latter's inherent uncertainty.

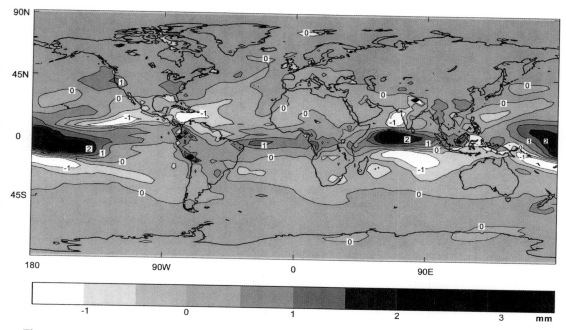

Figure 12   Change in annual precipitation in mm/day for the 2050s compared with the present day (from the U.K. Meteorological Office, 1997).

Inspection of the temperature predictions shown in **Figure 11** shows quite noticeable increases in annual average temperatures over the continental regions, particularly in South America, Africa, and India. In terms of population impact, India would be the region giving rise to most concern. In North America and southern Europe, changes of 2.5°C will probably lead to a marked increase in the frequency of periods of high summer heat.

A REVIEW OF KEY CONCLUSIONS FROM THE HADLEY CENTRE MODEL RESULTS

The impact of projected climate change on natural vegetation includes a reduction of tropical grassland and forest, but with a net expansion of temperate grassland and deciduous forest. Enhanced temperatures and CO_2 concentration could lead to a significant expansion of the biomass between 30° and 60°N latitude, as part of a potential substantial increase in the terrestrial carbon sink, currently estimated to be sequestering between 1 and 2 gigatonnes of carbon per year. The question of the degree and current distribution of terrestrial sequestration of carbon is today a very active research topic. The possibility of a very large terrestrial sink in North America is particularly intriguing (Fan et al., 1998).

Secondly, there is the key question of water resources. The important impact will be changes to annual runoff and recharge, which determine the resources available other than through expensive desalination. As a broad generalisation, although in a warmer world the rate of precipitation will increase, the models show **(Figure 12)** increases in equatorial and high latitude runoff, but decreases in midlatitudes. Couple this prediction with the fact that today a third of the global population lives in countries now experiencing moderate to severe water stress, where *stress* is defined as the ratio of withdrawal to renewable resources readily available (World Meteorological Organization, 1997) and the impact is likely to be substantial.

What is required above all else is a change in attitude, a recognition that in the future world the pattern of water distribution will have to change. Water provision must assume a higher priority in the allocation procedures for discretionary expenditure and in aid relief than is presently the case. Furthermore, users must become much more aware of the changes in patterns of usage that they will have to accept, if they are unwilling or unable to pay the price of provision.

In addition, it seems likely that changes in streamflow volumes and timing will lead to a greater frequency in flood conditions. These can be mitigated by investment in preventative measures, but the degree is limited by the pressure of population to occupy areas that are designed by nature to be flooded regularly.

Thirdly, food-production models are consistent in predicting enhanced yields in middle and high latitudes and reduced yields in low latitudes, but with strong regional differences. The Food and Agriculture Organisation (FAO) of the United Nations estimated that on standard assumptions and an unchanged climate, world cereal production could double from 1990 to 2050, hence keeping pace with requirements (FAO, 1987). The modeled impact of climate change is a 2% reduction and is therefore within the range of uncertainty. Although it is possible to see outcomes in which the impact of projected climate changes would increase the risk of hunger in particular regions, the issue of food supply is such an intensely political process that it is difficult to make any valid predictions.

Finally, in terms of human populations, the impact of sea-level rise on coastal zones will be determined by the mitigation efforts that are made. Most of the people at risk are concentrated in a few regions, particularly southern and southeast Asia, where the ability to fund the appropriate protection will be a problem. However, assuming the rise remains gradual (but see previous comments on west Antarctica), the long timescale means that there should be opportunities to develop the appropriate protective mechanisms.

All climate modeling work to date has concluded that the globe will continue to warm. There are no model predictions for a return to the cooler conditions that occurred at two intervals during the past two millennia. The historical record in the detail needed to make predictions comparable to the modeled outcomes for those applying to a warming world is really only available for northern Europe. The interested reader should refer to Lamb (1995).

There is no doubt that a 2°C drop in global temperatures would have a significant impact, particularly in temperate regions. If one consequence was a marked decline in spring and summer temperatures, as occurred during the "worst" years of the Little Ice Age, this would have a serious effect on food production in regions which have become major contributors to global output. With an ever-expanding world population, this could be catastrophic. However, the decadal and annual variability of the climate during periods of regional cooling, as shown by the historical records, is very marked, so broad generalisations are not easy to make without mathematical simulation.

CONCLUSIONS

The temperature history of the Quaternary and the record of past climates give an important context for a discussion of the impacts expected today from future changes. In particular, it provides a much broader perspective than one gets from a consideration of predictions based on climate models using data for the past 100 years only. Temperature and climate are constantly changing. It is absolutely impossible for mankind to stabilise the climate of the planet. There is no equilibrium state. At any point in time the temperature is either warming or cooling, and the historical record shows quite clearly the high magnitude of the changes.

The data also demonstrate very clearly the overall pattern of change and the very high degree of temperature variability on different timescales—millennial, centurial, decadal, and annual.

The highest temperature reached after the last glacial period occurred 6000 to 9000 years B.P (the postglacial thermal optimum). Temperatures are now on an irregular downward path in a manner quite similar to the latter part of the last interglacial. A key observation from the detailed records for the last 150,000 years is that once temperatures start to decline, successive recoveries are invariably lower than those reached at the preceding maximum. This would lead to the inference that the recovery over the last 200 years may not reach the level of the Medieval Warm Period of the twelfth century A.D.

The temperature increases projected from enhanced greenhouse gas concentrations are overlaid upon this high level of natural variability and will be counteracting the trend to global cooling. The rates of change projected for the next 100 years are not more rapid than those that have been experienced on half-century scales, but the magnitudes are significant. The uncertainties inherent in the modeling, particularly for projections on a regional scale, mean that notwithstanding the very high level of effort that has been devoted to the topic, predictions of outcomes are subject to wide margins of error.

The general messages that come through from all the impact studies are that temperature increases by themselves are much less of a concern than the likely changes in rainfall precipitation patterns. If these do turn out to be of the magnitude the current models predict, then not only will substantial investment be needed, but there will also have to be a significant change in political attitudes toward freshwater resources.

The issue of ecosystem response is difficult to judge, particularly if one accepts that there is a high degree of variability in the manner in which individuals in a species population react to change. In addition, active management of the landscape provides an opportunity for man to consciously manage floral and faunal shifts. However, this would require a change in our approach. Attempting to retain a preconceived concept of a state of prior equilibrium is false and bound to fail.

On the risk envelope, the most serious potential impact would come from a destabilisation of the west Antarctic ice sheet. Although this requires a temperature increase outside the range predicted, it remains a concern. Sea-level rise from thermal expansion of the oceans is very gradual and predictable, and there would be time to take action to mitigate the impacts.

REFERENCES CITED

Aaby, B., 1976, Cyclical climatic variations in climate over the past 5500 years reflected in raised bogs: Nature, 263, p. 281–4.

Alley, R. B., D. A. Meese, C. A. Shuman, A. J. Gow, K. C. Taylor, P. M. Grootes, J. W. C. White, H. Ram, E. D. Waddington, P. A. Mayewski, and C. A. Zielinski, 1993, Abrupt increase in Greenland snow accumulation at the end of the Younger Dryas event: Nature, 362 p. 527–9.

Atkinson, T. C., K. R. Briffa, and G. R. Coope, 1987, Seasonal temperatures in Britain during the past 22,000 years reconstructed using beetle remains: Nature, 325, p. 587–92.

Augustsson, T., and V. Ramanathan, 1997, A radiative–convective model study of the CO_2 climate problem: Journal of Atmospheric Sciences, v. 34, p. 448–51.

Bard, E., B. Hamelin, R. G. Fairbanks, and A. Zindler, 1990, Calibration of the ^{14}C timescale over the past 30,000 years using mass spectrometric U-Th ages from Barbados corals: Nature, 345, p. 405–9.

Bard, E., B. Hamelin, M. Arnold, L. Montaggioni, G. Cabioch, G. Faure, and F. Rougerie, 1996, Deglacial sea-level record from Tahiti corals and the timing of global meltwater discharge: Nature, 382, p. 241–4.

Broecker, W. S., 1994, Massive iceberg discharges as triggers for global climate change: Nature, 372, p. 421–4.

Bryson, R. A., and B. M. Goodman, 1980, Volcanic activity and climate changes: Science, 207, p. 1041–4.

Chapman, M. R., and N. J. Shackelton, 1998, Millennial-scale fluctuations in north Atlantic heat flux during the last 150,000 yrs.: Earth and Planetary Science Letters, 159, p. 57–70.

Cuffey, K. M., and G. D. Clow, 1997, Temperature, accumulation and icesheet elevation in central Greenland through the last deglacial transition: Journal of Geophysical Research, 102, no. C12, p. 26383–96.

Driscoll, N. W., and G. H. Haug, 1998, A short circuit in thermohaline circulation: A cause for northern hemisphere glaciation?: Science, 282, p. 436–8.

Fairbanks, R. G., 1989, A 17,000 year glacio–eustatic sea level record: influence of glacial melting rates on the Younger Dryas 'event' and deep ocean circulation: Nature, 342, p. 637–42.

Fan, S., M. Gloor, J. Mahlman, S. Pacala, J. Sarmiento, T. Takahashi, and P. Toms, 1998, A large terrestrial carbon risk in North America implied by atmospheric and oceanic carbon dioxide data and models: Science, 282, p. 442–6.

F. A. O., 1987, Fifth world food survey: Rome, UN Food and Agricultural Organisation: 75 p.

Gasse, F., and E. van Campo, 1994, Abrupt post-glacial climatic events in west Asia and north Africa monsoon domains: Earth and Planetary Science Letters, 126, p. 435–56.

Hoyle, F., 1996, The Great Greenhouse Controversy, in Energy and Environment, v. 7, no. 4, p. 349–357.

IPCC Working Group II, 1998, Special report on the regional impacts of climate change, for Intergovernmental Panel on Climate Change: Cambridge University Press, 527 p.

Kroon, D., W. E. N. Austin, M. R. Chapman, and G. M. Gaussen, 1997, Deglacial surface circulation changes in the northeastern Atlantic, temperature and salinity records of NW Scotland on a century scale: Paleoceanography, 12, p. 755–763.

Lamb, H. H., 1995, Climate, History & the Modern World, second edition, London, U.K., Routledge, 433 p.

Markgraf V., 1974, Palaeoclimatic evidence derived from timberline fluctuations. Colloque International CNRS, No. 219: Les Methodes Quantitatives d'Etude des Variations du Climat au Cours du Pleistocene, p. 67–77.

Mayewski, P. A., L. D. Whitlow, S. Twickler, M. S. Morrison, M. C. Grootes, P. M. Bond, G. C. Alley, R. B. Meese, D. A. Gow, A. J. Taylor, K. C. Ram, and M. Wumkes, 1994, Changes in atmospheric circulation and ocean ice cover over the North Atlantic during the last 41,000 years: Science, 263, p. 1747–51.

Mysak, L. A., D. K. Manak, and R. F. Marsden, 1990, Sea-ice anomalies observed in the Greenland and Labrador Seas during 1901–84 and their relation to an interdecadal Arctic climate cycle: Climate Dynamics, v. 5, p. 111–33.

Mysak, L. A., and S. A. Powers, 1991, Greenland sea-ice and salinity anomalies and interdecadal climate variability: Climatological Bulletin, v. 25, no. 2, p. 81–91.

Ritchie, J. C.,1984, Past and present vegetation of the far north-west of Canada. Toronto, University of Toronto Press, 251 p.

Roberts, N. ,1998, The Holocene. An Environmental History, second edition, Oxford, U.K., Blackwell, 316 p.

Shinn, E. A., R. N. Ginsberg, and R. M. Lloyd, 1965, Recent supratidal dolomite from Andros Island Bahamas: in Pray, L. C., and R. C. Murray, eds., Dolomitisation and Limestone genesis, a symposium: Society for Economic Paleontologists and Mineralogists Special Publication no. 13, p. 112–123.

Stott, P., 1998, Biogeography and ecology in crisis: the urgent need for a new metalanguage: Journal of Biogeography v. 25, p. 1–2.

U.K. Meteorological Office, 1997, Climate change and its impacts: a global perspective: The Met. Office, Bracknell, England.

Vaughan, D. G., and C. S. H. Doake, 1996, Recent atmospheric warming and retreat of ice shelves on Antarctic peninsula: Nature, 379, p. 328–30.

Warrick, R. A. and H. Oerlemans, 1990, Sea level rise: in Climate Change—the IPCC Scientific Assessment (eds. Houghton J. T. et al.) Cambridge University Press, 365 p.

Watson, R. T., M. C. Zinyonera, and R. H. Moss, 1996, Climate Change 1995. Impacts, Adaptions and Mitigation of Climate Change: Scientific Technical Analyses, for Intergovermental Panel on Climate Change, U.K., Cambridge University Press, 879 p.

Webb, III, T., 1991, The spectrum of temporal climatic variability. *In* R. S. Bradley (ed.) Global Change of the Past. Office of Interdisciplinary Earth Studies, Boulder Colorado, p. 61–81.

Webb, T., E. J. Cushing, and H. E. Wright, 1984, Holocene changes in the vegetation of the midwest: *in* H. E. Wright (ed.) Late Quaternary environments of the United States, v. 2, The Holocene, London, Longman, p. 142–65.

Wingham, D. J., A. J. Ridout, R. Schamoo, R. J. Arthem, and C. K. Shun, 1998, Antarctic elevation change from 1992 to 1996: Science, Oct. 16, p. 456–58.

World Meteorological Organisation, 1997, Comprehensive assessment of the freshwater resources of the world: Report to the U.N. Commission on Sustainable Development. Geneva, WHO, 33 p.

Zielinski, G. A., P. A. Mayewski, L. D. Meeker, S Whitow, M. S. Twickler, M. Morrison, D. A. Meese, A. J. Gow, and R. B. Alley, 1994, Record of volcanism since 7000 B.C. from the GISP2 Greenland ice core and implications for the volcano-climate system: Science, 262, p. 948–52.

Epilogue

The editors have tried to provide an overview of scientific studies that is both pertinent to the climate-change question and is available to the scientific community and policy makers. In addition, the papers in this volume provide a glimpse of the future of earth research, in which geologists will play an ever-increasing role in the formulation of public policy. As resources become more precious and populations more concentrated, conflicts between people and natural processes and people versus people will increase. Whereas the former role of the earth scientist was to locate resources for national and economic development, our mandate has expanded to include protecting people when they intervene in complex natural systems which they may not understand well.

We have not finished our work on the climate question. There remains preparation of a time-temperature chart of the last 600 million years, the time period we call the Phanerozoic, when life became abundant and evolved to its present level. It will likely take several years to prepare an authoritative and statistically robust geologic view of past climates, against which models of climate behavior and predictions for the future may be compared.

As part of preparing this book, it was necessary to keep abreast of current scientific literature. We have seen several major research studies involving oceanographic circulation and its potential impact published in international journals. Not long ago, solar influence on climate was thought to be inconsequential. Today, researchers are actively working to better understand these phenomena, and the relation between solar variability and the climate of our planet is becoming well documented. Measurements are now replacing theory, and they are providing the foundation for development of new theories.

In the past, decision makers have been somewhat insensitive to geologic reality. We hope we have at least partially demonstrated the need for broader scientific scope in those policy-making activities related to natural systems. If no effort is exerted to understand and consider the history of the earth and the dynamics of its natural systems, it is hopeless to try to understand the present and predict the future. Truly, James Hutton (1785) was right. The present is the key to the past, but we would add—*the past is the key to the future.*

Index

Acanthocaetetes wellsi, environmental reconstructions of, 140–147

Accelerator mass spectrometry (AMS) dating, 111, 156

Adaptability
of beetles to late-glacial climatic changes, 161
human, 351–352

Adsorption trapping, 290–291

Agricultural fertilizer application, and anthropogenic fluxes, 77–78

Alberta Basin, potential of, for CO_2 sequestration, 298

Amazon Basin, 104–110

AMS dating. *See* Accelerator mass spectrometry

Andes, ice cores from, 104–112

Antarctic ice cap, ice flux through, 352

Anthropogenic CO_2, sequestration of, 12, 285–301

Anthropogenic fluxes, 76–78

Atlantic Ocean
and heat distribution, 36
north
bottom-water temperature of, 122–125
flow and thermal structure, 125
sea-surface temperature of, 341–342

Atlantic Ocean conveyor circulation
alternate modes of operation, 88–91
during Pleistocene, 340, 342
See also Thermohaline circulation

Atmosphere-ocean
exchange flux, 78
obscure connection to climate changes, 91

Atmospheric water vapor, change in, experiment using, 318–319

Autocorrelation, of oxygen isotope ratios, 222–223

Autocorrelation function, 222–223

Baltic Sea
basin sediments, 235–237
classification by coast index, 241
coastline evolution, 237–243
digital elevation model, 234
geotectonic setting and hydrography, 233–235
sea-level changes, 231–232

Beetles
and climate change, 156–163
evidence of, 8–9
fossil distribution, 160
and global warming, 163–165
introduction to, 153–155

Biogeochemical cycles, C-N-P-S, 53–56
Bond, Gerard, 92
Borehole temperature measurements, 127–128, 132
Bottom water temperature (BWT), 8, 121
    examples using, 122–125
    impact of, on global climate, 134
    reconstructing history of, 128–131

Calcification, microbial
    effects of cell physiology on, 272–276
    effects of changing climate on, 267
    and photosynthesis, 268
    theoretical model for, 274
Calcium-carbonate crystals, 269
Carbon
    fluxes, 59–60
    partitioning, 57–62
Carbon cycle. *See* Global carbon cycle
Carbon dioxide
    anthropogenic sequestration of
        basin-scale criteria for, 291–295
        mechanisms for, 289–291
        overview, 285–287
        performance assessment of, 301
        public acceptance of, 301
        requirements for, 300–301
        site-specific criteria for, 295–296
        versus lime-mud deposition, 276–280
    concentration of CO_2
        cooling forces to counter, 317
        temperature, 169–171
    emissions, as main human forcings, 52
    geological traps for, identifying location and capacity of, 299
    and global warming, exploring through natural meteorological experiments, 317–323
    and human activities, 6–7, 12
    injectivity of, 296
    long-term fluctuations in, 178–183
        early-middle Eocene, 180–183
        late Neogene, 178–180
    and microbial calcification, 274–276
    paleoatmospheric, and stomatal frequency, 169–183

    relationship between, 171–172
        Stomatal Index, 9
    relevant properties of, 287–288
    rising, and global warming of past century, 323–324
    sequestered and injected, fate of, 300
    short-term fluctuations in, 173–178
    unwanted, disposal of, 12–13
Carbon equations, of model TOTEM, 72–73
Carbon-14 dating. *See* Accelerator mass spectrometry (AMS) dating
Carbon, nitrogen, phosphorus, sulfur (C-N-P-S)
    biogeochemical cycles, 53–57
    flux constants adopted in TOTEM, 70–71
    global reservoir masses, 67
    transfer fluxes, 68–69
Cavern trapping, 291
Cell, cyanobacterial, with calcium-carbonate crystal, 269
Cell physiology, effects of, on microbial calcification, 272–276
Cenozoic, reconstructing paleoatmospheric CO_2 of, 169–183
Central England temperature plot, 347–348
Climate
    clouds' effect on, 28–29
    direct measurement of, 204
    dynamic nature of, 338–339
    earth's recent history, 340
    impact parameters, 339
    predicting, from GISP2 data, 305–306
    solar forcing of, 26–30, 205
Climate change
    adaptation of beetles to late-glacial, 161
    anthropogenic, assumption of, 193–194
    anthropological record of, 203–204
    beetles' response to, 156–163
    during past two millennia, 346–350
    and food production, 356
    geologic record of, 195–202
    human adaptability to, 351–352
    impacts, 351–352
    indicated by coral reefs and tropical shorelines, 251–252

and microbial lime-mud production, overview, 267–268
model forecasts, 213
    limitations of, 214–215
and natural variability, 9–12
and natural vegetation, 355
overview, 1–4
Pleistocene, 340–342
rate and magnitude of past, overview, 193–195
records of, from Greenland ice core, 306–309
reflected by Baltic Sea sediments, 235–237
traditional models of, problems with, 314
and vertical crustal movements, 237–243
and water resources, 355–356
See also Climate drivers; Cooling; Global cooling; Global warming; Temperature, changes in
Climate controls, 2–7
in coastal locations, 232
Climate drivers, 4–7
CO_2 versus other, 351
geology, 1–2
and timescales, 2–3
Climatic record, 86–88
Cloud condensation nuclei (CCN), 325–326
Clouds
and solar energy, 28
cooling effect of, 324–326
CO_2. See Carbon dioxide
Coast index (ci), 231–232
classification of Baltic Sea coastal locations, 241
deriving, 240
Coastal locations
crustal-uplift and climate-controlled, 232
See also Shoreline accumulations
Coastal zone, and carbon partitioning, 60–62
Coastline evolution, 237–243
Coleoptera. See Beetles
Composite variable, on GISP2 data, 306–308

Computer modeling, and regional impact assessments, 352–355
Condensation levels, in tropics versus polar regions, 109
Conductivity-temperature-depth (CTD), 122
Cooling
effects of clouds, 324–326
in western Europe, potential for, 342
Cooling forces, to counter CO_2 buildup, 317
Coral reefs
Bahamas, 257–258
as indicator of climate/sea-level changes, 251–252
as potential environmental recorders, 138–140
and sclerosponges, 138–139
uplifted, "bathtub ring" on, 253–257
See also Key Largo Limestone
Crustal-uplift coastal locations, 232

Dasuopu Glacier, isotopic enrichment, 104, 106
Data, types of
alpine tree-limit, 197, 200–201
anthropological, on climatic fluctuation, 203–204
environmental, and skeletal characteristics, 144
hydrographic, 122
isotopic. See Oxygen isotopic records
relative and absolute, on sea-level changes, 243–245
subseafloor borehole temperature, 121–125, 127–128
See also Greenland Ice Sheet Project II (GISP2), Greenland Ice Core Project (GRIP)
Deep-sea water temperature, and oxygen isotope, 252–253
Dendrochronologic record, 200–201
Digital elevation model, of Baltic Sea area, 234
Drilling technology, and sequestered CO_2, 300

Dunde ice cap
  isotopic enrichment, 104, 106
  oxygen isotope ratio, 101–102, 106
Dust, vertical redistribution of, 319
Dynamical systems theory, 305
  and behavior of Holocene record, 315
  and simple nonlinear prediction,
    309–312

Earth
  equilibrium and nonequilibrium states
    of, 338–339
  recent climatic history, 340
Electrical field, solar wind and, 28
Electron-microprobe techniques, 142
Energy dispersive spectroscopy (EDS), 8,
  137, 140–143
Environmental data, syndepositional, and
  correlation with skeletal characteris-
  tics, 144
Environmental reconstructions, of A.
  Wellsi, 140–147
Eocene, early-middle, long-term fluctua-
  tions in CO_2, 180–183
Equations, of model TOTEM, 72–76
Equilibrium states, and nonequilibrium
  states, 338–339
Eustatic curves (esl), 237–240
Experiments, natural, exploring CO_2-
  induced global warming through,
  318–323
Extinction, beetle species, 154–155,
  158–159

Fertilizers, application of, 77–78
First-order climate controls, 2–4
Florida Keys, and sea-level changes,
  255–261
Food-production predictions, 356
Fossil-fuel burning
  and anthropogenic fluxes, 76
  carbon partitioning from, 57–62
  sequestering CO_2 emissions from,
    286–287
  See also Human activity
Fossils
  evidence of beetles' response to climate
    change, 156–163

leaf remains, stomatal frequency analy-
    sis on, 170–171
  Quaternary beetle, 155–156
Fourier analysis, 223–226
Fourth-order climate controls, 3, 6–7

General circulation model (GCM), 10
Geologic models, and global climate-ßΣ
    change models, 214
Geologic record, of climatic fluctuation,
    195–202
Geology, as climate driver, 1–2
Geotectonic setting, of Baltic Sea, 233–235
Geothermal regimes, and CO_2 sequestra-
    tion, 293–294
GISP2. See Greenland Ice Sheet Project II
Glaciers
  formation, 101
  tropical, melting of, 112–117
  Holocene fluctuations, 196–199
  late Pleistocene fluctuations, 196
  record of, 195–199
    anthropological, 203
    sedimentary, 199–200
  See also Global cooling, Ice caps, Ice
    cores, Little Ice Age
Global carbon cycle, 52–53
  future projections, 62–67
Global cooling
  greenhouse effect, 324–327
  versus global warming, 350
  from volcanic eruption, 345–346, 348
Global-equilibrium experiments, 320–321
Global warming
  beetles' response to, 163–165
  CO_2-induced, exploring through natur-
    al meteorological experiments,
    317–323
  cooling greenhouse effect, 324–327
  of past century, and rising CO_2,
    323–324
  predictions, 351
  of twentieth century, 114
  versus global cooling, 350
  See also Greenhouse effect; Ice masses,
    decreasing; Tropical glacier,
    melting of

Greenhouse effect
  and anthropogenic sequestration of
    CO_2, 286
  assumption of, 194
  of changing water vapor, 318–319, 323
  confirmation of, by Mars and Venus
    surface temperatures, 322
  cooling, 324–327
  expected magnitude, 84–85
Greenhouse events, and icehouse events,
  38–41
Greenland
  climatic record, 86–89
    anthropological, 204
  ice cap, ice flux through, 352
Greenland Ice Core Project (GRIP), 215
  oxygen isotopic record, 202, 205
Greenland ice cores
  oxygen isotopic records, 13
  small-scale climate changes indicated
    in, 196–199
  studies, 202
Greenland Ice Sheet Project II (GISP2), 10,
  198–199
  correlation with sunspot cycle, 24
  as Holocene record, 215–217
  ice flux through, 352
  long-term trend exhibited by, 213,
    217–221
  oxygen isotopic record, 202, 205
  predicting climate from data, overview,
    305–306
  rapid rate of change demonstrated, 343
  records of climate change from,
    306–309
GRIP. See Greenland Ice Core Project
Guliya ice cap, isotopic enrichment,
  104, 106

Hadley Centre analyses, 354
  key conclusions from, 355–356
Holocene record, 30–31, 215–216, 227–228,
  342–346
  Baltic Sea Basin sediments, 236
  Baltic Sea coastline changes, 237
  "climatic crises," 345
  and dynamical systems theory, 315

glacier fluctuations, 196–199
  leaf fossil's contradiction of, 176–178
  and long-term trends in temperature,
    217–221
  melting glaciers, 203
  persistence of conditions, 221–223
  possible flaw in interpreting, 207
  recurrent patterns, 223–226
  rise in sea level, 346
  tree-line elevation, 343, 345
  variations in precipitation, 344
Huascarán, ice cores from, 106–112
Human activity
  as cause of global warming, question
    of, 193
  as fourth-order climate driver, 6–7
  major pertubations due to, 52
  and natural variability, 14
  response of beetle to, 163–165
  effect of
    on beetle species, 154–155
    on beetles' survival, 154
    on carbon-dioxide levels, 6–7, 12
    on climate and sea-level changes,
      261
  See also Fossil-fuel burning
Human adaptability, 351–352
Hydrocarbon potential, and basin
  maturity, 294
Hydrodynamic regimes, and CO_2 sequestration, 293–294
Hydrodynamic trapping, 289, 291
Hydrographic regime, in Straits of
  Florida, 125–127
Hydrography
  of Baltic Sea, 233–235
  data from Middle Atlantic Bight,
    131–133
  limited data, 122
Hydrological model of moisture trans-
  port, 108–109

Ice caps
  Antarctic and Greenland, ice flux
    through, 352
  Dunde, 100–104, 106
  Guliya, 104, 106

Quelccaya, 111–112
  melting of, 114–116
Sajama, 111–112
  See also Glaciers, Ice cores
Ice cores
  Andes, 104–112
  global array of, 100–101
  Greenland studies, 202
    oxygen isotopic records, 13
  Holocene record
    Greenland, 215–221
    leaf fossil's contradiction of, 176–178
  Huascarán, 106–112
  locations of recovery, 100
  recurrence of patterns, 223–226
  Tibetan Plateau, 101–104
  See also Glaciers, Greenland Ice Core
      Project, Greenland Ice Sheet
      Project II
Ice masses, decreasing, 116
  See also Glaciers, melting
Icehouse events, and greenhouse events,
    38–41
Industrial Revolution, and long-term
    temperature trends, 218, 220
Infrastructure, surface, for sedimentary
    basins, 294–295
Injection technology, and sequestered
    CO_2, 300
Instrumental records, limitations of, 138
International Panel on Climate Change
    (IPCC), 338
    1995 report, 339
    Regional Impacts Special Report, 352,
      354–355
IPCC. *See* International Panel on Climate
    Change
Isotope. *See* Oxygen isotope
Isotopic data. *See* Oxygen isotopic records

Key Largo Limestone, 257
Key West, surface air-temperature data,
    131, 133

Landmasses
  distribution of, and effect on climate,
    37–39, 41–47

movement of. *See* Vertical crustal
    movements
  paleoreconstruction of, 43–47
  See also Coastal locations
Land-use activities
  changes in, 52
  and anthropogenic fluxes, 76–77
  CO_2 emissions and carbon partitioning,
    58
  effect of, on CO_2, magnitude of, 60
  planning for, and sea-level changes,
    231–232
  TOTEM and, 56
  See also Human activity
Leaf fossils
  and short-term fluctuations in CO_2,
    173–178
  stomatal frequency analysis on,
    170–171
Leetmaa Pacific Ocean Analysis, 144,
    146–147
  deposition of, versus atmospheric CO_2
    fluctuations, 276–278
  microbial precipitation of, 268–272
  overview of production, 268–268
Lindzen, Richard, 84
Little Ice Age
  current recovery from, 347
  and long-term temperature trends, 218
Lorenz system, 308–309, 314–315
  prediction scheme in reconstructed
    space for, 311
  prediction for trace of, 312
  x-coordinate of, 310

Magnetism, solar, 21–22
Major forcings and model analysis, 53–57
Marine photosynthetic flux equation,
    scaling factors, 79
Marine sediments
  and anthropogenic CO_2, 60–61
  as source of paleoenvironmental
    data, 138
Mars, temperature of, and greenhouse
    effect, 322
Mean stomatal density and index,
    181–182

Meteorological experiments, natural, exploring CO_2-induced global warming through, 317–323

Microbial lime-mud production, overview, 267–268

Middle Atlantic Bight, hydrographic data from, 131–133

Migration, of beetles, 159–163

Milankovitch theory, 252–253

Mineral trapping, 290

Models
  computer, results of, and regional impact assessments, 352–355
  digital elevation, of Baltic Sea area, 234
  general circulation, 10
  global climate change
    and geologic models, 214
    problems and limitations with, 214–215, 314
  hydrological, of moisture transport, 108–109
  ocean conveyor-belt, 88–91, 253
  standard solar, 20–21
  TOTEM, 53–56, 70–76
  See also Dynamical systems theory, Lorenz system

Natural variability, and records of change, 9–12

NEAPACC. See North-East Atlantic Paleogeography and Climate Change Project

Neogene, late, long-term fluctuations in CO_2, 178–180

Nitrogen equations, of model TOTEM, 73–74

Nonequilibrium states, and equilibrium states, 338–339

North Dakota, Devil's Lake, hydrograph of fluctuations, 205–206

North-East Atlantic Paleogeography and Climate Change Project (NEAPACC), 340

Ocean(s)
  changes in. See Sea-level changes, Surface ocean–deep ocean waters exchange

deep. See Bottom-water temperature; Deep-sea water temperature, and oxygen isotope; Surface ocean–deep ocean waters exchange

distribution of. See Tectonic geometry
  paleoreconstruction of, 43–47
  temperature
    limited hydrographic data on, 122
    subseafloor borehole data, 121–125, 127–128
    See also Bottom-water temperature; Deep-sea water temperature, and oxygen isotope
  thermal structure of, changes in, 122
  thermohaline circulation, 36–37, 62–67. See also Surface ocean–deep ocean waters exchange
  See also Atlantic Ocean, Baltic Sea

Ocean conveyor-belt model, 88–91, 253

Ocean Drilling Program (ODP), 8, 125, 127

ODP. See Ocean Drilling Program

Optimal eigenstate method, 226

Optimal eigenstate method, 226–228

Oxygen isotope ratio, Dunde ice cap, 101–102, 106

Oxygen isotope
  and deep-sea water temperature, 252–253
  and Greenland's ice core, 13
  and sea-level changes, 254

Oxygen isotope ratio
  distribution of, 313
  Dunde ice cap, 101–102, 106
  GISP2 record, 306–308

Oxygen isotopic records
  Andes of South America, 104–112
  introduction to, 100–101
  from low-latitude, high-altitude glaciers, overview, 100–101
  as methodology, 7
  paleontologic, 201–202
  Tibetan Plateau, 101–104

Palynologic record, 200

Phosphorus equations, of model TOTEM, 74–75

Photosynthesis, and microbial calcification process, 268, 274
Plant productivity, CO_2-induced, 326–327
Pleistocene
    climate variability, 10, 340–342
        and future predictions, 337
    late, glaciar fluctuations, 196
Policy drivers, 12–14
Power spectrums, 223–225
    during Holocene, 344
    microbial, of lime mud, 268–272
Predictions
    climate-change model forecasts, 213
    contradicting, 327
    examples of, 312–314
    food-production, 356
    from GISP2 data
        for interval prior to Holocene, 312
        overview, 305–306
    global warming, 351
    from principal component analysis, 313–314
    of sea-level changes, 245–246
    simple nonlinear, 309–312
    vegetation, 355
    water resources, 355–356
Principal component analysis
    on GISP2 data, 306–308
    predictions from, 313–314
Projections, global carbon cycle, 62–66

Quaternary beetle fossils, 155–156
Quaternary Period
    beetle fossils from, 155–156
    temperature changes during, 340
Quelccaya ice cap, 111–112
    recent warming of, 114–116
Radiative perturbations, and near-surface air temperature
    natural experiments, 318–323
    overview, 317–318
Redfield ratios, 79
Regional impact assessments, and computer modeling, 352–355
Relative sea level (rsl) curves, 237-238
Runs analysis, to examine trends, 221

Sajama ice-core records, 111–112
Salinity, and temperature, in Leetmaa Pacific Ocean Analysis, 145–150
Salinity-temperature-depth (STD), 122
Sargasso Sea, surface temperature, 201–202
Sclerosponges, as potential environmental recorders, 8, 138–140
Sea level
    changes in, 11–12, 231–232
        Fairbanks curve, 255–256
        Florida Keys, 255–261
        indicated by coral reefs and tropical shorelines, 251–252
            versus deep-sea sediments, 253
        and oxygen isotope stages, 254
    lowering, 259
    lowstands, 254–255
    predicting future, 245–246
        IPCC, 339
    relative sea-level curves, 237–238
    rising, 259–261, 352–353
        during Holocene, 346
        impact on human populations, 356
Sea-surface temperatures, 31
    during past 15,000 years, 343
Second-order climate controls, 2–5
Sedimentary basins
    Amazon, 104–110
    Baltic Sea, 235–237
    main types of, 292
    maturity of, and hydrocarbon potential, 294
    mechanisms and type of flow in, 290
    sequestering CO_2 in, site-specific criteria for, 295–296
    surface infrastructure, 294–295
    See also Lime mud, Marine sediments
Sedimentary record, of cold events, 199–200
Sediments
    deep-sea, versus coral reefs and shoreline accumulations, 253
    marine, and anthropogenic CO_2, 60–61
    as source of paleoenvironmental data, 138
Sequestration efficiency, 296

Sequestration medium, in-situ conditions of, 300
Sequestration, of anthropogenic CO_2, 12, 285–301
  applicability, 296–299
  basin-scale criteria for, 291–295
  characterization of medium for, 300
  current practice of, 299
  mechanisms for, 289–291
  overview, 285–287
  performance assessment of, 301
  public acceptance of, 301
  requirements for, 300–301
  site-specific criteria for, 295–296
Sequestration efficiency, 296
Sequestration medium, in-situ conditions of, 300
Sewage disposal
  and anthropogenic fluxes, 78
  into aquatic systems, 52
Shoreline accumulations, as indicator of climate/sea-level changes, 251–252
Simple nonlinear prediction, 309–312
Skeletal carbonate, procedure for analyzing, in *Acanthocaetetes wellsi*, 140–143
Solar energy, origin of, 21
Solar forcing, of earth's climate, 26–30, 205
Solar magnetism, 21–22
Solar radiation
  absorption, cycle of, and surface air temperature, 319–320
  variations in, 22–25
Solar wind, 28–29
Solubility trapping, 289, 291
Sorgenfrei-Tornquist Zone (STZ), 233
Speciation, beetles', 156–158
stasis in, 157–158
Spectral analysis, 223–226
Spectral transfer function approach, 144–147
Standard solar model, 20–21
Stars, sunlike, variability in, 26
Stomatal frequency, and atmospheric CO_2, relationship between, 171–172
Stomatal frequency analysis, on fossil leaf remains, 170–171

Stomatal Index (SI)
  and carbon dioxide, 9
  mean, 181–182
Straits of Florida, hydrographic regime in, 125–127
Subseafloor borehole temperature data
  introduction, 121–125
  measurements, 127–128
Sulfur equations, of model TOTEM, 75–76
Sunlike stars, variability in, 26
Sunspots, 22–25
Surface-air temperature
  near, to predict CO_2-induced warming, 317–318
  and solar radiation absorption, 319–320
Surface ocean–deep ocean waters exchange, 79
System transfer function. *See* Spectral transfer function approach

Tectonic geometry, 37–47
Tectonic setting, for sequestering anthropogenic CO_2, 292–293
Temperature
  ancient, methods of estimating, 7–9
  changes in
    continual fluctuations, 206
    during Tertiary, 195
    global average, from 1860 to 1998, 348–349
    melting of Quelccaya ice cap, 114–116
    perceived impact of, 350
    See also Climate change, Global warming
  comparing observations of, 131–132
  deep-sea water
    and oxygen isotope, 252–253
    See also Bottom-water temperature
  history of, from bottom waters, 128–130
  IPCC prediction on, 339
  long-term trends in, 217–221
  ocean, limited hydrographic data on, 122
  persistence of conditions, 221–223

and salinity, in Leetmaa Pacific Ocean
    Analysis, 145–150
sea-surface
    during past 15,000 years, 343
    from North Atlantic core, 341–342
    Sargasso Sea, 201–202
surface air
    data from Key West, 131, 133
    near, used to predict CO_2-induced
        warming, 317–318
    and solar radiation absorption,
        319–320
    true, from drilling operations, 127
    Venus and Mars, 322
    of water around Straits of Florida, 126
Terrestrial photosynthetic flux equation,
    scaling factors, 78
Terrestrial-Ocean-aTmosphere-Ecosystem
    Model (TOTEM), 53–56
    equations, 72–76
    flux constants in, 70–71
Tertiary, temperature fluctuation
    during, 195
Thermohaline circulation, 36–37, 62–67
    Atlantic Ocean's alternate patterns,
        89–91
    projected effects of, 92
    See also Ocean conveyor-belt model
Third-order climate controls, 2–3, 5
Tibetan Plateau, ice-core records, 101–104
Tidal cycles, 13
Time-temperature curve, 207
Time-temperature chart, 341
Tornquist-Teisseyre Zone (TTZ), 233
    inactivity of, 244–245
Total solar irradiance (TSI), 5, 19, 22–27,
    29–30, 32
TOTEM. See Terrestrial-Ocean-
    aTmosphere-Ecosystem Model
Trace-element chemistry, 139
Trans-European Fault (TEF), crustal block
    near, 245
Trapping, types of, of CO_2, 289–291
Tree-line elevation, during Holocene,
    343, 345

Tree-limit data, 197, 200–201
Tree-ring widths, 347
Trends
    atmospheric CO_2 fluctuations versus
        lime-mud deposition, 276–278
    temperature, Holocene record, 217–221
Tropical glacier
    Andes of South America, 104–112
    melting of, 112–117
Tropical shorelines, as indicator of climate
    changes, 251–252

Ultraviolet radiation, 25, 30

Vapor, change in
    experiment using, 318–319
    over tropical oceans, 323
Vegetation, impact of climate change
    on, 355
Venus, temperature of, and greenhouse
    effect, 322
Vertical crustal movements
    climate change and, 237–243
    recent data, 243–245
    in southern Baltic Sea, 241–243
Volcanic eruption, global cooling from,
    345–346, 348

Water resources
    fresh, adequate supplies of, 352
    question of, 355–356
Water-sampling temperature probe
    (WSTP), 127
Water vapor
    change in
        experiment using, 318–319
        over tropical oceans, 323
    and tropical hydrological system, 116
Well-completion technology, and
    sequestered CO_2, 300
Whitings, 270–271
Winds, solar, and earth's electrical field, 28
World Ocean Circulation Experiment
    (WOCE), 122